TSUNAMIS

TSUNAMIS

Editor

Wallace Dawison

scitus
academics

Tsunamis

Edited by **Wallace Dawison**

This Book is Printed on Acid-Free Paper. ♾

ISBN: 9781681178851
Library of Congress Control Number: 2017948912

SCITUS Academics LLC,
616, Corporate Way, Suite 2, 4766,
Valley Cottage, NY 10989

www.scitusacademics.com

Notice

Printed in United States of America.

Preface

A tsunami is a series of waves most frequently caused by an earthquake under the sea floor. Tsunamis can cause great loss of life and property damage in coastal areas. Very large tsunamis can cause damage to coastal regions thousands of miles away from the earthquake that caused them. In recent years, world has experienced few mega-tsunamis, which have caused extensive loss of life and properties. The most destructive ones were in December 2004 in Sumatra causing more than 290,000 death and March 2011 in Japan, creating a nuclear accident. The event of 2004 has triggered many global initiatives, such as a new tsunami detection system, more detail coastal modeling, tsunami compatible coastal developments, integrated approach for regional early warning system (EWS) and public educations, awareness and preparedness. The recent advances in tsunami study has lead the scientist for better understanding of the cause of tsunami, its propagation and most importantly the system of warning ahead of tsunami arrival to the vulnerable coastal area. In addition, now the secondary sources for strengthening the devastating effect of the tsunami, for example, the splay faulting, landslide caused by the parent earthquake sources, have been advancing.

This book 'Tsunamis' shows the state-of-the-art of tsunami research at different levels. The content of the book consists of information on tsunami propagation from the open sea to the coast and coastal tsunami warning system. It includes tsunami propagation, preliminary methods for evaluating the suitability of radar sites for tsunami detection using simulated tsunami velocities and factors affecting tsunami detectability are discussed and methods for reducing the false alarms are described. It also includes the effects of tsunami hazard on the coastal environment and infrastructure. The various contributions cover an array of themes that span from geological evidence to post-trauma human care, encompassing pre-tsunami analyses and modeling to post-tsunami management and preparedness techniques. The chapters are contributed by renowned experts involved in tsunami study for the purpose of covering its different aspects from generation to warning system to be applied in tsunami risk reduction.

Table of Contents

CHAPTER 1

Tsunami-Driven Debris Motion and Loads: A Critical Review

Ioan Nistor[1]*, Nils Goseberg[1,2] and Jacob Stolle[1]

[1]*Department of Civil Engineering, University of Ottawa, Ottawa, ON, Canada*

[2]*Ludwig-Franzius-Institute for Hydraulic, Estuarine and Coastal Engineering, Leibniz Universität of Hannover, Hannover, Germany*

ABSTRACT

Recent natural disasters, such as the 2004 Indian Ocean and 2011 Tohoku Tsunami, exhibited the importance of tsunami-resistant infrastructure in high-risk coastal areas. The failure of critical infrastructure in tsunami-stricken communities has led to a recent emphasis on extreme loading conditions associated with tsunami events. One of the critical loads identified by previous research was debris loads. Debris is defined as solid objects entrained within the inundating flows and can range from construction materials to shipping vessels. The emphasis of tsunami loading has led to recent progression in the understanding of debris loads and effects, particularly in evaluating the impact of a single debris piece on a structure. The following paper reviews state-of-the-art research in tsunami-driven debris motion and loads and identifies future directions of research into debris loads and effects to aid in the design of tsunami-resistant infrastructure.

Keywords: tsunami, debris, impact loads, debris damming, coastal engineering, hydraulic engineering

INTRODUCTION

Tsunamis are among the most destructive and deadly natural disasters. Several recent events, such as the 2004 Indian Ocean Tsunami, the 2010 Chilean Tsunami, and the 2011 Tohoku Tsunami, have emphasized the importance of studying tsunami-induced loading conditions. The failure of critical infrastructure (Yeh et al., 2014) and the lack of a clear understanding of the tsunami-induced loading conditions highlighted the deficiency of current building design in tsunami-prone areas (Taubenböck et al., 2013). In the aftermath of tsunami events, reviews of current building standards have clearly shown that existing standards do not properly account for, or in some cases explicitly address, tsunami loads and effects (Palermo et al., 2009). These findings have led to an increased emphasis on the need to understand tsunami flow conditions and associated loads by researchers, engineers, and policy makers in an attempt to design tsunami-resilient infrastructure. In North America, this effort resulted in the addition of a new chapter in the upcoming ASCE *Minimum Design Loads for Buildings and Other Structures* (ASCE 7, 2016) for the design of structures in tsunami-prone areas of the United States: West Coast, Alaska, and Hawaii (Chock, 2016).

Tsunami engineering research has primarily focused on hydrodynamic loading, as generally field observations have identified hydrodynamic conditions as the critical load (Charvet et al., 2014). Additionally, information regarding critical flow features, such as maximum inundation and run-up, is available from post-tsunami field surveys. However, many field observations of tsunami-affected built areas have shown that debris loads and effects can play an important role in structural failure (Yeh et al., 2013). Until recently, existing guidelines have conservatively addressed debris impact recommending that all structures be designed for the impact of a single object (FEMA, 2012). Similarly, in research, the focus has been on quantifying the load of a single debris impact (Haehnel and Daly, 2004; Matsutomi, 2009; Aghl et al., 2015). Another aspect in the design of tsunami-resilient infrastructure is the identification of critical areas for debris impact. This aspect has been significantly more difficult due to the random nature of debris motion (Matsutomi, 2009) and to the wide range of variables that affect debris motion (Naito et al., 2014).

The objective of this review is to (1) evaluate the current state-of-the-art research into tsunami-driven debris motion and loads, (2) indicate areas of research needs, (3) highlight results from a collaborative research effort by the University of Ottawa, Canada, the University of Hannover, Germany, and Waseda University, Japan, to develop new experimental methods to evaluate debris motion, and, finally, (4) evaluate debris impact loads on structures. These objectives have been reached by conducting a comprehensive literature review and further drawing conclusions regarding research gaps and outlining the methodological requirements and facilities to address current and future research needs.

LITERATURE REVIEW

Debris Transport

The focus of post-tsunami forensic engineering surveys of affected coastal communities has primarily evaluated the hydraulic conditions of the tsunami inflow, such as inundation depth, flow velocity, and spatial inundation limits (Saatcioglu et al., 2005; Borrero et al., 2006; Fritz et al., 2006). However, information regarding debris impact cannot easily be identified from the field data, aside from the fact that debris impact may have occurred. Critical information such as the flow conditions, debris velocity, and debris orientation at the time of impact cannot be determined without a video or images of the impact (Charvet et al., 2014). Therefore, very little information regarding debris dynamics within extreme hydrodynamic flows can be derived directly from field impact sites in the aftermath of a disaster.

Debris motion has been equally challenging to evaluate in the field, as the type of debris can vary significantly and few studies have performed a comprehensive survey of debris in the aftermath of a tsunami event. Any rigid or deformable objects entrained within the inundating flow, such as construction materials, vehicles, or shipping vessels, all common to many coastal communities, can become debris (Naito et al., 2014). Additionally, multiple potential debris sources make the identification of the source of debris challenging, particularly in the post-event site surveys. The work of Naito et al. (2014) was among the first field survey to examine overall transport of the debris after the 2011 Tohoku Tsunami. Naito et al. (2014) examined the position of displaced shipping containers and vessels after the 2011 Tohoku Tsunami as they originated from a clear source (the port area) for which substantial documentation of their position before the tsunami existed. From the survey, Naito et al. (2014) were able to develop a conservative estimate of the maximum spreading angle of the shipping vessels in relation to their original location.

Based on the identified sources and subsequent field investigations, Naito et al. (2014) proposed a method to determine the maximum spreading area of the debris. Their method considered the origin of the debris as the geometric center of the debris source (black dot in Figure 1). The debris was assumed to propagate in a shore normal direction (dashed line) and conservatively estimated to propagate within a ±22.5° spreading sector in the inundation direction to the onshore extent of the tsunami inundation limit. From the point if maximum debris displacement, a similar ±22.5° spreading sector was considered as a potential motion area a result of the outgoing flow. Based on the limited observations collected from field data, the majority of debris fell well within the proposed spreading angle.

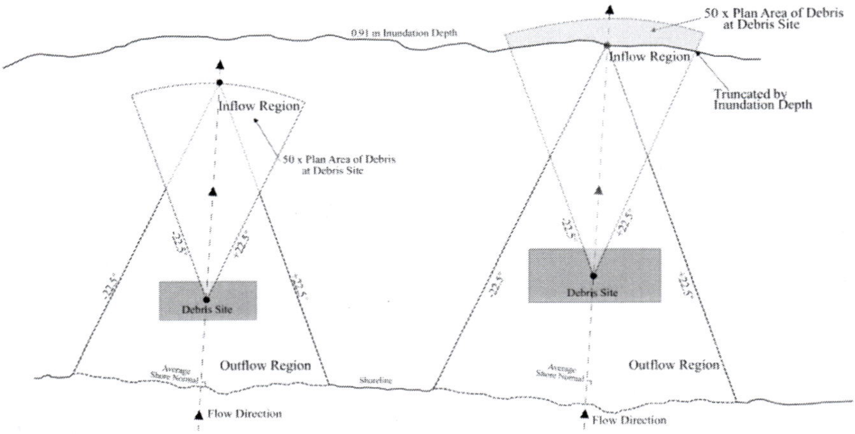

FIGURE 1. Identification of area for potential debris impact [adapted from **Naito et al. (2014)**].

The maximum displacement limit of the debris was calculated based on the debris concentration, which was defined as the plane area of the debris at the source divided by the spreading area of the debris. A conservative estimate was made to establish a debris concentration of 2% that would contain the majority of the debris. The debris limits for the inundating flow are defined by the ±22.5° sector area defining a spreading area that gave a debris concentration of 2% (50 times the plane area of the debris), termed the *inflow region*. However, if the spreading area encroached on an area where the inundation depth is less than 0.91 m (the "prescribed" floatation threshold for debris), this area is not considered as it was expected that debris would ground and no longer propagate forward (red area in Figure 1). As a result, the maximum inundation limit of the debris would be truncated. Additionally, from the center of the inundation limit, a return spreading area can be again determined using the ±22.5° conical area from the inundation limit toward the shoreline. The second area represents the potential spreading sector as a result of the outflow and is termed the *outflow region*. Based on the method outlined by Naito et al. (2014), any critical structures located within the two spreading areas (indicated by the light gray area in Figure 1) should be designed for debris impact. A detailed design example using this method can be found in Naito et al. (2016).

While the method of Naito et al. (2014) provided a conservative approach to evaluating the debris impact potential, several assumptions are associated with their proposed method. The topography of the spreading area has a substantial influence on the flow conditions and therefore the debris motion. Naito et al. (2016) indicated that, in cases with large topographical features that would divert flow, the spreading area should be shifted to account for these irregularities. Naito et al. (2014) also indicated that buildings could act as obstacles in the path of debris motion. In the case of industrial areas, large reinforced concrete buildings would act as a barrier to debris motion as long as the inundation depths were less than 0.91 m than the top of the building. However, in cases where the inundation depths' area greater than 0.91 m above the maximum height of the building or when the surrounding buildings are likely to be destroyed by the inundating flow (wooden structures), the buildings should not be considered as obstacles to the debris. Additionally, the method only examined one type of debris whereas debris are of wide variety of sizes and properties, such as buoyancy, that would influence their propagation distance.

To provide a larger dataset examining the spreading area of debris in tsunami-like flow conditions, Nistor et al. (2016) examined the transport of multiple scaled-down shipping containers in controlled laboratory conditions, over a flat horizontal topography and with no obstacles. The study found that the spreading angle of the debris increased as the number of debris increased. This was attributed to the inter-debris collisions and flow perturbations caused by the debris within the flow. Nevertheless, the motion trajectories of all debris occurred well within the spreading angle of ±22.5° proposed by Naito et al. (2014). Nistor et al. (2016) determined that as the number of debris increased, the longitudinal displacement of the debris decreased. The latter conclusion counters the method proposed by Naito et al. (2014), which suggested that as the number of debris increased, the spreading area also increased. Additionally, Nistor et al. (2016) noted that the debris tended to propagate as an agglomeration which counters assumptions, made in the FEMA P646 (FEMA, 2012), that the likelihood of multiple debris impacts occurring is unlikely.

Goseberg et al. (2016b) built upon the study of Nistor et al. (2016) by including a scaled-down built environment, to act as obstacles to the propagating debris. Their study found that the obstacles acted as a macro-roughness feature for both the debris and inundating flow, resulting thus in significantly shorter longitudinal displacements of the debris. However, the obstacles appeared to have no influence on the spreading angle and the debris once again fell well within the ±22.5° spreading angle proposed by Naito et al. (2014). Due to scaling issues related to debris transport in a scaled experimental environment, further work is needed to properly understand the momentum transfer for debris–debris and debris–fluid interactions to possibly amend the method proposed by Naito et al. (2014).

There are still multiple challenges that must be overcome to properly model debris motion in a reduced scale experimental setting. The foremost issue is the scaling of tsunami flow conditions (Madsen et al., 2008; Rossetto et al., 2011; Goseberg et al., 2013). Related to debris transport, the motion of debris was shown to be a highly variable process (Bocchiola et al., 2006; Matsutomi, 2009). Hence, extensive information is required to obtain meaningful results. To retrieve the necessary information, tracking of the debris' transient motion requires experimental methods that do not influence the debris motion while providing high-quality data regarding the debris' position, orientation, acceleration, and velocity.

Braudrick and Grant (2000) examined the entrainment of individual large woody debris (LWD) in steady flow conditions. Experiments were performed to test a simple entrainment model of a single piece of LWD based on a balance of forces as depicted in Figure 2. The original model considered the LWD as a smooth cylinder lying on a smooth bed; however, the debris cross-sectional geometry may also be rectangular. The initial movement of the LWD was by sliding, though Braudrick and Grant (2000) noted that the initial movement tends to be more complex with significant pivoting involved. Figure 2 outlines the basic force balance used in the model considering a flow downstream with a channel of slope θ. The gravity force is the effective weight ($W_{\text{eff}} = F_g - F_{\text{Buoyant}}$) of the debris. The friction force ($F_f = F_N\mu$) acts in the upstream direction and is a function of the normal force (F_N) and the friction coefficient (μ) between the bed and the LWD. The drag force (F_d) is a function of the water velocity, flow depth, drag coefficient, and angle that the log is traveling in relation to the flow direction. The model performed reasonably well under experimental scrutiny though the pivoting, which was not captured by the model, was an important aspect of the LWD motion.

Imamura et al. (2008) experimentally evaluated the transport of boulders in a dam-break flow and developed a simple model for estimating their motion. Imamura et al. (2008) determined that the boulders tended to be transported by saltation or rolling initially (within the higher velocity flow). When flow velocity began to decrease, as the bore reached the point of maximum inundation and as the bore receded, the boulders would be transported by sliding. Their study also found that the debris orientation affected the motion of the boulder. The boulder would always pivot to have the long axis perpendicular to the flow direction, using some of the hydrodynamic energy, resulting thus in the boulder aligning with the long axis perpendicular to the flow and having thus a greater displacement under inflow conditions.

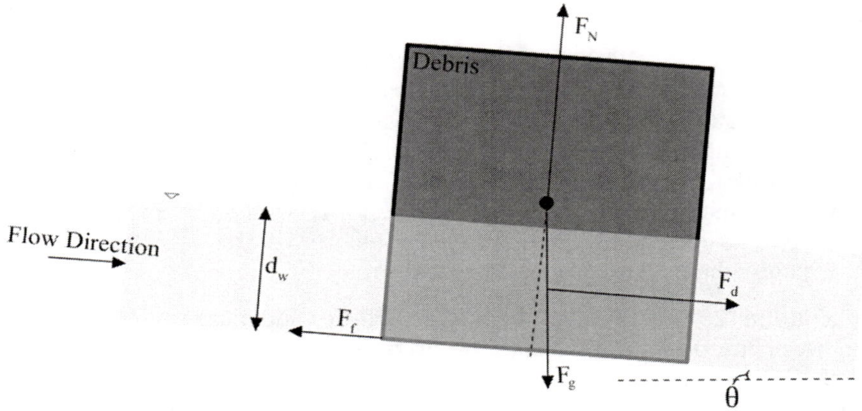

FIGURE 2. DEBRIS ENTRAINMENT FORCE BALANCE [ADAPTED FROM **Braudrick and Grant (2000)**].

The model developed by Imamura et al. (2008) was based on a force balance of the boulder in contact with the ground. The forces to be considered for the boulder transport are the hydraulic force, the frictional force, and the component of the gravitational force on the slope. The balance of these forces resulted in:

$$\rho_s k_r d^3 X'' = \frac{1}{2} C_D \rho_f (U - v)\,|U - u| \left(k_r d^2\right) + C_M \rho_f U' \left(kd^3\right) - (C_M - 1)\,\rho_f u' \left(k_r d^3\right) - k_r F_b - k_r F_g, \qquad (1)$$

$$F_b = \frac{\mu \left(\rho_s - \rho_f\right) k_r d^3 g \cos\theta\, X'}{|X'|}, \qquad (2)$$

$$F_g = \left(\rho_s - \rho_f\right) kd^3 g \sin\theta, \qquad (3)$$

where ρ_s is the density of the boulder, k_r is the ratio between the long axis and short axis of the boulder, $d$is the length of the short axis, u' is the acceleration of the boulder, C_D is the drag coefficient, ρ_f is the density of the fluid, U is the current velocity, u is the velocity of the boulder, θ is the angle of topographic slope, μ is the coefficient of friction, and C_M is the mass coefficient. This model tends to underestimate the propagation of the boulders, likely because the friction was always considered (even when saltation or rolling occurred) and the model does not consider the initial pivoting of the boulder, influencing thus the exposed surface area of the debris.

Matsutomi et al. (2008) examined the correlation between debris concentration and debris velocity as well as the hydrodynamic conditions in a dam-break flow. Debris concentration was expressed as the void ratio (1 – plan area of debris/area of flume bed). An increase in the debris concentration resulted in increased flow resistance in the bore front. The flow resistance in turn increased bore depth as well as decreased debris velocity and bore front propagation. The debris velocity was found to be always less than or equal to the bore front velocity.

Matsutomi (2009) evaluated the motion of driftwood pieces in steady-state, high-velocity flow to determine the probability of the driftwood colliding with a structure. The probability of impact was determined based on the lateral diffusion (*y*-direction) of the driftwood as the driftwood propagated downstream, therefore assuming that the structure would be in the center of the flume. The Gaussian probability distribution of the driftwood location (K_y) in the *x*- and *y*-direction was expressed as:

$$K_y(x, y) = \frac{1}{\sqrt{2\pi\overline{\delta_y}}} \exp\left(-\frac{(y-\overline{y})^2}{2\overline{\delta_y}^2}\right), \tag{4}$$

where δy¯¯¯δy¯ is the variance as a function of *x* (flow direction), roughly expressed as:

$$\frac{\overline{\delta_y}^2(x)}{L_w^2} = \frac{\frac{1}{n}\sum_{i=1}^{n}(y-\overline{y})}{L_w^2} = a\left(\frac{x}{L_w}\right)^b, \tag{5}$$

where L_w is the length of the driftwood, *a* and *b* are a function of the debris' physical properties and geometry. Extensive work is still needed in the classification of the *a* and *b* coefficients due to the variety of debris that are potentially present during a tsunami (Naito et al., 2014). The mere utilization of wooden debris with uniform draft restrains the validity of such approach in relation with different material debris.

As a preliminary investigation of debris motion in tsunami-like flow conditions, Yao et al. (2014) evaluated the motion of scaled-down coastal house models in flow conditions developed by the shoaling and subsequent breaking of a solitary wave over a sloping bed. Their study examined the maximum debris inundation compared to the limit of the maximum flow inundation. The debris initially propagated within the overflow of the bore due to high flow velocities and water depths. As the overflow approached maximum inundation, the flow velocity and depth decreased, resulting in the debris contacting the bed and slowing down, falling thus behind the advancing bore front. In the case of smaller offshore waves, the debris fell behind the bore front earlier due to an earlier contact with the bed. For the most part, the debris were unaffected by the receding flow, except when they had grounded significantly earlier and were thus much closer to the coastline. The receding wave, in this case, pulled the debris further seaward than their initial position.

Recent improvements in non-intrusive laboratory techniques for tracking debris motion have allowed for more a more detailed evaluation of the intermediate variables important for debris motion, such as changes in their orientation during displacement as well as their velocity (Goseberg et al., 2016a; Stolle et al., 2016). Rueben et al. (2014) used a novel camera-based tracking algorithm to assess the repeatability of the motion of square boxes on a sloped bed. The camera-based tracking algorithm was developed to identify a pattern of dots painted on the top of each box. The number of dots drawn on the top of each box was the identifier of each box as well as the orientation. The algorithm performed well for experiments conducted at larger scale comparing to smaller scale experiments for which the algorithm failed to identify the pattern on the top of each box.

Rueben et al. (2014) found the motion of the debris was generally 1-D and repeatable during the onshore direction phase. However, once a debris reached its maximum inundation, it would then ground and be subsequently washed seaward by the receding flow. The trajectory of the debris in the receding stage was significantly more variable, possibly due to eddies induced by the grounded debris within the returning flow.

Rueben et al. (2014) additionally examined the effect of multiple debris and fixed obstacles on debris motion. For multiple debris, the motion remained qualitatively similar to that of the single debris case. However, the presence of more debris led to the grounding point being significantly closer to the shore and the peak onshore velocity also occurred later. Alternatively, in offshore direction, peak velocities were similar to the case of single debris experiments. This is likely due to less disturbance induced in the local flow fields by debris as they had already dispersed. To examine the effect of obstacles, a fixed box was placed in front of the debris to initiate a forced rotation. The rotation of the debris resulted in a significantly more random trajectory and grounding point. The obstacle also reduced the peak onshore velocity of the debris, likely as a result of the energy lost during their rotation.

Shafiei et al. (2016) used a sensor-based tracking system, which recorded the accelerations of debris within dam-break flow conditions. The acceleration was then integrated over time to obtain debris velocity. Using a force balance and based on the assumption that debris entrainment begins after the leading edge of the bore passes the debris and that average stream-wise velocity behind the bore is constant, the following equation can be derived regarding the velocity profile (u) of the debris propagation:

$$u(t) = U - \left(\frac{C_d \rho_f A_d}{2 m_d} t + \frac{1}{U} \right)^{-1} \tag{6}$$

where U is the bore velocity, C_d is the drag coefficient, ρ_f is the density of fluid, A_d is the area of the debris projected to the incoming bore direction, m_d is the mass of the debris, and t is the debris travel time. The study showed that the debris velocity matched well to the proposed model and indicated the limit of Eq. 6 is the bore velocity.

Debris Impact

The primary approach to the modeling of debris loads has been on the impact force of a single debris on a structure (FEMA, 2012). As a result, the focus of current tsunami guidelines solely considers the impact of single debris; the likely occurrence of multiple, simultaneous debris impacts has not yet been evaluated. Nouri et al. (2010) examined the impact of a single wooden log on a structure. The debris impact resulted in an increase of the peak forces acting on the structure from 250 to 650 N. It was noted that while the debris impact load can be large, the dynamics of the impact are significantly different than those of the hydrodynamic loads, which are sustained. Nouri et al. (2010) also examined the impact duration of the debris and found that the impact duration was constant regardless of variation in log mass and velocity.

The majority of debris impact research has gone into determining maximum debris impact forces, as the maximum force will conservatively be included in the future design guidelines. The determination of the debris impact load has been based on the one-degree-of-freedom model proposed by Haehnel and Daly (2002) (Figure 3). Where the structure is considered to be rigid, the impact zone is considered to have a stiffness (k_i), the debris propagating at velocity (u_d) with a stiffness (k_d) impacts the zone causing a net displacement of x_0. Due to the relatively short duration of the impact, the damping has generally been ignored.

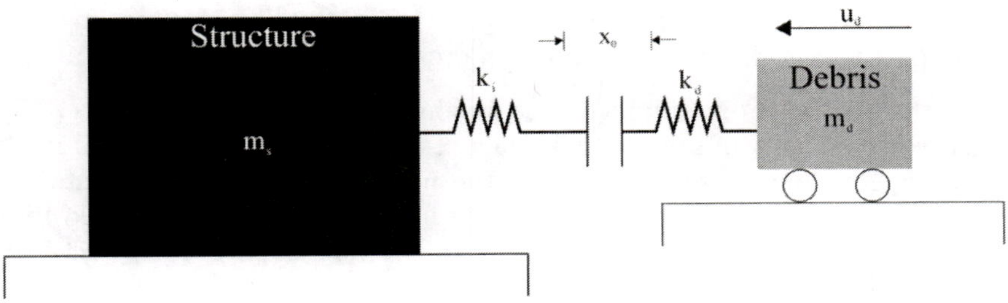

FIGURE 3. SCHEMATIC OF THE ONE-DEGREE-OF-FREEDOM MODEL PROPOSED BY **Haehnel and Daly (2002).**

There are several methods of solving for the maximum impact force, the most common approach being the contact-stiffness method, which is the approach used by the FEMA P646 (FEMA, 2012). The basis for this prescription is given by a debris modeled as a log impacting a rigid structure (Haehnel and Daly, 2004). Maximum impact force is thus expressed as:

$$F_{i,\max} = u\sqrt{k\left(m_d + C_M m_f\right)}, \tag{7}$$

where *u* is the velocity of the debris, *k* is the effective contact stiffness $\left(k = \frac{1}{k_i} + \frac{1}{k_d}\right)$, is the mass of the debris, C_M is the added mass coefficient (dependent on debris geometry and density), and m_f is the mass of displaced fluid. In considering the maximum impact force, Matskevitch (1997), in a study of ice impacts, included a reduction in the impact force as a result of the eccentricity (*e*) and Haehnel and Daly (2004) considered the obliqueness (β) of the impact.

$$e = \frac{1}{\sqrt{1 + \left(\frac{\varepsilon_0}{r_i}\right)\left(1 + \mu\left(\frac{r_0}{\varepsilon_0}\right)\right)}}, \tag{8}$$

$$\beta = \sin\varphi, \tag{9}$$

where ε_0 is the distance from the center of gravity of the debris to the point of impact, r_i is the radius of gyration of the debris, μ is the coefficient of friction between the debris and structure, r_0 is the radius of the log, and φ is the angle of impact relative to the log surface. Haehnel and Daly (2004) examined the effects of impact eccentricity, the distance from the impact axis to the center of gravity of the debris, and obliqueness, angle of the impact face to the structural face, on the impact force.

From the analysis of the impact force, Haehnel and Daly (2004) derived Eq. 10. The eccentricity of the debris impact results in a decrease in the impact force proportional to the distance of the impact away from the center of gravity of the debris. The obliqueness of impact generally decreased in a sine function from the 90° impact (long axis perpendicular to the structure). However, for a 0° angle of impact, the impact force was maximum. Although eccentricity resulted in a decrease in the maximum impact force, torque around the vertical axis of a structural member might be a side-product and no guidance exists at present to address such an additional load.

$$F_{i,\max} = e\beta u\sqrt{k\left(m_D + C_D m_f\right)}. \tag{10}$$

Riggs et al. (2014) evaluated the added mass coefficient by examining the impact of an aluminum box on a structure in-air and in-water. The in-water tests were performed by connecting an aluminum specimen to guidewires within a flume. The flume used a long wave generator to generate a wave that propagated the aluminum specimen along the guidewire toward the structure. The in-air tests were performed by connecting the aluminum specimen to a pendulum system that accelerated the specimen to a velocity matching the velocity in the in-water tests. The study found very little difference in the peak impact force between the in-air and in-water tests. Moreover, Riggs et al. (2014) determined that the difference in impact force between the in-air and in-water tests was unaffected by the debris impact velocity. Based on these findings, Riggs et al. (2014) did not support the use of the added mass coefficient in the evaluation of debris impact force. However, Shafiei et al. (2016) performed a similar study examining the effect of the added mass coefficient with denser debris and found the peak impact force to be up to 1.5 times greater in-water than in-air. More research is needed to determine if the added mass coefficient is necessary in the design for debris impact at full scale.

Aghl et al. (2014) expanded upon the work by Riggs et al. (2014) to develop a 1-D bar model that accurately estimated the impulse demand of a debris impact event. The model considered the structure to be rigid, therefore the impact force which was fully dependent on the properties of the debris. The impact force (F) was derived from the 1-D wave equation, assuming that the debris responds uniaxially to impact (Eq. 11). The impact force equation was very similar to that of the contact-stiffness approach (Eq. 7), without the added mass of the fluid. For the 1-D wave equation to be correctly evaluated, the stiffness of the debris must be considered as the equivalent stiffness of a 1-D bar (k_d) (Eq. 12).

$$F = u\sqrt{k_d m_d}, \tag{11}$$

$$k_d = \frac{EA_d}{L_d}, \tag{12}$$

where u is the impact velocity, m_d is the mass of the debris, E is the Young's modulus, A_d is the cross-sectional area of the debris, and L_d is the length of the debris. The derived formula also resulted in a constant impact force for the duration of the elastic impact, resulting in a rectangular impact pulse with a duration of t_d.

$$t_d = 2\sqrt{\frac{m_d}{k_d}}. \tag{13}$$

Aghl et al. (2014) evaluated the 1-D bar model with in-air experiments performed by accelerating debris (at full scale) to impact structures using a pendulum system. The experiments were performed with a wood pole, a steel tube, and shipping containers. The peak impact force was demonstrated to be within 5% for all the impact experiments when compared to the model equation. The impact duration was also well predicted in the cases of direct or close to direct impact of the shipping container. However, as the container impacted at the corners, the impact was not elastic, as was modeled, resulting in a significant difference between the analytical and experimental impact duration. Each debris impact force–time history had a characteristic impulse shape for the different debris types: half-sine (wood pole), rectangular (steel tube), and trapezoidal (shipping container). For all experiments, the impulse (area under the curve) for the impact model was less than the experimental results, indicating that the impact model is a conservative estimate of the impulse.

Ikeno et al. (2016) performed large-scale experiments evaluating the impact of large wood logs using a dam-break hydraulic boundary condition as well as in-air testing. The study compared the results of both test conditions to available debris impact models. The authors noted that the impact force using a hydraulic boundary condition had a significantly lower impact force than the in-air experiments as well as the impact models. As a result, they considered the potential effects of water cushioning between the wood-log and the impact surface. Additionally, Ikeno et al. (2016) considered the effect of oblique collisions using the transformation of kinetic energy to rotational energy during oblique debris impacts by defining a reduction coefficient λ:

$$\lambda = \frac{1 + \left(\frac{\varepsilon_0}{r}\right)^2 \cos^2\theta}{1 + \left(\frac{\varepsilon_0}{r}\right)^2},$$

(14)

where θ is the collision angle between the debris face and the impact surface. The authors noted that the improved equation still overpredicted the experimental results, though it did not reproduce the reduction in force observed in the experimental data.

The ASCE7–Chapter 6: Tsunami Loads and Effects is the first guideline which approaches the design of coastal structures in tsunami-prone areas in mandatory language (Chock, 2016). The code uses the approach outlined by Aghl et al. (2014) for calculating the maximum impact force and duration. In the event that the structure being designed is in close proximity to debris sources with large debris, such as shipping containers and shipping vessels, the code also uses the approach proposed by Naito et al. (2014)to determine if the structure must be designed for the larger debris impact loads.

Debris Damming

Debris entrained within the flow can result in additional loads and effects on structures, particularly when the debris forms a "dam" in front of the structure or between columns, referred to as debris damming or accumulation (Robertson et al., 2007). The hydrodynamic force on a structure can be estimated by using a similar equation to the drag force (F_D) (Bremm et al., 2015):

$$F_D = \frac{1}{2} C_D \rho_f A U^2,$$

(15)

where C_D is the drag coefficient, ρ_f is the density of the fluid, A is the cross-sectional area, and U is the flow velocity. The debris dam created by the debris will increase the cross-sectional area (A), which will in turn increase the hydrodynamic force linearly (Yeh et al., 2014). Alternatively, the properties of the dam will influence the drag coefficient as well as the surrounding hydrodynamic conditions (Parola, 2000).

While very little research has gone into debris damming in tsunami flows, research has been done within river engineering. Debris damming, in river flows, has been extensively studied due to the buildup of debris at bridge piers, depending on the shape of the structure the drag coefficient may vary as a result of the dam (Parola, 2000). The dam creates a blockage, further restring the flow path, the blockage results in a large increase in the flow depth. The debris dams also can result in increased scour as they tend to redirect the flow pattern resulting in increased scour underneath the debris dam (Pagliara and Carnacina, 2013). Due to the more diverse nature of debris in tsunami flows and due to different obstacle settings found in tsunami inundation areas, further research is expected to clarify whether river engineering findings on debris damming can be applicable to tsunami engineering.

Numerical Modeling of Debris

The numerical modeling of fluid–structure interactions (FSI) is a common topic in several fields of engineering. One of the main issues is that the numerical model needs to be able to reproduce all the physically relevant scales that affect the FSI (Canelas et al., 2015). However, the physical scales that affect the interaction are often not clear, resulting in these models requiring a high level spatial and temporal resolution. This difficulty has led to attempts to simplify the FSI by providing one-way coupling between model elements: fluid–structure or fluid–air–structure. One-way solid–fluid coupling considers the solids as massless marker particles that moved unconstrained on the water surface (Wu et al., 2014). One-way fluid–solid coupling causes the solids to move in relation to the fluid without the solids affecting the fluid. The one-way coupling methods can provide good results if the scale of the interaction between the phases is disproportionate in one direction, however, for many cases this is not the case. Two-way dynamic coupling of fluid and solid numerical solvers have become increasingly popular as computational resources are able nowadays to handle the significant computational demands of the two-way models. However, the development is still very much in its early stages.

Wu et al. (2014) used the Navier–Stokes equations coupled with the VOF free-surface tracking technique and a large eddy simulation turbulence model to calculate the flow field around the solids. Additionally, Wu et al. (2014) used partial cell treatment (PCT) to locate the faces of the solids. The basic principle of the PCT is similar to the VOF method where each cell was assigned a value between 0 and 1 indicating the phases present in the cell. When the solid phase was present within the cell, the cell is assigned a porosity that reduces the effective volume of the cell, and in the case where the cell was completely solid, the cell was removed from the Navier–Stokes equation calculations. The calculated porosity was used to adjust fluxing quantities, such as momentum and mass. The motion of the floating body was determined using the discrete element method (DEM) to calculate the translation and rotation. The translation was calculated using Newton's equations of motion, and the rotation was calculated using Euler's equations. The forces on the solid were calculated by integrating the fluid pressures on the surface of the solid.

The model from Wu et al. (2014) was validated using two laboratory experiments: a positively buoyant box in a tank of water (0.15 m × 0.14 m × 0.14 m) and a negatively buoyant box (0.02 m × 0.02 m × 0.02 m) in a tank of water. The boxes were released in the tank, and the motion of the box was tracked using cameras. Generally, good comparison between the numerical and experimental results was observed, with a maximum difference in displacement of 0.0044 m over a 0.06 m water column. The difference was due to the numerical simulation showing the negatively buoyant box rebounding off the bottom of the tank whereas the experimental results showed no such phenomena.

Smoothed-particle hydrodynamics (SPH) has been increasingly used in the modeling of free-surface flows (Gomez-Gesteira et al., 2012) and has been shown to be able to simultaneously deal with multiple body dynamics (Amicarelli et al., 2015). However, few current models can capably handle the transport of moving bodies in free-surface flow. Primarily, the current models have modeled the moving bodies as a rigid body of moving particle simulations (MPS) fluid particles with an imposed rigidity (Canelas et al., 2013). While the models have achieved generally good results (Manenti et al., 2008; Rogers et al., 2009), the modeling of the interactions between the body and the fluid was not based on rigid body contact laws (Canelas et al., 2013).

Canelas et al. (2013) incorporated a DEM model where the interparticle rigid body forces are taken from rigid body contact laws. The boundary particles of the rigid body are taken as fluid particle allowing the DEM model to be coupled with the SPH model. The contact force between the fluid and rigid body particles is decomposed into a repulsion force, which also takes into consideration the deformation of the particle, and a damping force, which takes in consideration for energy lost during deformation. The coupled model was validated using a dam-break experiment entraining PVC cubes. While the model and the experimental results had qualitatively similar results, the study noted that that the bottom friction was not properly modeled resulting in the motion of the debris (particularly of the bottom box) to lag behind the experimental results. Moreover, due to the high computational cost associated with the coupled model resulted in the particle resolution being too low to properly resolve the water surface and bore profile.

Amicarelli et al. (2015) modeled the moving bodies using the MPS method. However, the authors used a modified boundary condition, originally proposed by Adami et al. (2012), for a more stable pressure gradient around the moving bodies. The solid–solid interactions were also adjusted by the addition of a coefficient to the repulsive boundary conditions proposed by Monaghan (2005) to better preserve global momentum and kinetic energy through the body–body interactions. The SPH model was validated using a 2D wedge falling into a water tank. The study compared the acceleration of the experimental and numerical model with generally good accuracy. Pressure fluctuations were common throughout the experimental runs, which is a commonly noted problem with SPH models (Gomez-Gesteira et al., 2012; St-Germain et al., 2013).

The numerical model used in Amicarelli et al. (2015) was also used to qualitatively examine 3-D test cases where debris interacted with a bore front and two sets of obstacles. The bore was developed using a dam-break, and the debris was tracked using cameras. The numerical model resulted in good results when comparing both the trajectory and the orientation of the debris was also relatively well reproduced. The presence of the obstacles resulted in the formation of recirculation zones in front of the obstacles furthest downstream, and the body dynamics were maintained through this highly non-linear zone.

Canelas et al. (2015) presented the most recent version of the DualSPHysics with its many improvements on current modeling of fluid–solid interactions. The primary difference in the model presented by Canelas et al. (2015) was the addition of a δ-SPH term to the continuity equation which helps in the interface description between the solid and fluid phase. The rapid change in the density at the interface results in the pressure and density fluctuations that can be seen in many SPH fluid–solid modeling. The SPH model showed promising results when examining the rising of a submerged positively buoyant cylinder. The SPH model compared well to the analytical solution for the linear acceleration when the Reynold's number was laminar and showed a noticeable shift in acceleration as the boundary layer transitioned to turbulent flow. The model also showed the stabilization of the density fields indicating that the δ-SPH term helping handle that particular common problem.

While the work by Canelas et al. (2015) and Amicarelli et al. (2015) showed a lot of promise in the SPH fluid–solid modeling, there are still many issues that have yet to be thoroughly evaluated. Amicarelli et al. (2015) presented a method of handling solid–solid interactions, yet, the contact mechanics were not fully evaluated. Canelas et al. (2013) coupled the SPH model with a DEM; however, the large computational cost associated with the coupling makes the use of the coupled model unfeasible for many studies. The addition of solid deformation and inelastic collisions would also greatly improve the applicability of the model throughout coastal and hydraulic engineering.

CRITICAL REVIEW AND FUTURE RESEARCH DIRECTIONS

The above presented literature review outlines the state-of-the-art knowledge on debris-induced loading and associated effects in the context of extreme hydrodynamic flows as arising from natural disasters, such as storm surges, tsunami, and flash floods. The current body of literature covers fundamental processes of impacts on vertical structures, either derived from experimental or analytical strategies. To date, this knowledge has started to improve guidelines and standards written in mandatory language, yet many aspects of the problem of debris exerting forces on or interacting with the build environment remain unclear.

The study of debris within the context of tsunami engineering has been difficult due to the random nature of debris transport (Matsutomi, 2009). Determining aspects of debris dynamics from post-event engineering survey has been challenging due to a lack of documentation regarding potential debris sources (Naito et al., 2014). Moreover, determining impact forces without the flow conditions at the time of impact makes drawing conclusions from impact sites equally challenging. Due to these challenges, along with the relative rare occurrence of tsunami events, available field data to be compared to experimental and numerical modeling results are limited to few post-tsunami forensic engineering surveys. Therefore, the determination of debris dynamics and impact loads has primarily been performed in an experimental setting.

The study of tsunami flows and their interaction with coastal infrastructure in a laboratory setting has fundamental scaling issues related to the period of a tsunami wave. Historically, the study of tsunami loads was performed using a broken solitary wave as the hydrodynamic boundary condition (Arnason et al., 2009). However, Madsen et al. (2008) showed that the spatial and temporal duration of a solitary wave was not on the same order of magnitude of observed tsunamis. Despite these observations, broken solitary waves have still be used to examine the near-shore impact of tsunami waves as well as incipient debris transport (Arnason et al., 2009; Chinnarasri et al., 2013; Yao et al., 2014; Nistor et al., 2016). Chanson (2006) indicated that the dam-break solution was a good representation of tsunami surge profile over a coastal plain. The application of the dam-break in an experimental setting has shown better results representing the period of a tsunami wave (Imamura et al., 2008; Al-Faesly et al., 2012; Shafiei et al., 2016), though large experimental facilities are needed to achieve these flow durations at an appropriate scale. Other techniques, such as the use of N-waves (Tadepalli and Synolakis, 1994), cnoidal waves (Synolakis et al., 1988), and pump-driven long wave generation (Goseberg et al., 2013), have been shown to better represent tsunami inundation temporal features in a variety of cases. However, these experiments were performed at small scales. In determining maximum debris loads, the hydrodynamic boundary conditions must be carefully considered within the experimental procedure; it is generally recommended to aim at scales as large as possible to accurately model debris impact processes (Chock, 2016).

The problem of experimental scales extends to debris dynamics and impact loads, where little research has been done to determine minimum scales at which experiments can be performed. Studies of debris transport have mentioned the effects of turbulent eddies on debris transport (Rueben et al., 2014; Yao et al., 2014). However, as with most coastal physical models, studies are often scaled using the Froude number. The scaling of a model using the Froude number often does not adequately scale turbulent length scales (She and Leveque, 1994), resulting in an unproportioned effect of turbulence on the debris transport. Another scaling concern for debris transport is the effect of viscosity on the transport of debris at small scales. Similar to the modeling of dam-breaks (Lauber and Hager, 1998), there is likely a scale at which the viscosity of the fluid will have a significant effect on debris transport, and hence, the scaling of physical properties using Froude would no longer be adequate. A determination of the minimum scales required would help evaluate the applicability of studies to debris transport problems.

Additionally, the scaling of the debris for impact loads must also be taken into consideration. Studies of impact loads have often focused on scaling the physical properties of the debris, such as length, width, height, and mass; however, little research has been performed in examining the scaling of mechanical properties, such as stiffness and elasticity, which is critical in determining maximum impact loads. The minimum scales at which these experiments can be performed would be dependent on the material as the study must consider the loads at which the debris maintains elastic properties. A determination of the minimum scales would help identify experimental facilities capable of modeling debris impact loads. Moreover, the scaling of the plastic region of impact needs to consider the effect of debris deformation on maximum impact forces, building upon the studies by Aghl et al. (2015). Other physical properties, such as the draft of the debris, also need to be considered, as was shown in Shafiei et al. (2016) where the draft of the debris had a significant effect on the added mass coefficient.

In overcoming scaling issues related to debris dynamics, the physical modeling of debris is essential to provide benchmarking data for the numerical models, particularly due to the lack of available field or prototype data. Numerical modeling has made recent strides in the development of two-way coupled fluid–solid interactions. The availability of high temporal and spatial resolution of the various drivers of debris transport and impact would allow for numerical models to be accurately validated and calibrated. In particular, the momentum transfer between the fluid and free-floating solids needs to be thoroughly evaluated to be applied to numerical models. The momentum transfer between the fluid and the solid objects, as well as the reciprocal effects, is critical in resolving the entrainment and transport of debris within extreme hydrodynamic events.

In the determination of debris impact loads and effects, the primary objective is to apply the findings of the research to propose accurate methods of determining design loads on structures in extreme hydrodynamic events. One of the difficult aspects in designing for extreme hydrodynamic events, like tsunamis, is determining the design conditions and maximum loads associated with such rare events. Commonly, the design conditions for tsunamis are taken from historical maximum tsunamis (Okada et al., 2005). However, as observed during the 2011 Tohoku Tsunami, historical maximums may not always provide a measure of maximum conditions (Esteban et al., 2015). As a result, recent efforts have been made to implement probabilistic tsunami hazard analysis and tsunami-resistant design (Chock, 2015).

Assessing debris impact is currently considered as a deterministic design condition (Chock, 2016), despite the motion of debris being a random process (Matsutomi, 2009). The FEMA P-55 and FEMA P-646 (2012) guidelines maintain a conservative approach where debris impact should always be considered. The upcoming ASCE7 Chapter 6 also maintains a conservative approach when referring to debris ubiquitous to coastal areas, such as hydro poles and concrete debris (Chock, 2016). However, for larger debris, the ASCE7 uses the empirical approach proposed by Naito et al. (2014), based on a limited data set. The limited data set was due to difficulties in assessing debris sources in the aftermath of the 2011 Tohoku Tsunami, limiting the data set to debris with clearly defined debris sources, such as shipping vessels and shipping containers.

To better design structures for debris impact within the probabilistic assessment of tsunami hazard and account for the random nature of debris motion, the probabilistic assessment of debris motion would improve the quantification of debris loads. Considering the effect of the proximity of the debris site, the debris' physical properties, local topography, surround obstructions, and potential flow conditions on debris dynamics and entrainment would also help assess the likelihood of debris impact, as well as identify the type of debris impacting a structure.

Additional consideration is needed to quantity of debris entrained within the flow. Currently, design guidelines only consider the possibility of single debris impacts. However, in a study of the transport of shipping containers in extreme hydrodynamic events, Nistor et al. (2016) determined that the debris often tended to propagate as an agglomeration. Based on this finding, an assessment of multiple debris impact needs to be studied to determine if building standards need to consider multiple debris impacts as the critical load. Additionally, the agglomeration of debris would also increase the risk of debris accumulating at the upstream face of structures, therefore increasing the risk of debris damming.

CONCLUSIONS

Several post-tsunami forensic field surveys over the past decade have led to increased awareness about loads associated with tsunami-induced coastal inundation. One of the loads identified from these field surveys is debris loads, where objects entrained within the flow can impact and accumulate onto structures, causing supplementary loads in addition to the previously considered hydraulic ones. Due to the random nature of debris motion and the relative rarity of tsunami events, the assessment of debris motion loads in the field has been limited. However, recent advancements in the determination of debris dynamics using an experimental setting have allowed for improvements in the methods available for the assessment of mechanisms of debris load as well as of the potential maximum impact loads. Based on an extensive literature review of debris dynamics in extreme hydrodynamic conditions, the following conclusions can be drawn:

1. An assessment of the lower bounds for experimental scales and conditions is needed to determine the type of experimental facilities necessary to accurately assess debris loading conditions. Consideration is needed in scaling of the debris mechanical properties, such as their stiffness and/or elasticity. This is particularly necessary when assessing their plastic deformation, as these properties vary widely and would have to be accounted for in scaled models.

2. As there are several means of reproducing hydrodynamic boundary conditions used in tsunami engineering, the hydrodynamic forcing factor in an experimental setting needs to consider what the critical condition is when assessing debris loads.

3. Rigorous, high-resolution studies are needed to provide detailed, tailored benchmark data sets to aid in the calibration and validation of numerical models developed for debris impact.

4. Fundamental research is needed on the various factors governing the actual debris entrainment and the solid–solid/solid–fluid

momentum transfer which are critical for validating numerical models.

5. Due to the random nature of debris motion and recent consideration to probabilistic tsunami hazards design, a probabilistic design approach is needed to assess the likelihood of debris loads occurring, which would be a function of the proximity of debris sources, debris properties, and surrounding environment.
6. It is important to evaluate of multiple impact events, as well as the increased likelihood of debris damming as they propagate as an agglomeration.

AUTHOR CONTRIBUTIONS

IN–lead author, leader of the research project, and wrote the critical discussion. NG–author, part of collaboration, and helped in the writing of both the review and critical discussion. JS–author, graduate student working on debris project, and helped in writing of the review and critical discussion.

CONFLICT OF INTEREST STATEMENT

The authors declare that the research was conducted in the absence of any commercial or financial relationships that could be construed as a potential conflict of interest.

FUNDING

The authors acknowledge the financial support by the Strategic Research Foundation Grant-aided Project for Private Universities (No. S1311028) from Japanese Ministry of Education and by Waseda University, of the NSERC Discovery Grant (Canada) and of the Kajima Foundation (Japan). NG acknowledges that this research was supported by a Marie Curie International Outgoing Fellowship within the 7th European Community Framework Programme (No. 622214).

REFERENCES

1. Adami, S., Hu, X., and Adams, N. (2012). A generalized wall boundary condition for smoothed particle hydrodynamics. *J. Comput. Phys.*231, 7057–7075.
2. Aghl, P., Naito, C., and Riggs, H. (2014). Full-scale experimental study of impact demands resulting from high mass, low velocity debris. *J. Struct. Eng.* 140, 04014006.
3. Aghl, P. P., Naito, C., and Riggs, H. (2015). Estimation of demands resulting from inelastic axial impact of steel debris. *Eng. Struct.* 82, 11–21.
4. Al-Faesly, T., Palermo, D., Nistor, I., and Cornett, A. (2012). Experimental modeling of extreme hydrodynamic forces on structural models. *Int. J. Protect. Struct.* 3, 477–506.
5. Amicarelli, A., Albano, R., Mirauda, D., Agate, G., Sole, A., and Guandalini, R. (2015). A smoothed particle hydrodynamics model for 3D solid body transport in free surface flows. *Comput. Fluids* 116, 205–228.
6. Arnason, H., Petroff, C., and Yeh, H. (2009). Tsunami bore impingement onto a vertical column. *J. Disaster Res.* 4, 391–403.
7. Bocchiola, D., Rulli, M., and Rosso, R. (2006). Transport of large woody debris in the presence of obstacles. *Geomorphology* 76, 166–178.
8. Borrero, J. C., Sieh, K., Chlieh, M., and Synolakis, C. E. (2006). Tsunami inundation modeling for western Sumatra. *Proc. Natl. Acad. Sci. U.S.A.* 103, 19673–19677.
9. Braudrick, C. A., and Grant, G. E. (2000). When do logs move in rivers? *Water Resour. Res.* 36, 571–583.
10. Bremm, G. C., Goseberg, N., Schlurmann, T., and Nistor, I. (2015). Long wave flow interaction with a single square structure on a sloping beach. *J. Mar. Sci. Eng.* 3, 821.
11. Canelas, R. B., Dominguez, J. M., Crespo, A. J., ómez-Gesteira, M. G., and Ferreira, R. M. (2015). A smooth particle hydrodynamics discretization for the modelling of free surface flows and rigid body dynamics. *Int. J. Numer. Methods Fluids* 78, 581–593.
12. Canelas, R. B., Ferreira, R. M., Crespo, A., and Domínguez, J. M. (2013). "A generalized SPH-DEM discretization for the modelling of complex multiphasic free surface flows," in *Proceedings of the 8th International SPHERIC Workshop*, Trondheim.
13. Chanson, H. (2006). Tsunami surges on dry coastal plains: application of dam break wave equations. *Coast. Eng. J.* 48, 355–370.
14. Charvet, I., Suppasri, A., and Imamura, F. (2014). Empirical fragility analysis of building damage caused by the 2011 Great East Japan tsunami in Ishinomaki city using ordinal regression, and influence of key geographical features. *Stoch. Environ. Res. Risk Assess.* 28, 1853–1867.
15. Chinnarasri, C., Thanasisathit, N., Ruangrassamee, A., Weesakul, S., and Lukkunaprasit, P. (2013). The impact of tsunami-induced bores on buildings. *P I Civil Eng-Mar. En.* 166, 14–24.
16. Chock, G. Y. (2015). "The ASCE 7 tsunami loads and effects design standard," in *Structures Congress 2015* (Portland, OR: ASCE), 1446–1456.
17. Chock, G. Y. (2016). Design for tsunami loads and effects in the ASCE 7-16 standard. *J. Struct. Eng.* 142, 04016093.
18. Esteban, M., Takagi, H., and Shibayama, T. (2015). *Handbook of Coastal Disaster Mitigation for Engineers and Planners*, Waltham, MA: Butterworth Heinemann, Elsevier.
19. FEMA. (2012). *P646 Guidelines for Design of Structure for Vertical Evacuation from Tsunamis*. Redwood City, CA: Federal Emergency Management Agency.
20. Fritz, H. M., Borrero, J. C., Synolakis, C. E., and Yoo, J. (2006). 2004 Indian Ocean tsunami flow velocity measurements from survivor videos. *Geophys. Res. Lett.* 33, L24605.

21. Gomez-Gesteira, M., Rogers, B. D., Crespo, A. J., Dalrymple, R., Narayanaswamy, M., and Dominguez, J. M. (2012). SPHysics – development of a free-surface fluid solver – part 1: theory and formulations. *Comput. Geosci.* 48, 289–299.
22. Goseberg, N., Nistor, I., Mikami, T., Shibayama, T., and Stolle, J. (2016a). Nonintrusive spatiotemporal smart debris tracking in turbulent flows with application to debris-laden tsunami inundation. *J. Hydraul. Eng.* 142, 04016058.
23. Goseberg, N., Stolle, J., Nistor, I., and Shibayama, T. (2016b). Experimental analysis of debris motion due the obstruction from fixed obstacles in tsunami-like flow conditions. *Coast. Eng.* 118, 35–49.
24. Goseberg, N., Wurpts, A., and Schlurmann, T. (2013). Laboratory-scale generation of tsunami and long waves. *Coast. Eng.* 79, 57–74.
25. Haehnel, R. B., and Daly, S. F. (2002). *Maximum Impact Force of Woody Debris on Floodplain Structures*. Hanover: USACE.
26. Haehnel, R. B., and Daly, S. F. (2004). Maximum impact force of woody debris on floodplain structures. *J. Hydraul. Eng.* 130, 112–120.
27. Ikeno, M., Takabatake, D., Kihara, N., Kaida, H., Miyagawa, Y., and Shibayama, A. (2016). Improvement of collision force formula for woody debris by airborne and hydraulic experiments. *Coast. Eng. J.* 58, 1640022.
28. Imamura, F., Goto, K., and Ohkubo, S. (2008). A numerical model for the transport of a boulder by tsunami. *J. Geophys. Res. Oceans*197, 113.
29. Lauber, G., and Hager, W. H. (1998). Experiments to dambreak wave: horizontal channel. *J. Hydraul. Eng.* 36, 291–307.
30. Madsen, P. A., Fuhrman, D. R., and Schäffer, H. A. (2008). On the solitary wave paradigm for tsunamis. *J. Geophys. Res. Oceans* 113, 1–22.
31. Manenti, S., Panizzo, A., Ruol, P., and Martinelli, L. (2008). "SPH simulation of a floating body forced by regular waves," in *Proceedings of 3rd SPHERIC Workshop* (Lausanne: Swiss Federal Institute of Technology), 38–41.
32. Matskevitch, D. (1997). Eccentric impact of an ice feature: linearized model. *Cold Reg. Sci. Technol.* 25, 159–171.
33. Matsutomi, H. (2009). Method for estimating collision force of driftwood accompanying tsunami inundation flow. *J. Disaster Res.* 4, 435–440.
34. Matsutomi, H., Fujii, M., and Yamaguchi, T. (2008). "Experiments and development of a model on the inundated flow with floating bodies," in *Proceedings of the 31st International Conference on Coastal Engineering* (Hamburg: ASCE), 1458–1470.
35. Monaghan, J. J. (2005). Smoothed particle hydrodynamics. *Rep. Progr. Phys.* 68, 1703.
36. Naito, C., Cercone, C., Riggs, H. R., and Cox, D. (2014). Procedure for site assessment of the potential for tsunami debris impact. *J. Waterw. Port Coast. Ocean Eng.* 140, 223–232.
37. Naito, C., Riggs, H., Wei, Y., and Cercone, C. (2016). Shipping-container impact assessment for tsunamis. *J. Waterw. Port Coast. Ocean Eng.* 142, 05016003.
38. Nistor, I., Goseberg, N., Mikami, T., Shibayama, T., Stolle, J., Nakamura, R., et al. (2016). Hydraulic experiments on debris dynamics over a horizontal plane. *J. Waterw. Port Coast. Ocean Eng.* 04016022.
39. Nouri, Y., Nistor, I., Palermo, D., and Cornett, A. (2010). Experimental investigation of tsunami impact on free standing structures. *Coast. Eng. J.* 52, 43–70.
40. Okada, T., Sugano, T., Ishikawa, T., Ohgi, T., Takai, S., and Hamabe, C. (2005). *Structural Design Method of Buildings for Tsunami Resistance (Proposed)*. Tokyo: Building Technology Research Institute, Building Centre for Japan, 15.
41. Pagliara, S., and Carnacina, I. (2013). Bridge pier flow field in the presence of debris accumulation. *P I Civil Eng-Water Manag.* 187–198.
42. Palermo, D., Nistor, I., Nouri, Y., and Cornett, A. (2009). Tsunami loading of near-shoreline structures: a primer. *Can. J. Civ. Eng.* 36, 1804–1815.
43. Parola, A. C. (2000). *Debris Forces on Highway Bridges*. Washington, DC: Transportation Research Board.

44. Riggs, H., Cox, D., Naito, C., Kobayashi, M., Aghl, P. P., Ko, H.-S., et al. (2014). Experimental and analytical study of water-driven debris impact forces on structures. *J. Offshore Mech. Arctic Eng.* 136, 041603.
45. Robertson, I., Riggs, H. R., Yim, S. C., and Young, Y. L. (2007). Lessons from Hurricane Katrina storm surge on bridges and buildings. *J. Waterw. Port Coast. Ocean Eng.* 133, 463–483.
46. Rogers, B. D., Dalrymple, R. A., and Stansby, P. K. (2009). "SPH modeling of floating bodies in the surf zone," in *Proc. 31st Int. Conf. Coastal Eng. 2008*. 204–215.
47. Rossetto, T., Allsop, W., Charvet, I., and Robinson, D. I. (2011). Physical modelling of tsunami using a new pneumatic wave generator. *Coast. Eng.* 58, 517–527.
48. Rueben, M., Cox, D., Holman, R., Shin, S., and Stanley, J. (2014). Optical measurements of tsunami inundation and debris movement in a large-scale wave basin. *J. Waterw. Port Coast. Ocean Eng.* 141, 04014029.
49. Saatcioglu, M., Ghobarah, A., and Nistor, I. (2005). Effects of the December 26, 2004 Sumatra earthquake and tsunami on physical infrastructure. *ISET J. Earthq. Technol.* 42, 79–94.
50. Shafiei, S., Melville, B. W., Shamseldin, A. Y., Beskhyroun, S., and Adams, K. N. (2016). Measurements of tsunami-borne debris impact on structures using an embedded accelerometer. *J. Hydraul. Res.* 54, 435–449.
51. She, Z.-S., and Leveque, E. (1994). Universal scaling laws in fully developed turbulence. *Phys. Rev. Lett.* 72, 336.
52. St-Germain, P., Nistor, I., Townsend, R., and Shibayama, T. (2013). Smoothed-particle hydrodynamics numerical modeling of structures impacted by tsunami bores. *J. Waterw. Port Coast. Ocean Eng.* 140, 66–81.
53. Stolle, J., Nistor, I., and Goseberg, N. (2016). Optical tracking of floating shipping containers in a high-velocity flow. *Coast. Eng. J.* 58, 1650005.
54. Synolakis, C. E., Deb, M. K., and Skjelbreia, J. E. (1988). The anomalous behavior of the runup of cnoidal waves. *Phys. Fluids* 31, 3–5.
55. Tadepalli, S., and Synolakis, C. E. (1994). The run-up of N-waves on sloping beaches. *Proc. R. Soc. Lond. A Math. Phys. Eng. Sci.* 445, 99–112.
56. Taubenböck, H., Goseberg, N., Lämmel, G., Setiadi, N., Schlurmann, T., Nagel, K., et al. (2013). Risk reduction at the "last-mile": an attempt to turn science into action by the example of Padang, Indonesia. *Nat. Hazards* 65, 915–945.
57. Wu, T.-R., Chu, C.-R., Huang, C.-J., Wang, C.-Y., Chien, S.-Y., and Chen, M.-Z. (2014). A two-way coupled simulation of moving solids in free-surface flows. *Comput. Fluids* 100, 347–355.
58. Yao, Y., Huang, Z., Lo, E. Y. M., and Shen, H.-T. (2014). A preliminary laboratory study of motion of floating debris generated by solitary waves running up a beach. *J. Earthq. Tsunami* 8, 1440006.
59. Yeh, H., Barbosa, A. R., Ko, H., and Cawley, J. G. (2014). Tsunami loadings on structures: review and analysis. *Coast. Eng. Proc.* 34, 1–13.
60. Yeh, H., Sato, S., and Tajima, Y. (2013). The 11 March 2011 East Japan earthquake and tsunami: tsunami effects on coastal infrastructure and buildings. *Pure Appl. Geophys.* 170, 1019–1031.

CHAPTER 2

A New Approach to Short-Term Tsunami Forecasting

Yury Korolev[1]

[1] *Institute of Marine Geology and Geophysics, Russia*

1. INTRODUCTION

A short-term tsunami forecast and effective tsunami warning is a key problem of the tsunami service. At present, the conventional method for short-term tsunami forecasting is based on seismological information only (earthquake magnitude, time of the main shock and epicentre location). An earthquake magnitude which exceeds the established threshold value, which is different for different tsunamigenic zones, typically results in the issuing of a tsunami warning (Gusiakov, 2011). This approach, based on the "magnitude-geographical principle", is straightforward and rather effective: at least it ensures few tsunami omissions. However, this method cannot provide sufficiently accurate estimates of tsunami wave heights, especially for specific coastal areas. Firstly, this is because the dependence of the tsunami intensity on the earthquake magnitude is far from being deterministic (Gusiakov, 2011): strong earthquakes can produce weak tsunamis (or no tsunamis at all) and, vice versa, sometimes destructive tsunamis are induced by relatively weak earthquakes. Secondly, because actual tsunami run-ups are strongly variable along the coast (c.f. MacInnes et al., 2009), the tsunami alarm can be justified for one coastal site, but false for another. Moreover, the magnitude-geographical warning method does not allow evaluation of the duration of the tsunami alarm or prediction of the arrival time of the largest tsunami wave (not necessarily the first, which is quite typical for tsunami events (c.f. Rabinovich et al., 2008)). Consequently, up to 75% of tsunami warnings turn out to be false

(Gusiakov, 2011; Poplavsky et al., 1997). False tsunami warnings cause anxiety among emergency managers, government officials and the business community. These false alarms result in actual financial losses, sometimes significant, including losses due to production downtime and expenses for emergency evacuation procedures and navigation of vessels to the open sea (NOAA Tsunami, 2004).

According to the modern concept, a tsunami alarm should be declared in good time and only selectively for exposed coastal areas where the incoming tsunami will be a real threat. Each tsunami alarm should be accompanied by trustworthy information on the expected tsunami arrival time, wave heights along the coast, their frequencies and the duration for dangerous oscillations to remain in effect. However, based merely on the seismological data such information cannot be obtained. Meanwhile, emergency managers and local officials are in urgent need the effective operational tools that can provide reliable tsunami forecasts as guidance for rapid, critical decisions in which lives and property are at stake (Titov et al., 2005).

From this point of view, direct tsunami monitoring in the open-ocean by DART[1] - stations can significantly improve the situation (Titov et al., 2005; Titov,2009). These deep-ocean real-time tsunameters have been developed by the Pacific Marine Environmental Laboratory (PMEL/NOAA) for the early detection, measurement and real-time reporting of far-field tsunamis. These stations have been deployed at strategic locations throughout the Pacific Ocean near the main seismically active regions with a history of tsunami generation (Mofjeld, 2009). DART tsunami monitoring systems were expected to become the backbone of the modern Tsunami Warning System (TWS) in the Pacific Ocean.

The principal methodology of the tsunami forecast based on open ocean sea level measurements was developed during 1987-2005 (Chubarov & Shokin, 1995; Chung, et al., 1995; Korolev & Poplavsky, 1995; Poplavsky et al., 1997; Satake, 1987; Titov et al., 2005; Voronina & Tcheverda, 1998; Wei et al., 2003). However, only recent advances in tsunami measurement and numerical modelling technology have made it possible to create effective tsunami forecasting systems. The tsunami forecasting technology elaborated at PMEL is based on the integration of real-time DART measurements and state-of-the-art numerical modelling (Titov et al., 2005). This technology was applied in an operative regime for several major events during 2005-2011 and was demonstrated to be highly efficient (Japan (East Coast of Honshu) Tsunami, 2011; Tang et al., 2008; Titov, 2009; Wei et al., 2008; Yamazaki et al., 2006). The maximum wave heights and other parameters of arriving tsunamis for several US sites were predicted and were later found to be in good agreement with the actual in situ observed waves (Titov, 2009). The method described in (Tsushima et al., 2011) is similar to the above method.

An inversion method proposed in (Satake, 1987) also gave good results (Fujii & Satake, 2008, 2011), however, determination of the area of a seismic centre on aftershock area for one day after the main shock is required for computation. The latter does not allow applying them in an operating mode.

It should be noted, however, that the PMEL technology is based on a *pre-computed Propagation Database* to compute a quick preliminary forecast of the ocean-wide propagation of the tsunami as a linear combination of unit sources selected to represent the initial earthquake parameters. Once the actual tsunami wave reaches a DART station, inversion is performed to adjust the slip distribution of the selected unit sources. The Propagation Database is a key component of the operational tsunami forecasting system. At present, it includes a great number of pre-computed model runs for simulated earthquake sources in the Pacific, Atlantic and Indian Oceans (Titov, 2009; see also http://nctr.pmel.noaa.gov/propagation-database.html).

In the present study we suggest a simple method for short-term tsunami forecasting that does not require pre-computed tsunami waveforms. The method is based on a reciprocity principle (c.f. Rayleigh, 1945) and uses the only seismological information about co-ordinates of the earthquake epicentre and in situ tsunami data from one of the far-field open-ocean DART stations. This method can work in a real-time mode and, therefore, be effectively used for local tsunami warnings.

The method for a short-term tsunami forecast is based on Green's function for the wave equation possessing the fundamental property of symmetry. This property is well-known in acoustics and seismology as the reciprocity principle (Rayleigh, 1945). This principle is known as an asymptotic assumption.

The first use of this principle for long water waves was probably suggested by Loomis (Loomis, 1979). The reciprocity principle was found to be effective to optimize tsunami pre-computations. However, the effects of the ocean depth inhomogeneity had not been analysed previously. In view of large horizontal scales of tsunami sources, these effects can be significant. As a result of a study of the reciprocity principle and asymptotic conditions, the similarity conditions for reciprocal sources of long waves were obtained. Some applications of this principle to the tsunami research are considered in the current study. The numerical simulation was made using actual bathymetry. It was shown in numerical experiments that, with a quality sufficient for practical applications, the reciprocity principle works quite well, even when asymptotic conditions are not totally satisfied (i.e., when the characteristic time of the wave propagation between the reciprocity sources is only three times greater than the typical wave period and when the characteristic wave length is comparable with the size of the source area).

This paper presents a method of short-term tsunami forecasting using open-ocean sea level data from distant sites based on the reciprocity principle. The simple relationship (estimated transfer functions) enabled us to simulate tsunami waveforms for any selected oceanic point based only on the source location and sea level data from a remote reference site (Korolev, 2004, 2005). Information about the mechanism of the earthquake, the size of the centre or earthquake magnitude is not required.

This relationship (transfer function) consists of Laplace (Fourier) transforms. To obtain the target result, the inverse Laplace (Fourier) transform must be made. This problem is an ill-posed one. Two methods of inverse Laplace transform are described.

The method enables the transfer function to be evaluated during the event immediately after obtaining information about the earthquake epicentre location. The method enables us to compute expected tsunami waveforms in real-time mode and for any given ocean site. In contrast to the PMEL/NOAA technology, which is based on a pre-computed Propagation Database that needs updating for newly deployed open-ocean stations, the present method does not require pre-computed tsunami waveforms. The important advantage of this method is that it can be used irrespective of the actual source mechanism (seismic, submarine landslide or other phenomena).

The method was successfully applied to hindcast several recent tsunamis observed in the Pacific. The locations of the earthquake epicentres and the tsunami records from one of the NOAA DART sites were used as inputs for the modelling, while tsunami observations at other DART sites were used to verify and to estimate the effectiveness of this model.

The computed and observed tsunami waveforms for the regions of the Kuril Islands, Aleutian Islands and the US West Coast were in good agreement, satisfying the requirements of tsunami warning services (Korolev, 2011a, 2011b).

The good agreement is not only at points in open-ocean where DART stations are located, but also at points close to the coast of the Kuril Islands.

The proposed method can be considered as the basis for creating a program package that can be applied for early tsunami warnings for sites exposed to the threat of tsunamis. The method can be used for both regional and local tsunami warning services having access to the open-ocean (DART) data in a real-time mode. The result is a computational basis for decision making about issuing (confirmation) or cancelling a tsunami alarm. It is important that the method can be applied at newly developing tsunami warning centres.

2. RECIPROCITY PRINCIPLE

The reciprocity principle is a corollary of the properties of symmetry of Green's function for a linear wave equation. The Green's function is created, as the response of a medium on point-like (Dirac δ-function) disturbance. The reciprocity principle (relationship) expresses the same property of symmetry for sources with limited sizes. There are two aspects of this principle. A kinematic reciprocity principle says that the propagation time of a signal from point *A* to reciprocal point *B* is equal to the propagation time of a signal from point *B* to *A* and a dynamic reciprocity principle runs as follows: amplitudes of harmonic signals at points *B* (*A*) are identical, if the same source is located at *A* (*B*) (c.f. Rayleigh, 1945).

The validity of the reciprocity principle for an inhomogeneous medium with reflecting, absorbing and mixed (impedance) boundaries was proved later. It is valid for any area, including shadow zones and including the case of non-stationary waves (Chertock, 1970). The reciprocity principle is widely applied in hydroacoustics for hydrophone calibration (Uric, 1975).

Let us assume that a wave's process is described by the linear wave equation. Let, in the space limited by reflecting and/or absorbing boundaries, the harmonic Source **A** be located at point *A* with co-ordinates (xA,yAxA,yA), while the analogous harmonic Source **B** be at point *B* (xB,yBxB,yB). Let also the pressure amplitude on the radiating surface be *pA*, *QA=pASA*, where *SA* is the square of the radiating surface and *pA(B)* is pressure amplitude in point *B*. The same is true for Source **B**: the pressure amplitude on the radiating surface is *pB*, *QB=pBSB*, where *SB* is the square of the radiating surface and *pB(A)* is pressure amplitude at point *A*.

So the following reciprocity relationship (asymptotic equality) is valid (Brekhovskikh, 1960; Landau & Lifshitz, 1975)

$$p_A(B)\cdot Q_B = p_B(A)\cdot Q_A. \tag{1}$$

The obtained relationship is fair, if the following asymptotic conditions are valid:

a. the distance between the sources has to be significantly longer than the wave lengths;

b. horizontal scales of both sources areas have to be significantly shorter than the wave lengths.

In the case of an inhomogeneous medium, these conditions may be formulated in time terms:

a. a' the time of wave propagation between the sources should be significantly longer than typical wave periods;

b. b' the wave period should be significantly longer than typical time in the source.

Any connections of a source's parameters with parameters of a medium are not analysed. If for the acoustic processes, for example at ocean, the medium inhomogeneities are insignificant (variations of wave speed do not exceed 7%), then for tsunami waves they can give significant effects, as the speed of long waves differ by more than ten times (from 10 m/s on water depth 10 m up to 200 m/s on depth 4000 м). Considering large horizontal sizes of sources of a tsunami, the influence of these effects can be significant. Besides, the condition (b) can be defaulted, as the characteristic lengths of tsunami waves do not always exceed characteristic sizes of the centre. The problem arises as to whether or not the reciprocity principle for tsunami waves is valid. If yes, what are the conditions of validity and do they differ from conditions in acoustics? The conditions on sources should be investigated.

2.1. The Reciprocity Principle For Non-Stationary Long Waves

The deducing of the reciprocity relationship for tsunami waves is indicated in this paper with the purpose of showing analogies and differences from a traditional relationship. We follow the algorithm described in (Brekhovskikh, 1960; Landau & Lifshitz, 1975).

Let us assume that a wave's process is described by the system of linear shallow-water equations, or wave equation, with variable water depth (Stoker, 1957)

$$\frac{\partial^2 \zeta}{\partial t^2} - g\vec{\nabla} \cdot (D(x,y) \cdot \vec{\nabla} \zeta) = 0 \qquad (2)$$

where $\zeta=\zeta(t,x,y)$ – waveform, $D(x,y)$ – water depth, g – gravity acceleration, $\vec{\nabla}=(\frac{\partial}{\partial x},\frac{\partial}{\partial y})$ – – differential operator.

Let Source **A** be located within area *SA* with the centre at point *A* with co-ordinates *xA, yA,* while Source **B** be within area SBSB with the epicentre at *B* (co-ordinates *xB, yB*). For simplicity, the initial sea surface elevation is used as the wave source. *ζ(t,x,y)* is a waveform from Source **A**, *η(t,x,y)* is a waveform from Source **B**.

Initial conditions (*t*=0)

$$\zeta(0,x,y)=\zeta_0(x,y) \text{ in area } S_A \tag{3}$$

$$\eta(0,x,y)=\eta_0(x,y) \text{ in area } S_B \tag{4}$$

Boundary conditions

$$\zeta(t,x,y)=0 \qquad \text{on } \Gamma_1 \text{ (absorbing boundary)},$$

$$\zeta(t,x,y)=0 \qquad \text{on infinity } \Gamma_\infty, \tag{5}$$

$$u_n(t,x,y)=0 \qquad \text{on } \Gamma_2 \text{ (reflecting boundary)},$$

$$\alpha(x,y)u_n(t,x,y)+\beta(x,y)\zeta(t,x,y) \qquad \text{on } \Gamma_3 \text{ (mixed, or impedance, condition)}. \tag{6}$$

Here *un(t,x,y)* – normal component of mass velocity.

The same boundary conditions are applied to waveform *η* and mass velocity *wn* from Source **B**.

While tsunamis are non-stationary waves, Laplace transform is applied to equation (2) and boundary conditions (5) for a waveform from Source **A**

$$s^2\zeta(s,x,y)-g\vec{\nabla}(D(x,y)\cdot\vec{\nabla}\zeta(s,x,y))=s\zeta_0(x,y).$$

The same is applied to waveform from Source **B**

$$s^2\eta(s,x,y) - g\vec{\nabla}(D(x,y)\cdot\vec{\nabla}\eta(s,x,y)) = s\eta_0(x,y)$$

Multiplying the last but one equation by *η(s,x,y)* and the last equation by *ζ(s,x,y)* and subtracting one from another we obtain

$$g\zeta(s,x,y)\cdot\vec{\nabla}(D(x,y)\cdot\vec{\nabla}\eta(s,x,y)) - g\eta(s,x,y)\cdot\vec{\nabla}(D(x,y)\cdot\vec{\nabla}\zeta(s,x,y)) = \\ = s(\eta(s,x,y)\cdot\zeta_0(x,y) - \zeta(s,x,y)\cdot\eta_0(x,y))$$

This equation may be transformed to divergent form using relationship

$$f_1(x,y)\cdot\vec{\nabla}(D(x,y)\cdot\vec{\nabla}f_2(x,y)) = \vec{\nabla}(D(x,y)\cdot f_1(x,y)\cdot\vec{\nabla}f_2(x,y)) - D(x,y)\cdot\vec{\nabla}f_1(x,y)\cdot\vec{\nabla}f_2(x,y).$$

This yield

$$g\cdot\vec{\nabla}(D(x,y)\cdot\zeta(s,x,y)\cdot\vec{\nabla}\eta(s,x,y)) - g\cdot\vec{\nabla}(D(x,y)\cdot\eta(s,x,y)\cdot\vec{\nabla}\zeta(s,x,y)) = \\ = s\cdot(\eta(s,x,y)\cdot\zeta_0(x,y) - \zeta(s,x,y)\cdot\eta_0(x,y))$$

Then the integral is taken over space with boundaries *Γ=Γ∞+Γ1+Γ2+Γ3*. The integral of the left part of (6) is transformed into the integral along a contour *Γ=Γ∞+Γ1+Γ2+Γ3* (Korn & Korn, 1961):

$$g\cdot\oint_\Gamma D\cdot\left[\zeta\cdot\left(\frac{\partial\eta}{\partial x}dy - \frac{\partial\eta}{\partial y}dx\right) - \eta\cdot\left(\frac{\partial\zeta}{\partial x}dy - \frac{\partial\zeta}{\partial y}dx\right)\right] = \\ = -s\cdot\oint_\Gamma D\cdot(\zeta\cdot w_n - \eta\cdot u_n)\cdot dl$$

The obtained integral over *Γ∞* and *Γ1* is equal to 0 by virtue of the first and second conditions (5), the integral over *Γ2* is equal to 0 by virtue of the third condition (5). The integrand is transformed to impedance form and the integral over *Γ3* is equal to 0 by virtue of the fourth condition (5)

$$-s\cdot\int_{\Gamma_3} D\cdot(\zeta\cdot w_n-\eta\cdot u_n)\cdot dl=-s\cdot\int_{\Gamma_3}\frac{D}{\beta(x,y)}\cdot\left[\beta\cdot\zeta\cdot w_n+\alpha\cdot u_n\cdot w_n-\alpha\cdot w_n\cdot u_n-\beta\cdot\eta\cdot u_n\right]\cdot dl=$$

$$=-s\cdot\int_{\Gamma_3}\frac{D}{\beta}\cdot\left[(\alpha\cdot u_n+\beta\cdot\zeta)w_n-(\alpha\cdot w_n+\beta\cdot\eta)u_n\right]\cdot dl=0\,.$$

So, the integral of the left part of (6) is equal to 0. The integral of the right part (6) gives

$$\int_{S_A}\eta(s,x,y)\cdot\zeta_0(x,y)\cdot dS=\int_{S_B}\zeta(s,x,y)\cdot\eta_0(x,y)\cdot dS$$

Until now any suppositions, simplifying the problem, have not been done. Relationship (7) is the exact one.

The classical statement of the reciprocity principle requires the following conditions to be satisfied:

a. the distance between the sources has to be longer than the wave lengths (*x, y* scales of the source areas), i.e., the variation $\Delta\zeta$ should be significantly smaller than function $\zeta(s,x,y)$ within the area *SB*: $\Delta\zeta<<\zeta(s,x,y)$. This should also be true for the function $\eta(s,x,y)$;

a' an alternative asymptotic condition may be formulated in the time domain: typical wave periods should be significantly smaller than the time of wave propagation between the sources;

1. the horizontal (*x, y*) scales of both source areas have to be shorter than the wave lengths of the generated tsunami waves.

If conditions (a), (a') and (b) are satisfied, functions $\zeta(s,x,y)$ and $\eta(s,x,y)$ in (7) may be removed from the integrals and replaced by mean values at points *B* and *A*. Thus, an approximate (asymptotic) expression that follows from (7) is

.

$$\eta(s,x_A,y_A)\cdot\int_{S_A}\zeta_0(x,y)\cdot dxdy\approx\zeta(s,x_B,y_B)\cdot\int_{S_T}\eta_0(x,y)\cdot dxdy\,.$$

Designating

$\int\limits_{S_A} \zeta_0(x,y)\cdot dxdy = Q_A$ and $\int\limits_{S_B} \eta_0(x,y)\cdot dxdy = Q_B$, where Q_A and Q_B are the volumes of the initial disturbances, we can rewrite the last relationship as

$$\eta(s,x_A,y_A)\cdot Q_A \approx \zeta(s,x_B,y_B)\cdot Q_B$$

The form of relationship (8) matches with the form of traditional relation (1).

Due to the medium inhomogeneity (variable speed of wave propagation) and large horizontal source scales in comparison with the water depth, the initial free surface elevations should satisfy the following conditions based on common reasons of similarity and dimensions:

- the forms of the initial disturbances should be similar;
- the disturbance radii, RARA and RBRB

, are related as

$$\frac{R_A}{\sqrt{gD_A}} = \frac{R_B}{\sqrt{gD_B}},$$

where *DA* and *DB* are the water depths at points *A* and *B,* and *g* is the gravitational acceleration.

The solution of one problem confirms this relationship (see Appendix A).

The approximate (asymptotic) relationship (8) and the condition of similarity (9) is the *dynamic reciprocity principle* (relationship) for non-stationary long water waves (Poplavsky et al., 1997).

Additionally, if the disturbance volumes are identical ($aARA^2=aTRT^2$, i.e. $QA=QB$), then after the inverse Laplace transform, the waveforms will also match each other

$$\zeta_T(t,x_A,y_A) = \zeta_A(t,x_T,y_T).$$

2.2. The Analysis Of The Reciprocity Principle For Long Waves

It is of interest to identify whether relationship (10) is fair if conditions (a) and (b) are not satisfied, therefore, this will be the subject of the following study.

To verify this approach we made several numerical experiments. The numerical simulations based on actual bathymetry were conducted for the region of the Southern Kuril Islands. Source **1** was located in the Sea of Okhotsk northwest of Kunashir Island; Source **2** was located in the Yuzhno-Kurilsk Strait, between the Kunashir and Shikotan Islands; and Source **T** was located in the Pacific Ocean southeast of Shikotan Island. The computational domain for the numerical experiments indicating the location of the sources is shown in Fig. 1. The bathymetry grid for these experiments was chosen to have a spatial step

$$\Delta x = \Delta y = \textbf{100m.}$$

The parameters of the initial free surface elevations were chosen taking into account the above-mentioned conditions. They were the following:

Source 1: Max ampl. $a1 =$ 0.4 m, radius $R1 =$ 25 km, water depth in epicentre $D1$= 2418 m;

Source 2: Max ampl. $a2 =$ 15.6 m, radius $R2 =$ 4 km, water depth in epicentre $D2$ 63.6 m;

Source T: Max ampl. $aT =$ 0.49 m, radius $RT =$ 22.5 km, water depth in epicentre DT= 2030 m.

The results of these simulations are shown in Fig. 2. The correlation coefficient, ρ, was chosen as a criterion for comparison of the reciprocal waveforms.

The coincidence of waveforms according to (10) is quite good despite wave sheltering by a chain of islands. The typical wave periods are about 10-18 min; the time of wave propagation is 57 min between Sources **1** and **T**, 50 min between Sources **1** and **2**, and 40 min between Sources **2** and **T**. It is possible to make certain that wave lengths in the reciprocity source areas slightly exceed their sizes. For example, the wave length of the wave from Source **1** ($\lambda 12$) at Site **2** for the period of $T =$ 18 min and water depth $D2 =$ 63.6 m is

$$\lambda_{12} = T \cdot \sqrt{gD_2} \approx 27 \text{ km},$$

, while the diameter of Source **2** is 8 km.

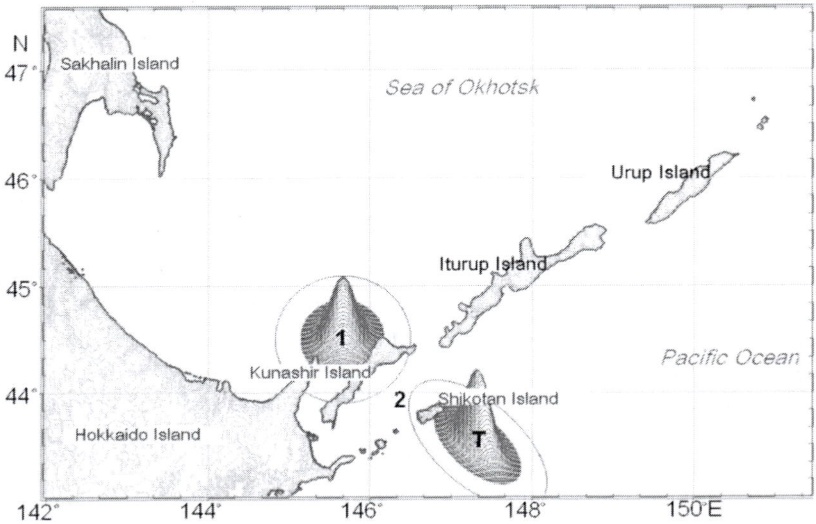

FIGURE 1. A map showing the region of the South Kuril Islands and a scheme of numerical experiments. Symbols 1, 2 and T indicate the epicentres of the sources. Initial sea surface elevations at Source 1 and Source T are pictured.

The wave length of the wave from Source **T** (λ_{T2}) at Site **2** for period of $T = 8$ min and water depth $D2 = 63.6$ m is

$$\lambda_{T2} = T \cdot \sqrt{gD_2} \approx 12 \text{ km},$$

while the diameter of Source **2** is 8 km.

Numerical experiments show that relationship (8) with condition (9) is true even if the time of wave propagation between sources is only 3 – 4 times longer than wave periods and if the wave lengths are comparable to or slightly longer than these sizes. As was found from our experiments, changing the radius of one of the sources by 25% does not significantly influence the output.

Additional experiments were done when Sources **1** and **2** were the same, but Source **T** was not a circle but an ellipse. The ellipse minor (b_2) and major (b_1) axes were chosen as (first)

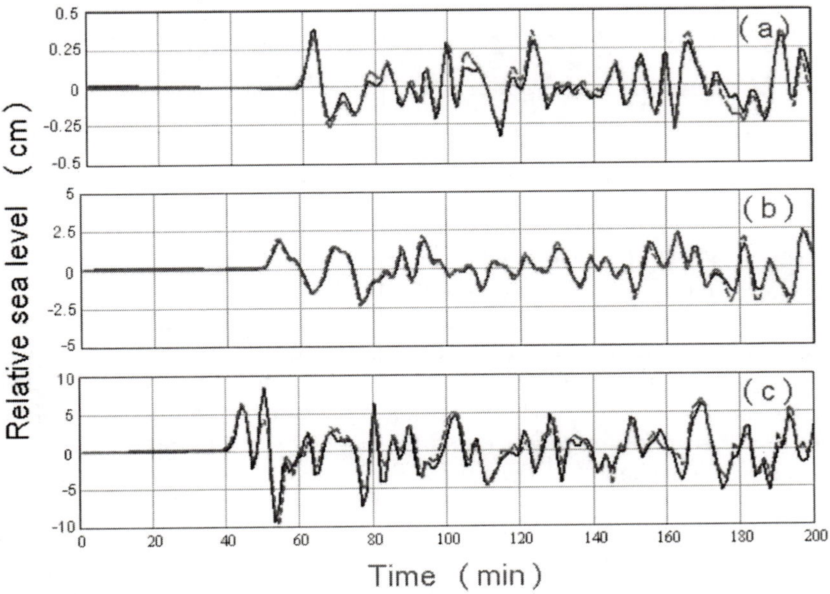

FIGURE 2. Illustration of the reciprocity relationships. (a) Waveforms at Site T due to Source 1 (blue solid line) and at Site 1 due to Source T (red dashed line); correlation coefficient ρ_{1T} = 0.96. (b) Waveforms at Site 2 due to Source 1 (blue solid line) and at Site 1 due to Source 2 (red dashed line); correlation coefficient ρ_{12} = 0.94. (c) Waveforms at Site T due to Source 2 (blue solid) and at Site 2 due to Source T (red dashed line) correlation coefficient ρ_{2T} = 0.91.

(b_1) = 63.6 km and (b_2) = 31.8 km $(b_2 / b_1 = 1/2)$; and (second) b1= 77.9 km and b2= 26.0 km (b2/b1b2/b1= 1/3); maximum amplitude a_T = 0.49 m and water depth at epicentre D_T = 2030 m (for both cases). We also changed the orientation of the major axis relative to true north. The condition of similarity of the source forms was not true, but it was accepted that

$$\frac{2R_{1(2)}}{\sqrt{gD_{1(2)}}} = \frac{\sqrt{b_1 \cdot b_2}}{\sqrt{gD_T}} \text{ and } 4a_1 R_1^2 = a_T b_1 b_2 .$$

The respective numerical results are shown in Fig. 3.

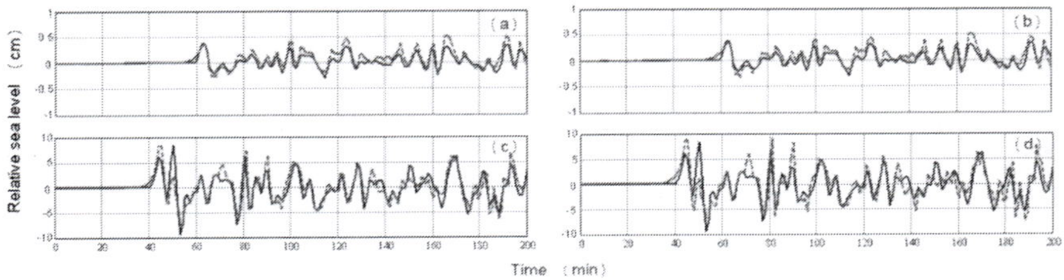

FIGURE 3. Fig. 3. Illustration of the reciprocity relationships. Elliptic Source T is elongated to the northwest (see **Fig.1)**. The ratio of minor/major axes is equal to 1/2 (left column) and to 1/3 (right column). Upper row: waveforms (a) at Site T due to Source 1 (blue solid line), correlation coefficients are $\rho_{1T} = 0.90$ and (b) at Site 1 due to Source T (red dashed line), $\rho_{T1} = 0.87$. Lower row: waveforms (c) at Site T due to Source 2 (blue solid line), correlation coefficients are $\rho_{2T} = 0.86$ and (d) at Site 2 due to Source T (red dashed line), $\rho_{T2} = 0.78$.

The correlation coefficients in all experiments are in the range of 0.67 to 0.90. The best result is in the case when the major axis was directed toward the northwest. This agreement is good enough and sufficient for any practical applications. Relationship (10) was found to be fair not only for circular sources, but also for elliptic sources, at least for the moderate ratios of axes' lengths.

3. THE METHOD OF SHORT-TERM TSUNAMI FORECASTING

The classical statement of the reciprocity principle requires conditions (a) and (b) (see the previous section), however, our numerical experiments demonstrate that it can still be applied without these strict contingencies. The results presented below could become the basis for the short-term tsunami forecasting.

3.1. Derivation Of Relationships For A Short-Term Tsunami Forecast

It was demonstrated (Korolev, 2004, 2005) that the derived reciprocity relationships could be efficiently applied for the problem of short-term (operative) tsunami forecasts based on sea level data from a remote open-ocean site. It is assumed that reciprocity relationship (8) with similarity condition (9) is true when one of the sources is natural, while the other source is artificial (numerical).

Let's tsunami Source **T**, the initial sea surface elevation, is located within the area *ST*. The total volume of the disturbed uplift is *QT*. Let us also assume that *M* is the site of a sea level gauge (i.e., DART site) and *A* is the site of interest (a point for tsunami forecasting). The problem is to compute the tsunami waveform at site *A* ("target site") using the sea level data at site *M* ("reference site"). Mathematically, this problem does not have a unique solution, however, it can be considered as a "consumer problem". Specific methods should be developed for solving this problem.

In this section, "waveform" means the Laplace (Fourier) transform (spectra) of the waveform. Let the reciprocal sources be located within areas with epicentres at *A* and *M*. Their disturbed volumes are *QA*and *QM*, while *ηA(s,xT,yT)* and *ηM(s,xT,yT)* are the waveforms at site *T* from reciprocal sources *SA* and *SM*. For the natural tsunami waveform *ζT* from tsunami Source **T** and for waveforms from the reciprocal sources in *SA* and *SM*, relationships (8) are

$$\zeta_T(s,A)\cdot Q_A = \eta_A(s,T)\cdot Q_T;$$
$$\zeta_T(s,M)\cdot Q_M = \eta_M(s,T)\cdot Q_T.$$

In these expressions we designated: *ζT(s,A)≡ζT(s,xA,yA)*, *ζT(s,M)≡ ζT(s,xM,yM)* and so on.

In addition, let's assume that the auxiliary sources are in the areas with the same epicentres at *A*, *M* and *T* (waveforms and disturbed volumes are designated by symbols with primes). For waveforms from these sources the reciprocity relationships are also true

$$\eta'_T(s,A)\cdot Q'_A = \eta'_A(s,T)\cdot Q'_T$$
$$\eta'_T(s,M)\cdot Q'_M = \eta'_M(s,T)\cdot Q'_T$$

As the auxiliary sources may be selected rather arbitrarily, in particular, *QA, QM* and *Q'A, Q'M* may be selected in such a way that *η'A* is equal to *ηA* and similarly for *η'M* and *ηM*. It is true if *QA* = *Q'A* and the same for Source **M**.

Eliminating unknown *QT* from the above-mentioned two sets of equations yields the following relationship

$$\frac{\zeta_T(s,A)}{\zeta_T(s,M)} = \frac{\eta_T(s,A)}{\eta_T(s,M)}$$

Let us assume that functions on the left side of the equation relate to one tsunami source and functions on the right side relate to another source, and that both sources have the same epicentre. Relationship (11) supports the well-known fact that spectra of any two tsunamis from different sources are similar when recorded at the same site, whereas spectra of any two tsunamis from the same source are dissimilar at different sites (Takahasi & Aida, 1961; Miller, 1972).

From (11) we can get the following main relationship

$$\zeta_T(s,A) = \zeta_T(s,M)\frac{\eta_T(s,A)}{\eta_T(s,M)}$$

Here *ζT(s,A)* is the target function (tsunami waveform at *A*); *ζT(s,M)* is the reference function (tsunami waveform at *M*, i.e., open-ocean DART data); and *ηT(s.M)* and *ηT(s,A)* are the numerical waveforms at *M* and *A* from the auxiliary source located at *ST*. All functions are Laplace transforms. The ratio on the right side of (12) plays the role of a transfer function. The transfer function allows computing a tsunami waveform at a target point on ocean level data at a reference site.

As the sea level data and computed data are digital ones, so the functions included in (12) are represented as numerical Laplace transformations or z-transformation, for example

$$\zeta(s,A) = \zeta(z,A) = \sum_{k=0}^{N} \zeta_k(A)e^{-ks\tau} = \sum_{k=0}^{N} \zeta_k(A)z^k$$

(see Appendix B), where *ζk(A)* – is the value of function *ζ(t,A)* for a moment *t=kτ*, *τ* – is a time sampling. Due to this, (12) may be rewritten as

$$\zeta(z,A) = \sum_{k=0}^{N} \zeta_k(A)\cdot z^k = \sum_{k=0}^{N_1} \zeta_k(M)\cdot z^k \frac{\sum_{k=0}^{N_2} \eta_k(A)\cdot z^k}{\sum_{k=0}^{N_3} \eta_k(M)\cdot z^k}$$

The arrival time of the forecasted tsunami *TA* at target point *A* can be estimated as *TA = T1 + T2-T3*, where *T1* - time of tsunami propagation to the point of registration *M*, T_2 - an arrival time of wave from an auxiliary source to point *A*, *T3* - an arrival time of wave from an auxiliary source to point *M*.

At the first stage, while the record obtained from the station DART has a short duration (for example, one half period of the first wave, *N1* = 10-20 minutes), the lengths of all power series in (13) should be equal: *N1 = N2 = N3 = N*, otherwise the result will be inadequate. But the approximate forecast is possible if *N3 = N1*, but *N2* is significantly more than *N1*. Such a forecast is approximate, but probably takes into account secondary waves. As the sea level information from the DART station at *M* is received, the duration and quality of the forecast is increased. The computations with variable duration of level data allow making the forecast in real-time mode (see Appendix B).

The practical interest is to predict tsunami waveforms of long duration adequately, which would take into account secondary waves reflected from island chains, submarine ridges and mainland coasts. These secondary waves have amplitudes which often significantly exceed amplitudes of first tsunami waves.

Such a tsunami forecast of long duration can be made on the limited information about a level. Often the tsunami in the ocean only has some first waves with significant amplitudes. If we neglect waves with small amplitude in a wave tail, the duration *N1* should be equal to the duration of the registered tsunami with significant waves and the duration *N2* could somehow be long, but durations *N3* and *N1*should be equal. The forecast duration *N* in this case is equal to *N2*. In this way secondary waves can be taken into account.

The moment of the preliminary forecast completion is determined by the time of tsunami propagation up to a point of registration and the time necessary for tsunami identification (the first half period of the first wave).

The division and the multiplication of polynomials in the right part of (13) could give the required result. Unfortunately, the direct application of the offered algorithm, gives, as a rule, a diverging result. This problem is an ill-posed one. Two methods of the solving this problem (regularization methods) permit to obtain a converging result and are described in Appendix B.

It should be noted that the generation mechanism of the actual tsunami is not important here. The method can be applied for short-term forecasting irrespective of the tsunami source, in particular for tsunamis produced by earthquakes, volcanic eruptions, submarine landslides or other phenomena. However, for the practical realization of the proposed method it is important to assume that the transfer function for the corresponding event can be constructed as the ratio of the Laplace transforms of the waveforms at appropriate sites generated by an initial free surface elevation. This elevation was supposed to be circular with a diameter of 50-100 km which is the characteristic transversal size of the earthquake source. There is no requirement to know the detailed seismological information about the source; it is enough to know the co-ordinates of the earthquake (or another source) epicentre. This will enable us, with accuracy sufficient for practical use, to create a short-term tsunami forecast for specific coastal sites.

In fact, the proposed method is a generalization of the methods suggested earlier (Chubarov & Shokin, 1995; Chung et al., 1995).

3.2. Dependence Of The Hindcast Quality On The Diameter Of The Auxiliary Source

There are questions about how the size of auxiliary sources affects the results of the computations.

Numerical experiments with auxiliary sources of different diameters (20, 30, 40, 50, 75 and 100 km) for sites in the ocean gave almost identical results.

3.3. Computing The Short-Term Tsunami Forecast

The algorithm for computing the tsunami waveform based on sea level data from remote sites includes the following steps:

1. deriving the co-ordinates of the earthquake (or other source) epicentre;
2. computing the tsunami waveforms for the reference site *M* and the target (forecasting) site *A*, i.e., functions $\eta T(s,M)$ and $\eta T(s,A)$. The auxiliary source is a circle with diameter of 50-100 km located at the centre of the actual tsunami source. The computation (evaluation of the transfer function) should be completed before the tsunami reaches the reference DART site (normally the site nearest to the source);
3. acquiring to the tsunami data from the appropriate reference DART station; eliminating tidal components from the corresponding record;
4. after recording and identifying tsunami waves at the reference DART station (Site *M*), the tsunami forecast can be made for the target site *A* based on relationship (12, 13). Similarly, in real-time

mode, tsunami forecasts can be provided for other target sites (e.g., for particular sites along the coast). In fact, the preliminary forecast can be given after recording the first tsunami semi-wave. Hereafter, as new and more complete information is received, this forecast can be corrected and improved;

5. the forecast results can be used by TWSs to assist in making their decision about declaring or cancelling the tsunami warning (alarm).

4. APPLICATION OF THE METHOD FOR SHORT-TERM TSUNAMI FORECASTING

Some preliminary research results published in 2005 (Korolev, 2005) demonstrated that the proposed method can be efficiently used for short-term tsunami forecasting. In particular, it was shown that the proposed method can work in a real-time mode and that the errors in the epicentre location only slightly affect the quality of the forecast.

Some new results were published recently by Korolev (2011a, 2011b) and some of these are represented below.

4.1. Preliminary Information

In the present study we apply the method of short-term tsunami forecasting to hindcast tsunami waveforms for some tsunamigenic earthquakes. Earthquakes occurred in 2006, 2007 and 2009 in the region of the Central Kuril Islands, eastward of Simushir Island, near the Chilean coast in 2010 and eastward of Honshu Island in 2011. Associated tsunamis were recorded by some tide gauges on the coasts of Russia and by a number of DART stations located along the Kuril Islands, Japan, the Aleutian Islands and the US West Coast (NDBC DART, n.d.).

Tsunami waveforms were simulated at the locations of the DART stations and near to the coast of Kuril Islands. The actual in situ tsunami data from the DART stations closest to the earthquake epicentres were used for the hindcast. The data from other DART stations, located further from the epicentre, were used to compare with the predictions and to verify the method. The correlation coefficient was chosen as criterion of coincidence. In the present study we used deep-ocean stations not affected by coastal resonant effects and surf beats. All computations were made based on the shallow-water numerical model (Poplavsky et al., 1997). We used the bathymetric data that were interpolated from the ETOPO2 global dataset (Smith & Sandwell, 1994).

No seismological information about the earthquake, except the epicentre co-ordinates, was used in the following computations.

The scheme of the research experiments, indicating the earthquake epicentre and positions of the DART stations, is shown in Fig. 4. The hindcasting results are represented below.

4.2. Simushir Tsunamis Of 2006, 2007 And 2009

The auxiliary source with an initial circular sea surface uplift of 75 km in diameter and with a maximum height of 10 m was chosen for transfer function creation in the cases of the 2006 and 2007 events. The selected source centre coincided with the earthquake epicentre.

Results were published in (Korolev, 2011b).

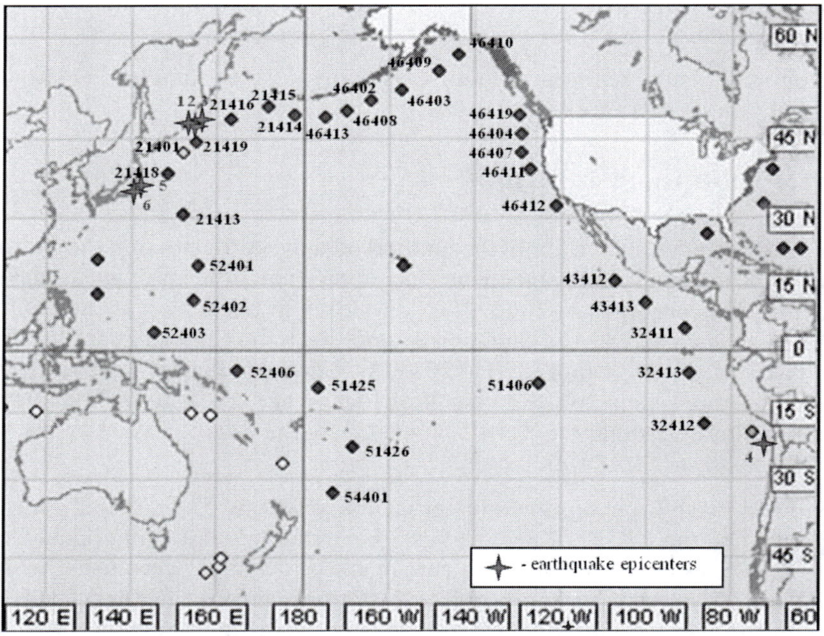

FIGURE 4. A map of the Pacific Ocean showing positions of DART stations (from NDBC DART, n.d.) and epicentres of some earthquakes.Digits denote: 1 – 2006 Simushir tsunami; 2 – 2007 Simushir tsunami; 3 – 2009 Simushir tsunami; 4 – 2010 Chile tsunami; 5 – 09.03.2011 tsunami; 6 – 2011 Great Tohoku tsunami.

4.2.1. The november 2006 simushir tsunami

The tsunami was generated by a major earthquake (M_w = 8.3) on November 15, 2006 at 11:14 UTC with the epicentre at 46.592ºN, 153.266ºE near the Central Kuril Islands (NOAA/WDC, n.d). Significant wave heights were observed along the coasts of the Kuril Islands (c.f. Rabinovich et al., 2008; McInnes et al., 2009).

The November 15, 2006 Simushir tsunami was accompanied by a big run-up at islands nearest to the earthquake centre, but small amplitudes in the northern and southern Kuril Islands, and at the same time caused considerable damage in Crescent City on the US West Coast (Horrillo et al., 2008; Kowalik et al., 2008).

A tsunami alarm was declared at 11:30 UTC by Sakhalin Tsunami Warning Center on the coast of all Kuril Islands and an evacuation of people from dangerous areas was done. Vessels were ordered to sail out into open sea. The alarm was cancelled at 13:19 UTC.

The proposed method was applied to simulate the operational tsunami forecast in the North Pacific Ocean along the Aleutian Islands and the US West Coast. To compute the tsunami waveforms, the data of DART stations 21414 and 46408 were used as the basis (see Fig. 4 for the DART positions). The ocean depth at the epicentre, based on the bathymetry grid readings, was 2803 m. Spherical co-ordinates with a spatial grid step of 3.8 km at latitude 40ºN were used.

Figure 5 shows hindcast results and the waveforms of the observed tsunami waves.

The computed waveforms match well the observed tsunami waveforms at DART stations 46402 and 46409 along the Aleutian Islands. Similar good agreement was observed also for DART stations 46419 and 46411 along the US West Coast. In all cases, the best agreement was for a few leading waves; the correlation coefficients for these waves were within the range ρρ = 0.70–0.85.

4.2.2. The january 2007 simushir tsunami

The tsunami was generated by a major earthquake (M_w = 8.1) on January 13, 2007 at 4:23 UTC with the epicentre at 46.243ºN, 154.524ºE located on the oceanic slope of the Central Kuril Trench (NOAA/WDC, n.d.). This tsunami was weaker than the 2006 Simushir tsunami, but was locally quite intensive (Rabinovich et al., 2008).

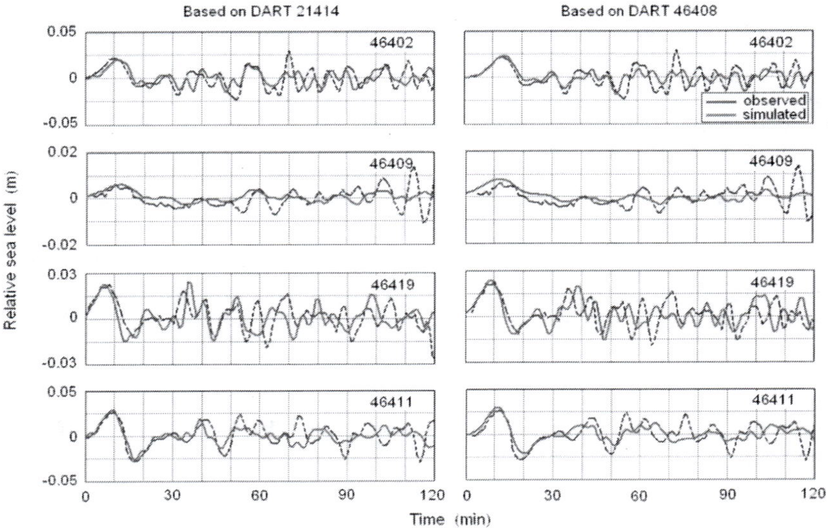

FIGURE 5. Computed and observed tsunami waveforms for the 2006 Simushir earthquake for DART sites in the North Pacific Ocean. Left column: using DART 21414 as the reference station; right column: using DART 46408 as the reference station.

The tsunami alarm was declared on the coast of the Kuril Islands at 4:36 UTC.

Tsunami waves were recorded with amplitudes at Yuzhno-Kurilsk 5 - 6 cm and at Malokurilskoe 36 cm. Due to the weak manifestation of the tsunami on the coasts of the Kuril Islands, Kamchatka and the Aleutian Islands, the tsunami alarm was cancelled at 7:49 UTC.

This tsunami was recorded by a smaller number of DART stations than the 2006 tsunami (c.f. Laverov et al., 2009), however, one of these stations was DART 21413, which was not in operation during the previous (2006) event. This station was located southward from the 2007 epicentre (690 nautical miles southeast from Tokyo). During the 2006 event the tsunami was only recorded by DART stations located to the east of the source. It is of additional interest to compare the results for open-ocean stations located in various directions from the source. Simulation of the operational tsunami forecast was done. The data of DART stations 21414 (east of the source) and 21413 (south of the source) were used as the basis (see Fig. 4 for the DART positions). The ocean depth at the epicentre was 6877 m. Spherical co-ordinates with a spatial grid step of 3.8 km at latitude 40ºN were used. The corresponding results are shown in Fig. 6.

In general, the agreement between computed and observed waveforms is satisfactory both for DART 46408 near the Aleutian Islands and for DARTs 46419 and 46412 along the US West Coast. Hindcasts based on data from both stations 21414 (eastward from the source) and 21413 (southward from the source) are of almost identical quality. For all cases, the computations show correctly the tsunami manifestation during the initial (negative) phase. The correlation coefficients were within the range ρρ = 0.50-0.72.

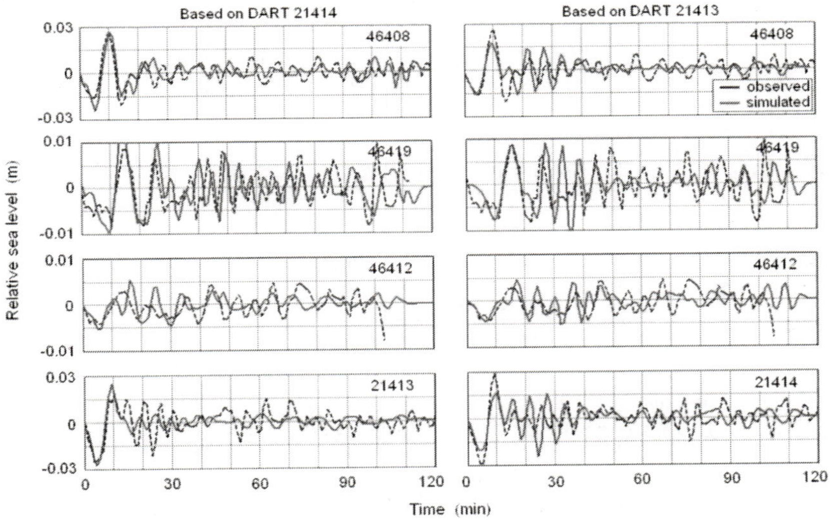

FIGURE 6. The same as in **Fig. 5**, but for the 2007 Simushir tsunami. Left column is for DART 21414 as the reference station; right column is for DART 21413 as the reference station.

4.2.3. The January 2009 Simushir Tsunami

This earthquake (M_w = 7.4) occurred on January 15, 2009 at 17:49 UTC approximately in the same region of the Central Kuril Islands as the 2006 and 2007 earthquakes, but it was weaker than the two previous earthquakes. The epicentre of the earthquake was located at 46.857ºN, 155.154ºE (NOAA/WDC, n.d.). The earthquake generated a tsunami that was recorded on the coast of the Kuril Islands and by a few DART stations, in particular by DART 21416 located 240 miles from the Kamchatka

Peninsula. The latter station was the nearest to the epicentre. A tsunami alarm was not announced during this event.

Figure 7 shows a more detailed map than in Fig. 4, indicating epicentres of earthquakes, DART positions in the northwest Pacific and target points on the coast of the Kuril Islands.

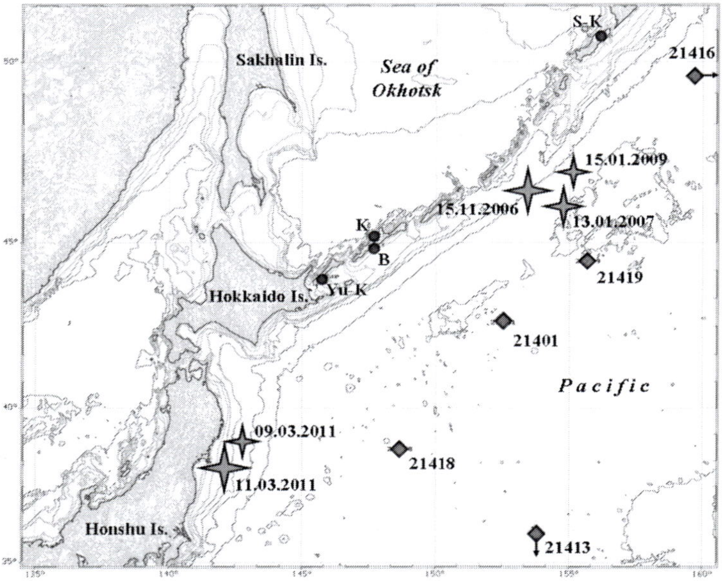

FIGURE 7. Map fragment of northwest Pacific. Conventional signs: S-K – Severo-Kurilsk, K - Kurilsk, B – Burevestnik (Kasatka Bay), Yu-K – Yuzhno-Kurilsk; asterisks – locations of earthquake epicentres (and dates), rhombuses – locations of DART stations.

The location of DART station 21416 is out of the scheme, the true position is to the east of that marked on the map at longitude 163° 30'E, the true site of DART station 21413 is to the south of that marked on the map at a point with latitude 30° 31'N.

The auxiliary source was taken as a circular uplift of the sea surface 50 km in diameter and maximum height of 8 m; the source centre coincided with the earthquake epicentre. This source was used as the input for the model. The ocean depth at the epicentre was 6645 m.

The computed and observed tsunami waveforms are shown on Fig. 8. The data from DART 21416 were used to hindcast the waveforms for DART

46408 and 46413, located eastward from the source, and for DART 21413, located southward from the source. Similar computations were also made for DARTs 46408, 46413 and 21416 based on the data from DART 21413.

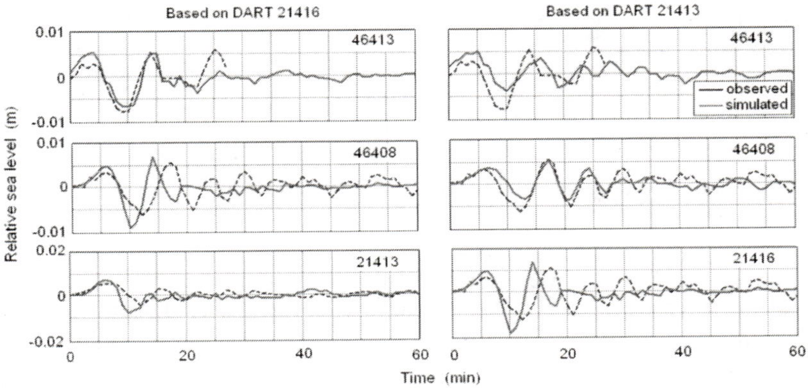

FIGURE 8. The same as in **Fig. 5**, but for the 2009 Simushir tsunami. Left column is for DART 21416 as the reference station; right column is for DART 21413 as the reference station.

The computed waveforms are in good agreement with the observed tsunami records for both the area of the Kamchatka Peninsula and the area of the Aleutian Islands, especially for the leading waves. The correlation coefficients were within the range ρρ = 0.75–0.80.

4.2.4. The 2009 simushir tsunami on the kuril islands

Previous results demonstrated the possibilities of the proposed method. More interesting is the answers the following questions. How can the proposed method produce a tsunami forecast near to the coast? Can the proposed method predict secondary waves of large amplitudes? How well can the proposed method estimate the duration of the alarm period?

The auxiliary source to create a transfer function was an axisymmetric free surface elevation with a diameter of 50 km and a maximum amplitude of 8 m. Water depth at the earthquake centre based on the bathymetry grid readings was equal to 6887 m. Mercator projection with a spatial grid step 1.1 km was used with latitude of scale 52º.

The data of the closest to the earthquake centre DART station 21416, located north-east of the source, were used for the simulation of the tsunami forecast on the coast of the Kuril Islands. The sea level recorded by this station is shown in Fig. 9.

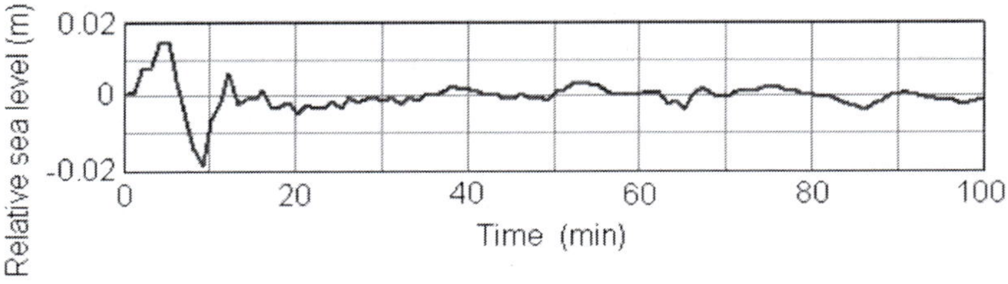

FIGURE 9. The sea level recorded by DART station 21416.

The January 15 tsunami was recorded by a tide gauge at Severo-Kurilsk, located north of the source. The tsunami waveform was computed at the point near Severo-Kurilsk in the strait at a distance of about 1.5 km offshore and at a depth of 20 m.

Results of simulation are represented in Fig. 10. This figure shows a tide gauge record at Severo-Kurilsk with a duration of 17 hours. The specific natural oscillations of the strait up to 19 hours on the sea level record on January 15 can be seen. Then the nature of the oscillations changed due to the tsunami. After 6 - 7 hours the specific mode of natural oscillations was restored, but with increased amplitude.

The blue line designates the recorded form of oscillations of a sea level at the port of Severo-Kurilsk, red - hindcast:

a. preliminary forecast at the point S-K (see Fig. 8) based on data of DART station 21416 with a duration of 20 min;

b. updated forecast at the same point based on data of DART station 21416 with a duration of 82 min.

On Fig. 10 the vertical solid line shows the time of the main shock (17:49 UTC), the vertical dashed line shows the moment of the forecast.

The preliminary estimated tsunami waveform based on 20 minute data of DART station 21416 is shown in Fig. 10 (a). The updated forecast using 82 minute level data of DART station21416 is shown in Fig. 10 (b).

Forecast, preliminary and adjusted, shows that the tsunami arrival is expected at 19:20, 90 min after the main shock. Waves with a greater amplitude followed the first wave are expected 1.5 hours later. The duration of the tsunami is estimated to be 6 - 7 hours.

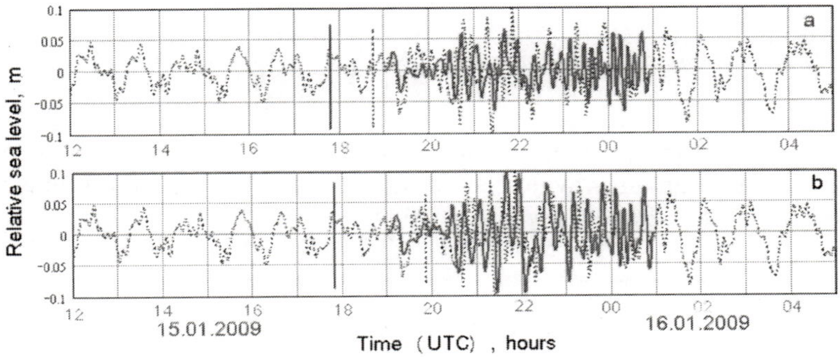

FIGURE 10. Results of the 2009 Simushir tsunami hindcast close to Severo-Kurilsk.

The computations necessary to create a transfer function are performed for 15 min, while the tsunami travel time to the point 21416 equals to 41 min, to Severo-Kurilsk - 75 min. Taking into account the time of tsunami identification (20 min), the preliminary tsunami forecast at Severo-Kurilsk may be given 30 minutes before the first wave arrives at this point.

The above example shows the principal possibility of tsunami forecast near to the coast of the Kuril Islands based on data of level stations, located on the ocean side of the Kuril-Kamchatka Trench.

4.3. 2010 Chile Tsunami

The tsunami was generated by earthquake (M_w = 8.8) which occurred on February 27, 2010 at 6:34 UTC near the Chilean coast. The co-ordinates of the earthquake epicentre are 36.122ºS, 72.898ºW (NOAA/WDC, n.d.). The tsunami was recorded by a number of DART stations in the Pacific.

4.3.1. Hindcast of 2010 chile tsunami in ocean

Post-event computations simulating the short-term tsunami forecast are based on data closest to the epicentre of the earthquake recorded at DART station 32412 over 30 minute duration (Fig. 11). An auxiliary source to create the transfer function was defined as a circular initial free surface elevation with a diameter of 100 km and a maximum height of 10 m with the centre coincident with the epicentre of the earthquake. The computations were performed in spherical co-ordinates with a spatial grid step 5 km at the equator.

The results of hindcasting the tsunami waveforms in the Pacific (see Fig. 4 for the positions of DART stations and tsunami epicentre) are represented in Fig. 12.

The results of the simulation show a good agreement between hindcasted and observed tsunami waveforms in the north and the west directions from the earthquake centre.

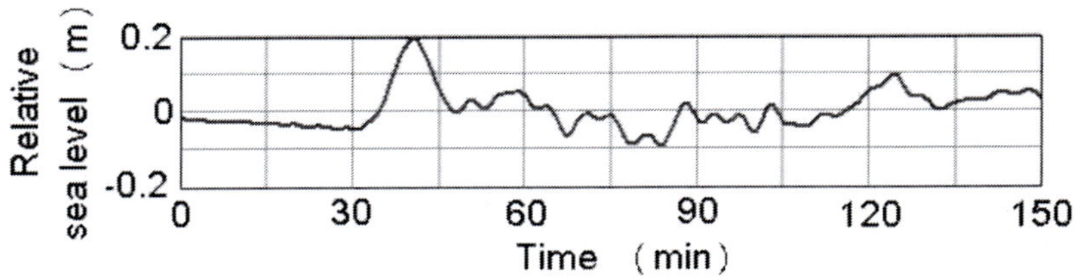

FIGURE 11. Sea level recorded by DART station 32412.

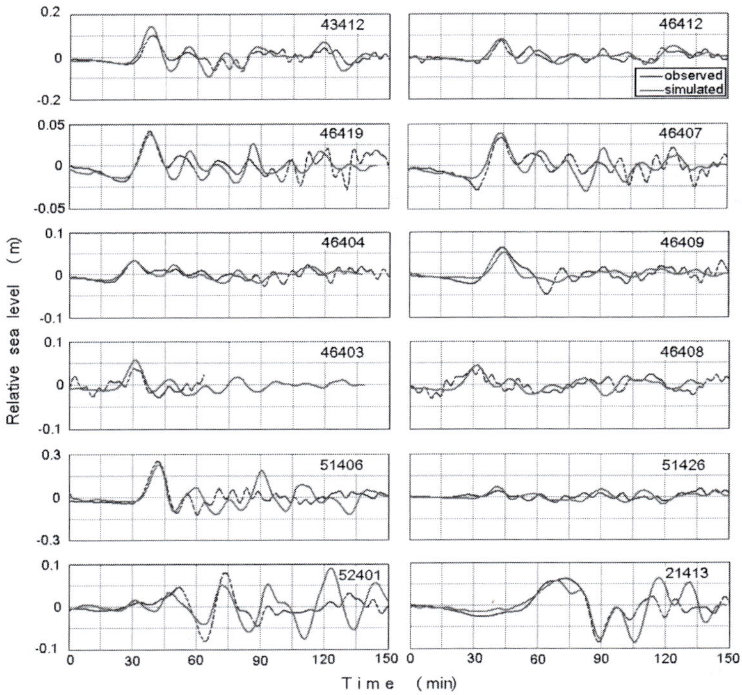

FIGURE 12. Results of hindcasting the 2010 Chile tsunami in the Pacific based on DART 32412 data.

4.3.2. Hindcast Of 2010 Chile Tsunami Near The Coast Of The Kuril Islands

The earthquake occurred on February 27 at 06:34 UTC. The tsunami alarm was declared by the Sakhalin Tsunami Warning Center on all the Kurile Islands on February 28 at 00:50 UTC (10:50 local time) 3 h 22 min before the first wave arrived at Severo-Kurilsk (Paramushir Island). The population of all settlements of the Kuril Islands was evacuated and ships were withdrawn to the open sea. The wave amplitudes for 4 hours did not exceed 0.3 m. The duration of the alarm was about 4 hours. Shortly after the alarm was cancelled (4 hours 8 minutes after the first wave arrival) a wave with an amplitude of 1 m was recorded. Once again the tsunami alarm had not been announced.

For tsunami prediction near the coast of the Kuril Islands, computations have been made from the source of the tsunami off the coast of Chile, initially using spherical co-ordinates with a spatial grid step 5 km at the equator. Then computations were made using a more detailed grid (left bottom corner 37° 51'N, 139° 26'E, right top corner 45° N, 180° E). The spatial grid step was 1 km on scale latitude 45°N. Part of the scheme is shown in Fig. 7.

The tsunami hindcasting is based on data of DART station 32412 with a 30 minute duration The hindcast result is represented in Fig. 13.

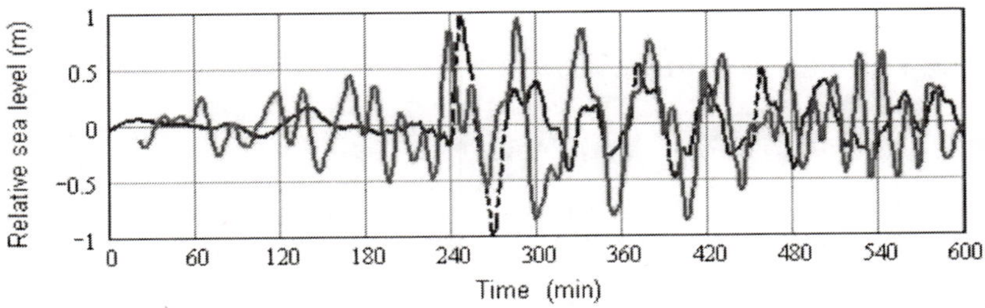

FIGURE 13. Chile tsunami hindcast at Severo-Kurilsk (point S-K on Fig. 7) using data of DART station 32412.

The simulated tsunami waveform coincides well with the form of the observed tsunami. The wave amplitudes during the first 4 hours do not exceed 0.5 m. After 4 hours the appearance of waves with increased amplitude up to 1 m is correctly predicted.

4.4. Tohoku Foreshock 2011

The weak tsunami M_w=7.2 which arose off east coast of Honshu Island on March 09, 2011 (epicentre co-ordinates 38.510°N, 142.792°E (NOAA/WDC, n.d.)) was recorded by a small number of DART stations. The tsunami hindcast using data of the 21418 DART station, nearest to the centre, was made in the area eastward of the Kuril Islands at points 21401, 21419 and in the Pacific Ocean at a point 21413 1300 km to the southeast of Tokyo (see Fig.7 for DART positions).

The auxiliary source was taken as a circular uplift of the sea surface 100 km in diameter and maximum height of 10 m; the source centre coincided with the earthquake epicentre. The ocean depth at the epicentre was 1374 m. We used the Mercator projection with a spatial grid step 5 km at latitude 45°N.

Represented in Fig. 14, the result shows a good coincidence between computed and actual tsunami waveforms.

The computations were executed for points in the open-ocean. The various edge, resonant effects do not influence tsunamis in the open-ocean. This result demonstrates the possibilities of the proposed method. Based on data of DART station 21418, located to the east of the earthquake epicentre, the tsunami waveforms are obtained to the northeast of the epicentre (points 21401 and 21419) and to the southeast (point 21413).

4.5. 2011 Great Tohoku Tsunami

4.5.1. 2011 Great Tohoku Earthquake And Tsunami

This earthquake with magnitude M_w = 9.0 off the northeast coast of Honshu Island occurred on March 11 at 5:46 UTC. The co-ordinates of the epicentre are 38.322ºN, 142.369ºE (NOAA/WDC, n.d.).

As a result of the earthquake a tsunami occurred. It was anomalously high and very destructive on the coast of Honshu Island. Against the backdrop of a long rise in sea level with a height of 2 metres there was an additional shorter rise, with increased amplitude up to 5 m 11 minutes after (Fig. 15).

The great tsunami resulted from a combination of crustal deformations of the ocean floor due to up-thrust tectonic motions, augmented by additional uplift due to the quake's slow and long rupturing process, as well as large coseismic lateral movements which compressed and deformed the compacted sediments. This mechanism is discussed in (Pararas-Carayannis, 2011).

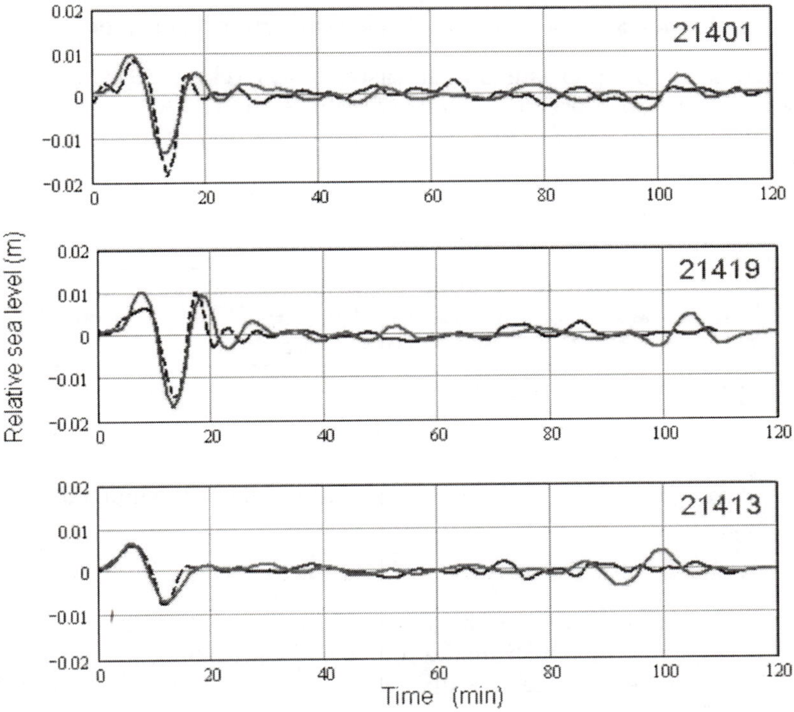

FIGURE 14. Hindcast of the March 9, 2011 tsunami in the Pacific.

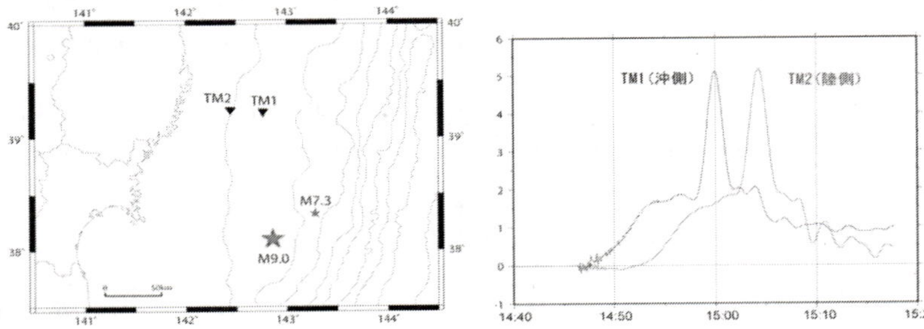

FIGURE 15. Registration of the 11.03.2011 tsunami near to the centre. Positions of earthquake epicentre with M = 9.0 and sea level stations TM1 and TM2 - at the left. The form of tsunami registered by stations - on the right (see **http://outreach.eri.u-tokyo.ac.jp/wordpress/wp-content/uploads /2011/03/TsunamiKamaishiMeter.jpg**).

4.5.2. The hindcast of 2011 great tohoku tsunami in the ocean

A simulation of operative tsunami forecasting for March 11, 2011 was made for points in the open-ocean. The data of DART station 21401 were used in the computations. The relative level of the ocean registered by station 21401 is shown in Fig. 16.

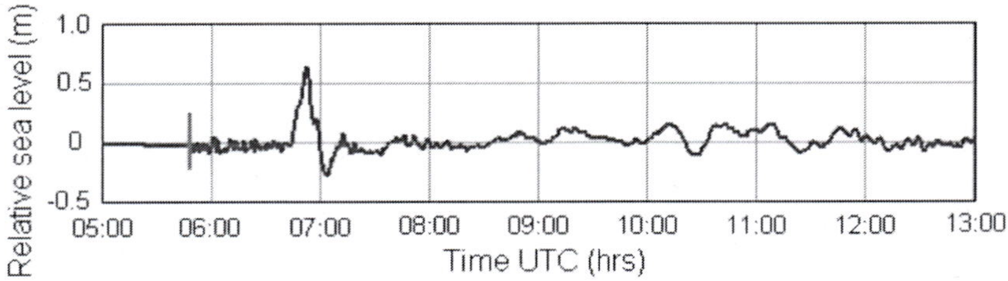

FIGURE 16. ea level recorded by DART station 21401 on March 11, 2011 (the tidal components are deleted), the vertical red line – the moment of the main shock.

To create the transfer function the auxiliary computations of the waveform at the corresponding points from the source in the form of an initial circular elevation of the free-surface with diameter 100 km and an amplitude 10 m are made. The centre of the source coincides with the epicentre of the earthquake, the depth of the ocean according to the bathymetric grid equals to 887 m. Computations were made using spherical co-ordinates with the spatial grid step 5 km at a scale latitude 0°. The computations were based on a 30 minute tsunami duration of the data from 21401 DART station.

Figure 17 demonstrates the high quality of the coincidence of simulated and observed tsunami waveforms both in the northeastern and southern directions from the source. Durations exceeding 30 min in obtained data from DART station 21401 used in the computations do not affect the quality of the forecast.

Analogous simulations have been performed using other means by Titov, Fujii and Satake (Japan (East Coast of Honshu) Tsunami, 2011; Fujii & Satake, 2011). The quality of the coincidence of simulated and observed tsunami waveforms in these works as well as in the presented work is comparable.

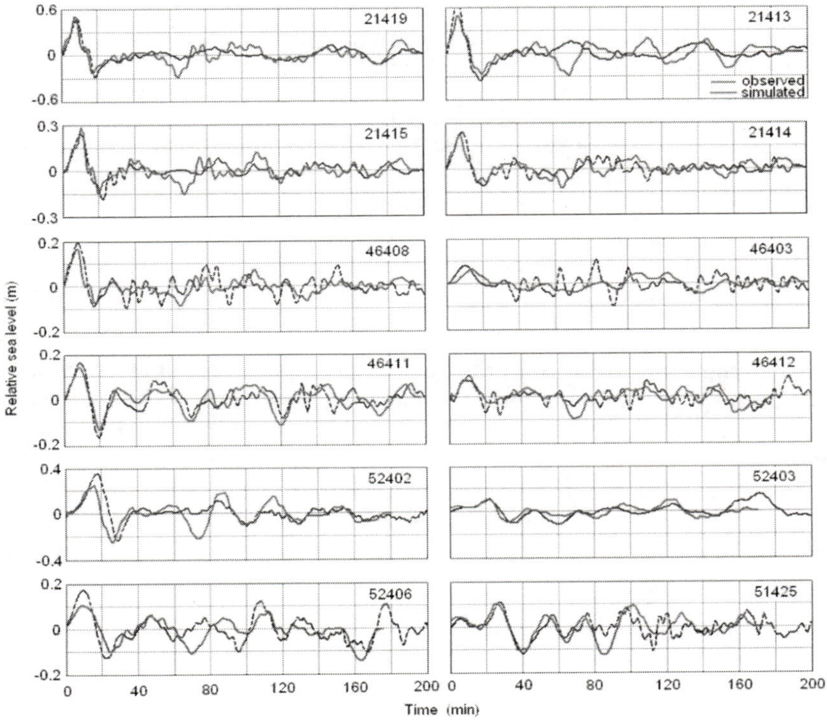

FIGURE 17. Hindcast of the March 11, 2011 Great Tohoku tsunami in the Pacific.

4.5.3. The hindcast of 2011 great tohoku tsunami near the coast of the kuril islands

This strong earthquake off the northeast coast of Honshu Island occurred on March 11 at 5:46 UTC.

The tsunami alarm was declared by the Sakhalin Tsunami Warning Center at 5:58 UTC (local time 15:58) on all Kuril Islands. The population deemed to be in the threatened areas was evacuated to safe places and vessels were recommended to sail out into the open sea.

At 6:43 deep-sea DART station 21401 (located southeast of Iturup Island) registered a tsunami wave with amplitude 67 cm.

The oscillations of sea level, caused by the tsunami, with amplitude of about 1 metre lasted until March 12, therefore, the duration of the tsunami alarm was about 20 hours. The tsunami alarm on all Kuril Islands was cancelled on March 12 at 2:20 UTC (local time 12:20).

The tsunami arrived at Severo-Kurilsk at 9:00. From information from vessels in the sea near Severo-Kurilsk the sea level under keel changed from 4.2 m up to 2.6 m, the swing was 1.6 m. Recorded wave amplitudes were about 0.3 m in Kurilsk. According to visual observations the amplitudes of the tsunami in Burevestnik were approximately 1 m. The tsunami reached Yuzhno-Kurilsk at 7:44, the recorded wave amplitudes were about 1 m. The tsunami alarm was declared 1.5 to 3 hours before the tsunami arrival at different points of the Kuril Islands.

To create the transfer function the auxiliary computations of the waveform at the corresponding points from the source in the form of an initial circular elevation of the free-surface with diameter 100 km and an amplitude 10 m were made. The sea depth at the epicentre is equal to 970.1 m. The centre of the source coincides with the epicentre of the earthquake. We used the Mercator projection with a spatial grid step 0.9 km at latitude 45°N. Computations based on data from DART station 21401 with tsunami duration 30 min were made.

The hindcast was made near the above-mentioned points S-K, K, B and Yu-K. The points are located at distances from the coast, respectively, 1.1, 0.8, 2.1 and 5.4 km. Water depths at these points are 16, 3.2, 30 and 29 m. Locations of these points are shown in Fig. 7.

The results of a hindcast of tsunami waveforms with a duration of 6 - 7 hours made for the above-mentioned points at the Kuril Islands are represented in Fig. 18.

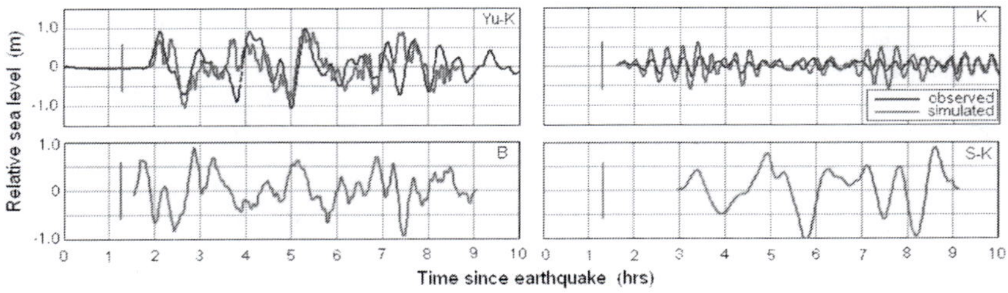

FIGURE 18. Tsunami hindcast at the points of the Kuril Islands. The vertical red line – the moment of forecast producing.

The simulated waveform at point Yu-K and the observed tsunami waveform at Yuzhno-Kurilsk are in good agreement. The structure of the tsunami, arrival times of all four waves and wave amplitudes coincided well. According to computations, maximal wave amplitudes should not exceed 1 m; that is in agreement with the observations.

The simulated tsunami waveform at Kurilsk (point K) demonstrates a brightly expressed batch wave structure. The computed wave periods coincide well with recorded periods. According to the forecast, the wave amplitudes should not exceed 0.5 metres.

According to the forecast for Burevestnik (point B), the wave amplitudes should not exceed 1 m. This result agrees with observational data: the wave amplitudes on visual observations were about 1 metre.

At Severo-Kurilsk (point S-K), according to the forecast, increase of amplitude of consequent waves almost up to 1 m is possible. The range of oscillations of a level should not exceed 1.8 - 2 metres. This result agrees well with observational data at Severo-Kurilsk. Based on information from vessels, the swing of sea level was equal to 1.6 m.

Unfortunately, the registration of a tsunami at Burevestnik and Severo-Kurilsk was not conducted. Nevertheless, the results of the computation are confirmed by visual observations and information from vessels in these points.

The time of auxiliary computations for transfer function creation did not exceed 25 min, while the tsunami travel time up to DART station 21401 is equal to 58 min. The auxiliary computations are completed before registration of a tsunami by station 21401. Taking into account the time of tsunami recognition at point 21401 (the first period with a duration of 19 min) the preliminary forecast could be given 78 - 80 min after the main shock of the earthquake. That is 2 hours prior to arrival of the first wave of a tsunami at point S-K, 46-48 minutes prior to arrival of a wave at Yu-K, 40 minutes before tsunami arrival at K and 20 minutes before arrival of the first wave of the tsunami at B. In the latter case the offered hydrophysical method appears to be ineffective.

5. CONCLUSION

The reciprocity principle, known in acoustics as an asymptotic assumption, is applied to describe non-stationary long ocean waves. The similarity conditions for reciprocal sources, not previously studied, are obtained. It is shown in numerical experiments that, with a quality sufficient for practical applications, the reciprocity principle works quite well, even when asymptotic conditions are not totally satisfied (i.e., when the characteristic time of the wave propagation between the sources is only three to five times greater than the typical wave period or when the characteristic wave length is only two to three times greater than the size of the source area).

The proposed method of computing the tsunami waveform from an unknown source using ocean level data at some points demonstrated its efficiency.

There were questions how the size of auxiliary sources affects the results of computations. Numerical experiments with auxiliary sources of different diameters (with a 20 – 100 km range) gave almost identical results.

There were also questions as to whether it is possible to effectively apply auxiliary axially symmetric sources to create a transfer function. The form of the source influences the directivity of the tsunami. The relief of ocean bottom has a major influence on the evolution of the waves. The use the observational data from the 2009 Simushir, 2010 Chile, 2011 weak Honshu and 2011 Great Honshu tsunamis gives very encouraging results. Despite the fact that the auxiliary source was axially symmetric, the tsunami forecast gives almost identical results based either on the data from reference DART stations located in the same direction as the target sites or from the stations located in other directions.

Despite its approximate character, the proposed method can provide a forecast of tsunami wave parameters for any ocean site with sufficiently high quality and provide real-time operative information for tsunami warning services. The results of this paper demonstrate that the proposed method satisfies the requirements of the warning services and can be effectively used for tsunami warnings.

The numerical experiments described in the paper were conducted for open-ocean target sites. For these sites the method works pretty well, however, the main purpose of TWSs is to provide reliable tsunami forecasts for coastal areas. For this purpose we made auxiliary computations (to evaluate transfer functions) using the detailed bathymetry grid (detailed enough to resolve small-scale resonant features of coastal topography). This allows us to provide the adequate tsunami forecast for the entire coastal zone exposed to arriving tsunami waves. This ensures a straightforward forecast, not only of the leading tsunami waves, but also the following tsunami waves which sometimes are much more intensive than the head waves. The practical interest is to predict tsunami waveforms of long duration, which would take into account secondary waves reflected from island chains, submarine ridges and mainland coasts. Examples of tsunami hindcasts of 2009, 2010, 2011 have shown that the forecast of long duration based on short information from DART stations is possible.

The proposed method of short-term tsunami forecasting uses only seismological information about the location of the earthquake epicentre and works independently of the mechanism of tsunami generation. Numerical experiments show that, despite the anomalous character of the 2011 Great Tohoku tsunami, the hindcast using only level data from DART 21401 without the additional seismological data gives an adequate result. The same example shows that the forecast, based on preliminary computations, is not always effective.

The advantage of the proposed method is that it does not require a pre-computed database of synthetic tsunami waveforms. This is especially important for areas where tsunami early warning systems are just being established and there are no pre-computed databases.

The time of the forecast producing is determined by the time of tsunami propagation up to a point of registration by the DART station and the time of tsunami identification (the first half-cycle, about 15 - 20 minutes). Such a forecast is effective for points for which the difference between the time of tsunami attack and the time of forecast producing is enough for an evacuation of the population and vessels from a threatened zone. For the Kuril Islands this time is equal, conditionally, to 30 minutes.

The comparison of results presented in this paper with some other results based on assimilation of DART data (Japan (East Coast of Honshu) Tsunami, 2011; Tang et al., 2008; Titov, 2009; Wei et al., 2008) shows that the forecast quality obtained by different methods is comparable.

Currently, post-event computations are performed using four non-related software, thence routine work takes a long time. This, unfortunately, does not allow forecasting in real-time.

NOTES

[1] - DART = Deep-ocean Assessment and Reporting of Tsunami is a deep-sea tsunami monitoring system by the US National Oceanic and Atmospheric Administration (NOAA).

REFERENCES

1. L. M. Brekhovskikh, 1960Waves in Layered Media, Academic Press, New York.
2. G. Chertock, 1970Transient flexural vibrations on ship-like structures exposed to underwater explosions. J. Acoust. Soc. America, 482170180

3. L. B. Chubarov, Yu. I. Shokin, 1995Mathematical modeling in mitigating the hazardous effect of tsunami waves in the ocean. A priori analysis and timely on-line forecast. Sci. Tsunami Hazards, 13N0. 1, 2744
4. J. Y. Chung, S. D. Kim, V. V. Ivanov, 1995Tsunami wave hindcasting in the Japan Sea, In: Tsunami: Progress in Prediction, Disaster Prevention and Warning, edited by Y. Tsuchiya and N. Shuto, 8598Kluwer Academic Publishers, Dordrecht, Netherlands
5. G. Doetsch, 1967Anleitung zum praktischen gebrauch der Laplace-transformation und der z-transformation. R. Oldenbourg, Munchen, Wien
6. Y. Fujii, K. Satake, 2008Tsunami sources of the November 2006 and January 2007 Great Kuril Earthquakes. Bull. Seism. Soc. Amer., 98315591571
7. Y. Fujii, K. Satake, 2011Off Tohoku-Pacific Tsunami on March 11, 2011. Available from: http://iisee.kenken.go.jp/staff/fujii/OffTohokuPacific2011/tsunami.html.
8. V. K. Gusiakov, 2011Relationship of tsunami intensity to source earthquake magnitude as retrieved from historical data. Pure Appl. Geophys., 16820332041
9. J. Horrillo, W. Knight, Z. Kowalik, 2008The Kuril Islands Tsunami of November 2006. Part II: Impact at Crescent City by local enhancement. J. Geophys. Res., 113C01021
10. Japan (East Coast of Honshu) Tsunami, March 11,2011Main Event Page. (2011). In: NOAA Center for Tsunami Research, Currently accessed, Available from: http://nctr.pmel.noaa.gov/honshu20110311/images/comp_plots/20110311_darts_comp.png
11. G. A. Korn, T. M. Korn, 1961Mathematical handbook for scientists and engineers. Definitions, theorems and formulas for reference and review. McGraw-Hill Book Company, Inc., New York, Toronto, London
12. Y. P. Korolev, 2004Tsunami numerical modeling for short-term forecasting using data of remote level gauges. Oceanology, 44346352
13. Y. Korolev, 2005The method of the early tsunami warning using data of remote level gauges, Proc. of 22nd Int. Tsunami Symp., Chania, Crete Island, Greece, 27-29 June 2005, 113119
14. Y. P. Korolev, 2011aRetrospective Short-Term Forecast of the 1996 Andreanov (Aleutian Islands) Tsunami. Oceanology, 51385393
15. Y. P. Korolev, 2011bAn approximate method of short-term tsunami forecast and the hindcasting of some recent events. Nat. Hazards Earth Syst. Sci., 1130813091doi:nhess-11-3081-2011. Available at: www.nat-hazards-earth-syst-sci.net/11/3081/2011/
16. Y. P. Korolev, A. A. Poplavsky, 1995A method of tsunami estimation for short-range forecast service, In: IUGG, XXI General Assembly, Boulder, Colorado, A335
17. Z. Kowalik, J. Horrillo, W. Knight, T. Logan, 2008The Kuril Islands Tsunami of November 2006. Part I: Impact at Crescent City by distant scattering. J. Geophys. Res., 113C01020
18. L. D. Landau, E. M. Lifshitz, 1975Fluid Mechanics: 6Course of Theoretical Physics). Pergamon Press, Oxford, New York, Toronto, Sydney, Paris, Braunschweig
19. N. P. Laverov, L. I. Lobkovsky, B. W. Levin, A. B. Rabinovich, E. A. Kulikov, I. V. Fine, R. E. Thomson, 2009The Kuril tsunamis of November 15, 2006, and January 13, 2007: Two trans-Pacific events. Transactions (Doklady) of the Russian Academy of Sciences, Earth Science Section, 4263386392
20. M. A. Lavrentiev, B. V. Shabat, 1987Methods of the theory of the functions of complex variables, (in Russian), Nauka, Moscow
21. H. G. Loomis, 1979Tsunami prediction using the reciprocal property of Green's functions. Marine Geodesy, 22739
22. Innes. B. T. Mac, T. K. Pinegina, J. Bourgeois, N. G. Razhigaeva, V. M. Kaistrenko, E. A. Kravchunovskaya, 2009Field survey and geological effects of the 15 November 2006 Kuril tsunami in the middle Kuril Islands., Pure Appl. Geophys., 166936
23. G. R. Miller, 1972Relative spectra of tsunamis. Hawaii Inst. Geophys. HIG-728Honolulu, 7 p.
24. H. O. Mofjeld, 2009Tsunami measurements, In: The Sea, 15edited by E.N. Bernard and A.R. Robinson, 201235Harvard University Press, Cambridge

25. NDBC DART Program. (n.d.). In: National Data Buoy Center, Currently accessed, Available from: http://www.ndbc.noaa.gov/dart.shtml
26. NOAA/WDC Global Historical Tsunami Database at NGDC.n.d.). In: NOAA National Geophysical Data Center, Currently accessed, Available from: http://www.ngdc.noaa.gov/hazard/tsu_db.shtml
27. NOAA Tsunami Warning System Receives High Marks.2004In: NOAA Magazine on-line, 30.03.2012, Available from: http://www. magazine. noaa.gov/stories/mag153.htm
28. G. Pararas-Carayannis, 2011Tsunamigenic source mechanism and efficiency of the March 11, 2011 Sanriku Earthquake in Japan. Science of Tsunami Hazards, 302126152
29. A. A. Poplavsky, V. N. Khramushin, K. I. Nepop, Y. P. Korolev, 1997The Operative Tsunami Prediction on the Sea Coasts of the Far East, (in Russian), DVO RAN, Yuzhno-Sakhalinsk, Russia
30. A. B. Rabinovich, L. I. Lobkovsky, I. V. Fine, R. E. Thomson, T. N. Ivelskaya, E. A. Kulikov, 2008Near-source observations and modeling of the Kuril Islands tsunamis of 15 November 2006 and 13 January 2007. Advances in Geosciences, 141105116
31. Lord. . J. W. Rayleigh, Strutt, 1945The Theory of Sound, 2Dover Publications, New York
32. K. Satake, 1987Inversion of tsunami waveform for the estimation of a fault heterogeneity: method and numerical experiments. J. Phys. Earth, 353241254
33. W. H. F. Smith, D. T. Sandwell, 1994Bathymetric prediction from dense satellite altimetry and sparse shipboard bathymetry. J. Geophys. Res., 9921803824
34. J. J. Stoker, 1957Water waves. The mathematical theory with applications. Interscience Publishers, New York, London
35. R. Takahasi, I. Aida, 1961Studies on the spectrum of tsunami. Bull. Earthq. Res. Inst., 39523535
36. L. Tang, V. V. Titov, Y. Wei, H. O. Mofjeld, M. Spillane, D. Arcas, E. N. Bernard, C. Chamberlin, E. Gica, J. Newman, 2008Tsunami forecast analysis for the May 2006 Tonga tsunami. J. Geophys. Res., 113C12015,
37. A. N. Tikhonov, V. Y. Arsenin, 1977Solutions of ill-posed problems (F. John, translation editor), V. H. Winston and Sons (distributed by Wiley, New York)
38. V. V. Titov, F. I. González, E. N. Bernard, M. C. Eble, H. O. Mofjeld, J. C. Newman, A. J. Venturato, 2005Real-time tsunami forecasting: challenges and solutions. Natural Hazards, 3513541
39. V. V. Titov, 2009Tsunami forecasting, In: The Sea, 15edited by E.N. Bernard and A.R. Robinson, 367396Harvard University Press, Cambridge
40. H. Tsushima, K. Hirata, Y. Hayashi, Y. Tanioka, K. Kimura, S. Sakai, M. Shinohara, T. Kanazawa, R. Hino, K. Maeda, 2011Near-field tsunami forecasting using offshore tsunami data from the 2011 off the Pacific coast of Tohoku Earthquake. Earth Planets Space, 63821826
41. R. J. Uric, 1975Principles of Underwater Sound, McGraw-Hill Book Company, New York
42. T. A. Voronina, V. A. Tcheverda, 1998Reconstruction of tsunami initial form via level oscillation. Bull. Novosibirsk Comp. Center, Ser. Mathematical Methods in Geophysics, 4127136
43. Y. Wei, K. F. Cheung, G. D. Curtis, C. S. Mc Creery, 2003Inverse algorithm for tsunami forecasts. J. Waterw., Ports, Coastal Ocean Eng., ASCE, 12926069
44. Y. Wei, E. Bernard, L. Tang, R. Weiss, V. Titov, C. Moore, M. Spillane, M. Hopkins, U. Kânoğlu, 2008Real-time experimental forecast of the Peruvian tsunami of August 2007 for U.S. coastlines. Geophys. Res. Lett., 35L04609,
45. Y. Yamazaki, Y. Wei, K. F. Cheung, G. D. Curtis, 2006Forecast of tsunamis from the Japan-Kuril-Kamchatka source region. Natural Hazards, 383411435

CHAPTER 3

Integration of Spatial Analysis for Tsunami Inundation and Impact Assessment

Abu Bakar Sambah[1,2], Fusanori Miura[1]

[1]*Graduate School of Science and Engineering, Yamaguchi University, Ube, Japan*

[2]*Fisheries and Marine Science Faculty, Brawijaya University, Malang, Indonesia*

ABSTRACT

Disaster mitigation and reconstruction plan due to tsunami can be implemented with various actions. An integration of spatial analysis through Geographical Information System (GIS) application and multi-criteria analysis through Analytical Hierarchy Process (AHP) is one of the methods for tsunami inundation and impact assessment. In this study, vulnerability, inundation and impact assessment due to tsunami hazard in Ofunato city, Iwate Prefecture, Japan was carried out. Appropriate input parameters were derived from Digital Elevation Model data, and satellite remote sensing and field data were analyzed through GIS. We applied the parameter of elevation and slope created from of Aster GDEM version 2, coastline distance created from vector map of the study area and vegetation density created from ALOS ANVIR-2 image. We applied AHP process for weighting the parameter through pair-wise comparison using five iterations of normalized matrix. Five classes of vulnerability were defined and analyzed for tsunami inundation mapping. We used weighted overlay through spatial analyst in GIS to create the final map of tsunami vulnerability. The assessment results indicate that 7.39 square kilometer of the study area was under the high vulnerability zone due to tsunami, and 8.13 square kilometer of building area was under the inundation area. Our result showed good agreement with the observed data and historical map. The result presented here can aid as preliminary information for the coastal zone management related to disaster mitigation and for the evacuation process and management strategy during disaster.

Keywords: Spatial Analysis; GIS; AHP; Tsunami

1. INTRODUCTION

Natural disaster including tsunami is the natural phenomena which is difficult to prevent. An initial assessment for vulnerability mapping due to disaster is one analysis that can be applied to reduce its impact. Tsunami vulnerability assessment is essential to disaster planning in the term of mitigation. It can provide preliminary information that important for tsunami disaster risk management plans. This includes pre-planning appropriate response activities in order to minimize the impact of disaster and all possibilities that will happen, and preparing and mitigating for the future events [1]. It can include mapping of evacuation routes and evacuation building, tsunami barriers construction, disaster risk management and regulation and disaster education for public.

In order to prepare those activities for effective implementation, the availability of reliable and accurate information concerning the spatial and temporal characteristics and impact of potentially damaging due to tsunami at different scales of magnitude is needed. Thus, a tsunami vulnerability analysis should be developed based on many parameters as possible to get a realistic description of vulnerability area in both spatial and temporal.

In addition to the hazard probability, exposure, and capacity measures, vulnerability is one of the parameters in determining disaster risk [2-4]. It can be defined as the degree to which a person, community or a system is likely to experience harm due to an exposure to an external stress. Vulnerability is described as a set of conditions and processes resulting from physical, social, economic and environmental factors that increase the susceptibility of a community to the impact of hazards [5,6].

The application of remote sensing and GIS made the possibilities in order to map the distribution of damage area due to disaster and to assess the vulnerability area. GIS through the spatial multi-criteria analysis helps in making a priority related to the decision making process using geo-reference data. In addition to the decision maker's preferences related to a parameter evaluation, spatial multi-criteria analysis needs both information on criterion attributes and the geographical references [7,8]. Some previous work have applied GIS method for tsunami mapping and overlaid with official land-use map [9], and have analyzed tsunami risk using multi-scenario approach [10], and have analyzed the vulnerability using remote sensing data and integrated analysis using GIS to the physical built-up infrastructure, i.e. buildings, and identified the inundation area based on the contour and the highest recorded of tsunami event related to the building vulnerability and human vulnerability [1,11].

The aims of this study are to map the vulnerability area due to tsunami disaster and to assess its impact by mapping the possible inundation area due to tsunami by using GIS. The vulnerability was calculated based on elevation, slope, coastline distance and vegetation density. In terms of vegetation density, we tried to apply Soil Adjusted Vegetation Index (SAVI) instead of vegetation mapping using Normalized Difference Vegetation Index (NDVI). In addition, we also applied the Analytical Hierarchy Process for estimating weights of the parameters.

2. MATERIALS AND METHODS

2.1. Data

The elevation data used in this study was obtained from The ASTER Global Digital Elevation Model (ASTER GDEM) version 2. The Advanced Space-borne Thermal Emission and Reflection Radiometer (ASTER) GDEM is a joint product developed and made available to the public by the Ministry of Economy, Trade, and Industry (METI) of Japan and the United States National Aeronautics and Space Administration (NASA). It is generated from data collected from a space-borne earth observing optical instrument, namely ASTER.

The ASTER GDEM covers the entire land surface of the Earth at high resolution. Version 2 of the ASTER GDEM is employing an advanced algorithm to improve global digital elevation resolution and elevation accuracy and reprocessing a total of 1.5 million scene data including additional 250,000 scenes acquired after the previous release. The data are posted on a 1 arc-second (approximately 30 m at the equator) grid [12].

We analyzed ALOS AVNIR-2 image for vegetation density mapping. Moreover, we used run up and inundation information of the 2011 Tohoku

Earthquake which is recorded in the report book of detailed map of the impacts of the 2011 Japan tsunami in Miyagi Prefecture [13]. The step of analysis starts from data collection, surface analysis of DEM data, vector data processing, vegetation density mapping, AHP process and ends with raster overlay processing through GIS approach (Figure 1).

2.2. Study Area

The study area was Ofunato city in Iwate prefecture, Japan (Figure 2). This area is one of the areas affected by The 2011 Tohoku earthquake tsunami, which is the fourth huge earthquake that caused tsunami since 1900, besides the 1960 Chile Earthquake tsunami, the 1964 Alaska earthquake tsunami and the 2004 Sumatra earthquake tsunami [14].

The significant damage to city in previous event was in 1896, 1933 and 1960. In 2013, the city has an estimated population of 39,136. The total area is 323.28 square kilometer. Population residing in inundated area during the 2011 Tohoku Earthquake tsunami was 19,073. Ofunato faces the Pacific Ocean. Outside its bay, the warm and cold ocean currents meet. In this city, the prefectural government provides hazard mapping to the municipal government who then work with community groups to develop the evacuation maps and identify suitable places for refuge. In other hand there is limited illustration of evacuation routes on the maps [15].

2.3. Spatial Analysis

2.3.1. Elevation

Elevation data is one of the principal datasets required for the model to generate vulnerability and inundation of the tsunami. In order to derive a set of parameters that describe the physical vulnerability a digital elevation model was generated by elevation map using ASTER GDEM version 2 data. This data was downloaded from ASTER GDEM website (http://gdem.ersdac.jspacesystems.or.jp/).

We classified elevation into five classes of vulnerability based on the height of the surface. It describes that low elevation (in meter) of the surface will have high vulnerability to tsunami wave, as shown in Table 1. Vulnerability mapping based on elevation data describe that most of coastal area is in the slightly high class of vulnerability (Figure 3(a)).

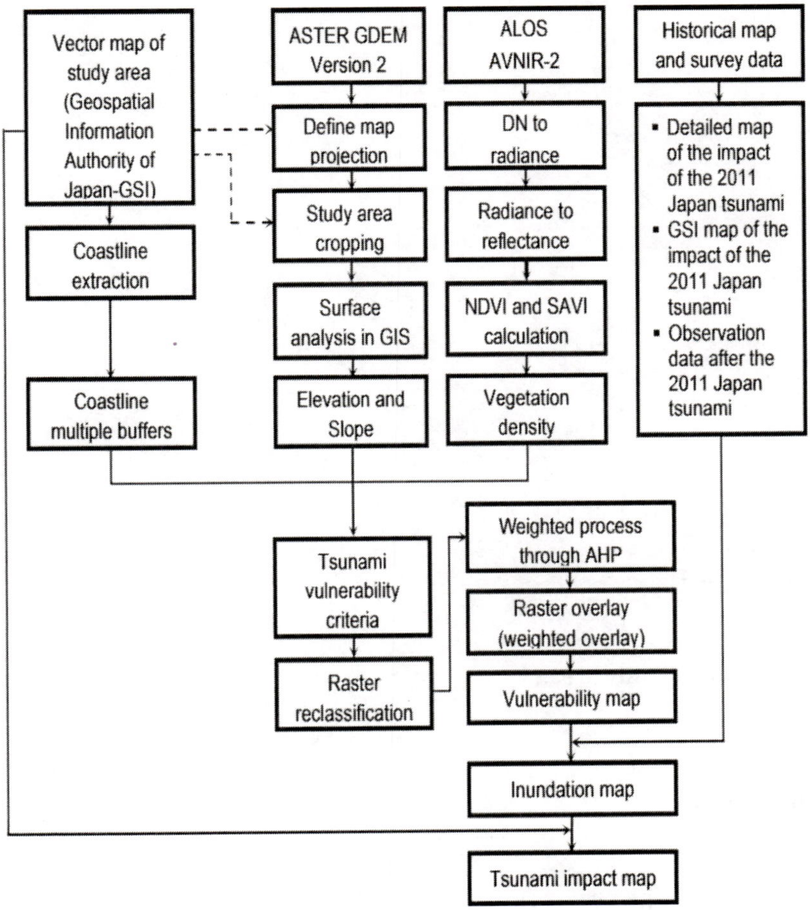

Figure 1. General framework of the study.

2.3.2. Slope

Slope is the rate of maximum change in z-value from each cell of the image. We created slope map using the third-order finite difference method proposed by [16]. Slope percentages range from 0 to near infinity. A flat surface is 0 percent, and as the surface becomes more vertical, the slope percent becomes increasingly larger. Slope map was created from ASTER GDEM version 2. We classified slope map into five classes according to the value of tsunami vulnerability as described in Table 1. We applied slope classification based on tsunami vulnerability class through reclassify process to generate the vulnerability map (Figure 3(b)). High vulnerability areas based on slope was identified in the urban area close to the river in the southeast part of Ofunato.

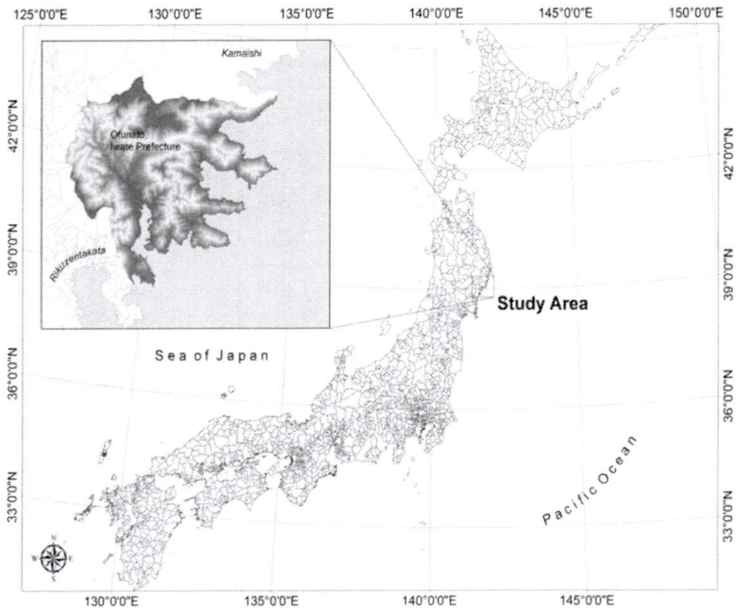

Figure 2. Map of study area.

Table 1. Vulnerability class based on elevation [17], slope [18], and coastline distance.

Elevation (meter)	Slope (percentage)	Coastline distance (meter)	Vulnerability class
<5	0 - 2	0 - 603.50	High
5 - 10	2 - 6	603.50 - 1451.76	Slightly high
10 - 15	6 - 13	1451.7 - 2453.60	Medium
15 - 20	13 - 20	2453.60 - 3572.06	Slightly low
>20	>20	>3572.06	Low

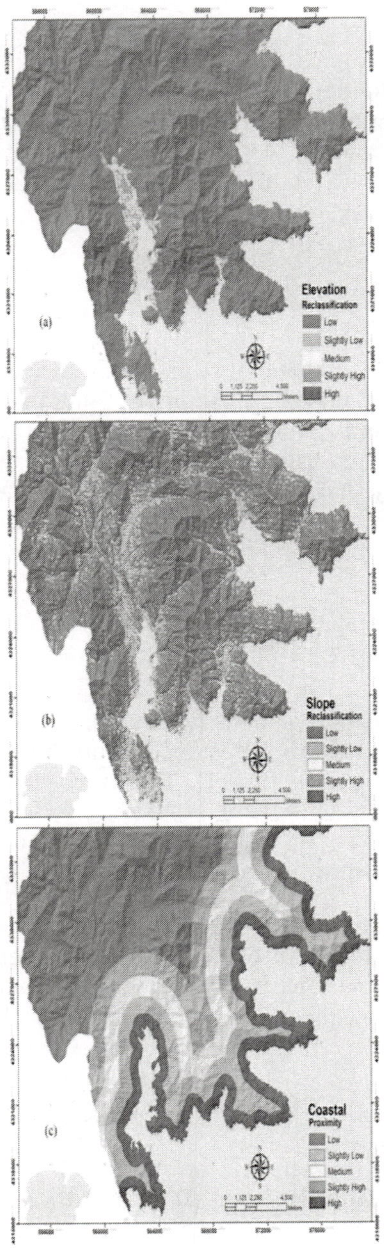

Figure 3. Tsunami vulnerability map based on the parameter of (a) Elevation; (b) Slope; and (c) Coastline distance.

2.3.3. Coastline Distance

The calculation of the distance from coastline to the land using multi-buffering was done to divide area into five classes of vulnerability. The buffering distance was set based on the possibility range of the tsunami to reach the land. The distance is depend on the historical report of the maximum run up in the area of study, and is expressed by Equation (1) [9,19]:

$$\log X_{max} = \log 1400 + \frac{4}{3}\log\left(\frac{Y_o}{10}\right) \quad (1)$$

X_{max} represents the maximum reach of the tsunami over land, and Y_o is the tsunami height at the coast. Maximum run-up of Ofunato area based on The 2011 Earthquake Tsunami Joint Survey Group was 30.1 meter and minimum run up was 5.32. By using the algorithm above, we generated five range of coastline distance that vulnerable to the tsunami as shown in Table 1. We applied coastline distance in vector map of Ofunato through GIS processing to generate vulnerability map based on coastline distance (Figure 3(c)).

2.4. ALOS AVNIR-2 Image Processing

We tried to use SAVI for vegetation density mapping. The digital numbers of ALOS AVNIR-2 were converted to reflectance values before generating the synthetic NDVI and SAVI image [20,21]. The steps to create vegetation density map as follows.

2.4.1. Reflectance Calculation

Radiance convertion from digital number values is needed before generating reflectance. Calculation of radiance is the fundamental step in relating image data from multiple sensors and platforms into a common radiometric scale [22]. The general equation for DN to radiance conversion as in Equation (2) [23]

$$L_\lambda = G_{rescale} \times QCAL + B_{rescale} \quad (2)$$

L_l is spectral Radiance at the sensor's aperture (W/m^2/sr/µm), $G_{rescale}$ is rescaled gain, QCAL is Digital Number (DN), and $B_{rescale}$ is rescaled bias. Moreover, Equation (3) was used to convert radiance to reflectance [24]:

$$\rho_\lambda = \pi \times L_\lambda \times d^2 / ESUN_\lambda \times \cos\theta_s \quad (3)$$

ρ_l is unitless planetary reflectance, L_l is spectral radiance at the sensor's aperture, d^2 is earth-sun distance in astronomical units from nautical handbook, $ESUN_l$ is mean solar Exoatmospheric irradiances, and q_s is solar zenith angle in degrees.

2.4.2. NDVI Calculation

The principle behind NDVI is that channel 1 is in the red-light region of the electromagnetic spectrum where chlorophyll causes considerable absorption of incoming sunlight, whereas channel 2 is in the near-infrared region of the spectrum where a plant's spongy mesophyll leaf structure creates considerable reflectance. Consequently, dense vegetation shows up very strongly in the imagery where red spectrum reflectance is low and in contrast high near infrared reflectance [25-27].

NDVI is a measure of the difference in reflectance between these wavelength ranges that takes values between −1 and 1, with values 0.5 indicating dense vegetation and values < 0 indicating no vegetation including water. NDVI was estimated using Equation (4) [28]

$$\mathrm{NDVI} = \frac{(\mathrm{NIR} - \mathrm{VIS})}{(\mathrm{NIR} + \mathrm{VIS})} \tag{4}$$

NIR is Near Infra-Red band, and VIS is visible band of red band. ALOS AVNIR-2 band 3 is red and band 4 is NIR.

2.4.3. SAVI Calculation

SAVI is one of the algorithms developed to generate vegetation index by eliminating soil factor [29]. Although SAVI model is not significant in the intra-class analysis, in the similar spectral space, this model presents a reasonable performance in the characterization of forested and non-forested areas [21]. Atmospheric influences lower the NDVI of vegetated areas whereas dark or wet soil backgrounds increase the NDVI for given vegetation canopy [30]. SAVI was developed to minimize soil brightness, including shadow, influences found in the NDVI by accounting for first-order soil-vegetation spectral interactions as in Equation (5) [31].

$$\mathrm{SAVI} = \frac{(\mathrm{NIR} - R)}{(\mathrm{NIR} + R + L)} \times (1 + L) \tag{5}$$

NIR is Near Infra-Red band, R is red band, and L is soil calibration factor, in which L equal to 0.5 [32]. A correction factor (L) was used to minimize the secondary backscattering effect of canopy transmitted-soil background reflected radiation. L value of 1 was optimal in semiarid environments. SAVI map is shown in Figure 4(a).

SAVI map was used to map the vegetation coverage. The image was used to identify the density of vegetation and to map vulnerability area of tsunami. We assumed that this area is more vulnerable to tsunami attack than another area. Vegetation density of SAVI describes the minimum value was −0.22, while the maximum was 0.38. We reclassify SAVI value to generate the vegetation density map in five classes (low, slightly low, medium, slightly high and high density) as shown in Figure 4(b), which then will be classified based on tsunami vulnerability. We assumed that low density of vegetation has a high vulnerability to tsunami. The result of this classification as shown in Figure 4(c). The lowest value of vegetation index found in the coastal area of Ofunato where indicated as building, bare soil, rocky land and water body.

3. RESULTS

3.1. AHP Processing

All parameters were in grid cells which are then classified into five classes of vulnerability. For the vulnerability classes we used integer numbers of 1, 2, 3, 4, and 5 that represent low, slightly low, medium, slightly high and high vulnerability. We then overlay the entire parameter of tsunami vulnerability by using weighting and scoring system. Scoring is intended to assess the limiting factor on each parameter, while weighting is based on the dominant influence of these parameters in determining the class of tsunami vulnerability.

We applied AHP in assigning weights to each of the parameters. By applying pair-wise comparisons, AHP helps in creating a scaled set of preferences and describing the importance of each parameter relative to other parameter [33-35]. AHP allows the investigators in assessing the vulnerability related to the natural hazard. A number of 1 until 9 as shown in Table 2 is used in comparison scale to create pair-wise comparison matrix. This number is depends on the relative importance of each parameter. The relative weights of each parameter will be produced through pair-wise comparison as shown in Figure 5.

Table 2. Nine-point comparison scale by Saaty [37].

Intensity of importance	Definition
1	Equal importance
3	Weak importance of one over another
5	Essential or strong importance
7	Demonstrated importance
9	Absolute importance
2, 4, 6, 8	Intermediate value between the two adjacent judgments

The calculation of consistency level is needed prior to the spatial analysis. Consistency Ratio (CR) is a procedure for determining the index of consistency. It indicates the probability that the matrix judgments were randomly generated [33]. AHP tolerates inconsistency through the calculation of consistency ratio. Acceptable consistency ratio is less than or equal to 10 percent [36]. CR is the ratio between the consistency index (CI) and random consistency index (RI) and can be expressed using Equation (6).

$$\mathrm{CR} = \frac{\mathrm{CI}}{\mathrm{RI}}, \text{and } \mathrm{CI} = \frac{(\lambda_{\max} - N)}{(N-1)} \tag{6}$$

l_{max} represents the largest eigenvalue, and N the size of comparison matrix. In this study N = 4. l_{max} is calculated from the sum of all parameter which is then multiplied by its eigenvector. RI is based on the random consistency index (RI = 0.09 for four parameters).

Normalized matrix as shown in Figure 5 was calculated from the pair-wise comparison in five iterations, and it shows that CI was 0.057, while CR was 6.4%. The pair-wise comparison describes that elevation is the most important factors followed by slope, coastline distance, and vegetation density.

3.2. Spatial Analysis for Vulnerability Mapping

After creating the parameter maps, we applied weighted overlay through spatial analyst in GIS. The weighted overlay tool applied to solve multi criteria problems. It is a method for applying a common scale of values to diverse input parameters which have different importance for creating an integrated analysis. The general step of this process as shown in Figure 6(a).

A number of 1 (low vulnerability) to 5 (high vulnerability) was used to represent vulnerability classes. Then we calculated the statistics of vulnerability map. The vulnerability classes as shown in Table 3 was created using the subtraction of maximum and minimum value and divided by number of class. Subsequently tsunami vulnerability in five classes is mapped (Figure 6(b)).

3.3. Tsunami Inundation and Impact Assessment

We compared our result to the survey data from Joint Survey Group of the 2011 Tohoku Earthquake, the existing map of tsunami inundation area due to the 2011 Tohoku Earthquake published by Geographical Survey Institute (GSI) Japan (Figure 7), and detailed map of the impacts of the 2011 Japan tsunami, vol. 1: Aomori, Iwate and Miyagi Prefecture [13].

The comparison describes that most of inundation area was occurred in the area of high and slightly high vulnerability with an area of 1.10 square kilometer and 6.29 square kilometer. Based on these areas we estimate that total area of the possibility of inundation is 7.39 square kilometer (Figure 8(a)), while the inundation area due to due to 2011 Tohoku earthquake based on GSI report is 8 square kilometer. Maximum run-up of Ofunato area based on The 2011 Earthquake Tsunami Joint Survey Group was 30.1 meter and found in the area of Shirahama Sanrikucho Ryori. Minimum run up was 5.32 meter at the area of Sakaricho. This area is identified as the high vulnerability area. We estimated that inundation area will cover 8.13 square kilometer of building area along coastal area of Ofunato (Table 4 and Figure 8(b)).

Inundation area as shown in Figure 8(a) describe that the farthest area of inundation could reach until 3.5 kilometer from coastline. In this area, the elevation was 12 meter, while reflectance value was 0.13, and SAVI was 0.05. Elevation map created from ASTER GDEM version 2 describes that the highest elevation in the inundation area is 20.5 meter.

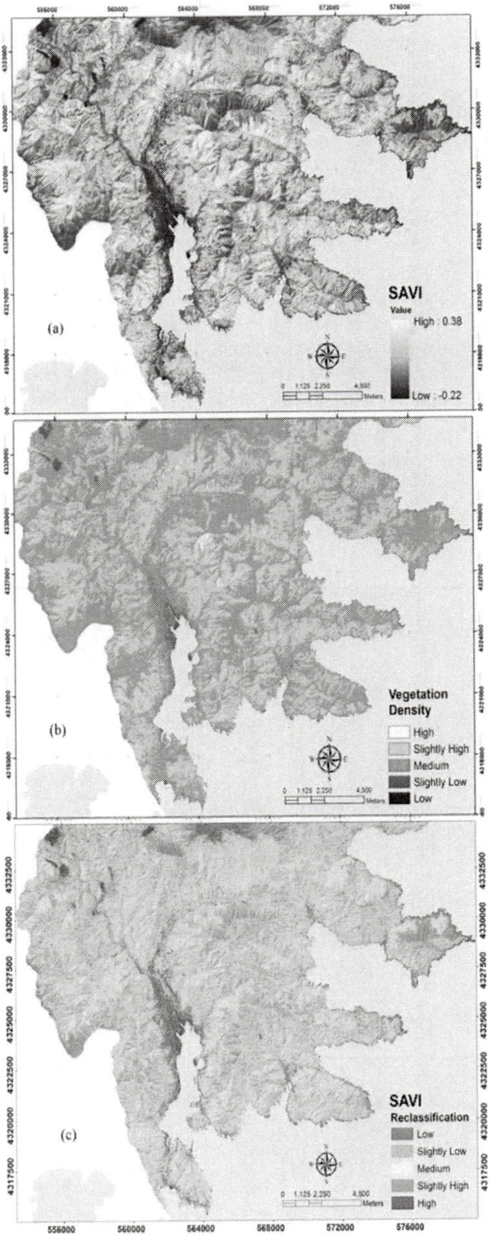

Figure 4. (a) SAVI map; (b) Vegetation density; and (c) Vulnerability map based on vegetation density.

Pair - wise comparison matrix

	Elevation	Slope	Coastline distance	Vegetation density
Elevation	1	2	3	3
Slope	1/2	1	2	2 1/2
Coastline distance	1/3	1/2	1	3
Vegetation density	1/3	0.4	1/3	1

Normalized matrix

Elevation	0.4615	0.5129	0.4737	0.3158
Slope	0.2308	0.2564	0.3148	0.2632
Coastline distance	0.1539	0.1282	0.1579	0.3158
Vegetation density	0.1539	0.1026	0.0526	0.1053

5th iteration

0.4595	0.4595	0.4595	0.4595	45.94%
0.2554	0.2554	0.2554	0.2554	25.53%
0.1671	0.1671	0.1671	0.1671	16.71%
0.1181	0.1181	0.1181	0.1181	11.81%

Figure 5. Pair-wise comparison and normalized matrix.

Table 3. Vulnerability classes.

Vulnerability Class	Area (square kilometer)	Percentage (%)
Low (1)	234.79	73.63
Slightly low (2)	71.10	22.30
Medium (3)	5.59	1.75
Slightly high (4)	6.29	1.97
High (5)	1.10	0.34

Table 4. Vulverability class of building area.

Vulnerability Class	5	4	3	2	1
	Inundation Area				
Building area (square kilometer)	3.35	4.78	6.25	1.71	6.01

4. DISCUSSION

Vulnerability is related to the capacity to reduce the impact of disaster and mitigation plan. It describes the potential area to be damaged by natural disasters. Class of vulnerability could be based on a physical parameter, such as elevation, slope, coastline distance, land cover and coastal shape. Moreover, inundation can be defined as the result of a tsunami traveling a long distance inland and is a horizontal measurement of the path of the tsunami. The analysis of satellite image, elevation data, and survey data followed by multi-criteria analysis through AHP and raster overlay in GIS processing can be used as the basic information for vulnerability mapping, inundation mapping, and impact assessment due to tsunami disaster.

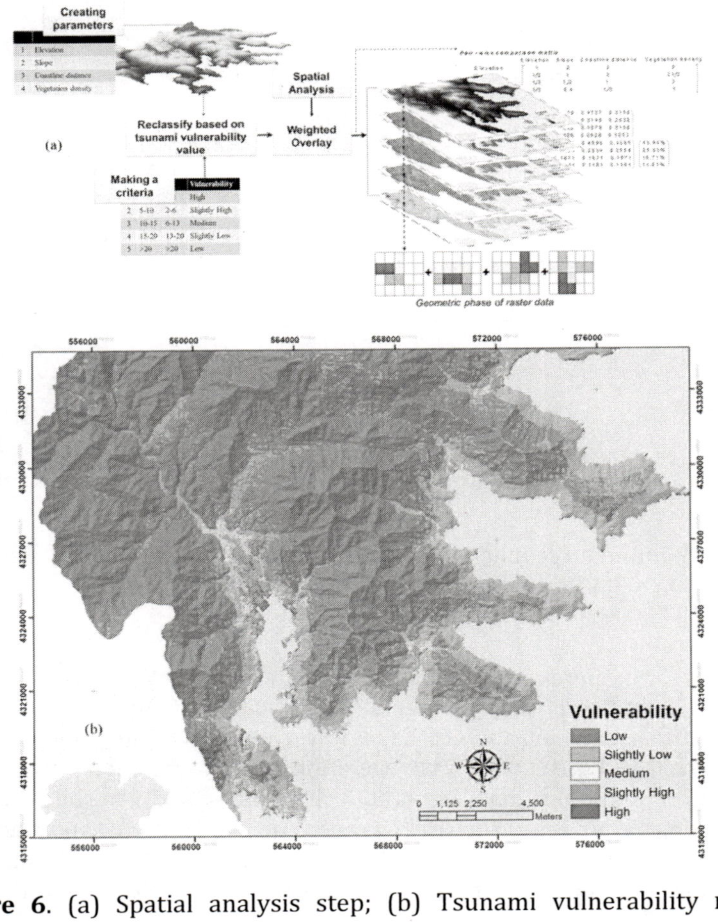

Figure 6. (a) Spatial analysis step; (b) Tsunami vulnerability map of Ofunato area.

The use of AHP method helps in the analysis of spatial multi-criteria where all of the parameters used in this study were calculated based on its weight factor to create vulnerability map. This study is a first attempt to assess tsunami vulnerability by using the parameter of SAVI besides elevation, slope and coastline distance, and apply AHP methods combining with raster overlay through GIS processing in Ofunato area, Japan. According to the result of this study, a simple method for inundation prediction can be a valuable step for carrying out a prelimnary tsunami vulnerability mapping and impact assessment when the high resolution of DEM data and detailed topographic data is not available.

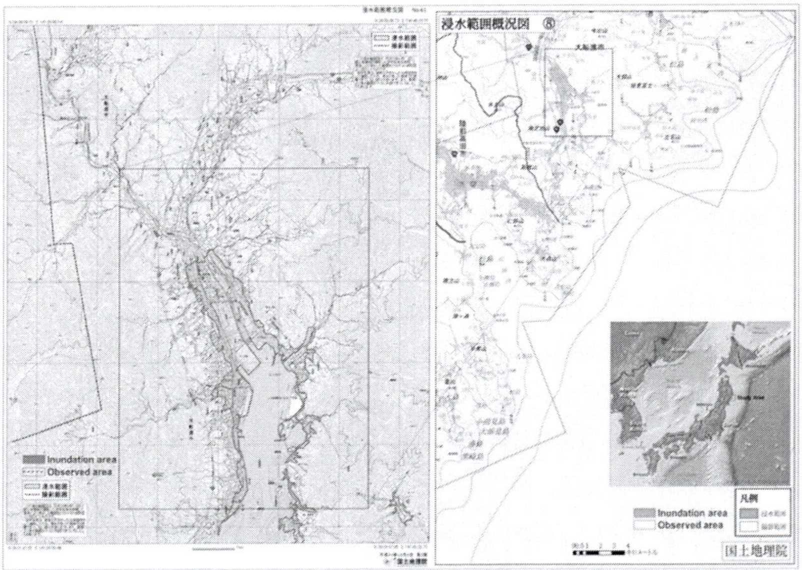

Figure 7. Inundation map of Ofunato during the 2011 Japan Tsunami, published by GSI [38] (modified).

Elevation is the parameter that has the highest weight and 2.31% of the area that identified as slightly high and high class of tsunami vulnerability found in the coastal area which has a lower elevation. Most of the area is highly developed areas with low vegetation density. Tsunami vulnerability map and inundation map generated in this study can be used for determining a priority for risk prevention, mitigation, and land-use planning to reduce the tsunami risk. Particularly, in the slightly high and high vulnerability area that can be identified as possible inundation area describe that 11.85% of building area has potential consequences of a tsunami impact.

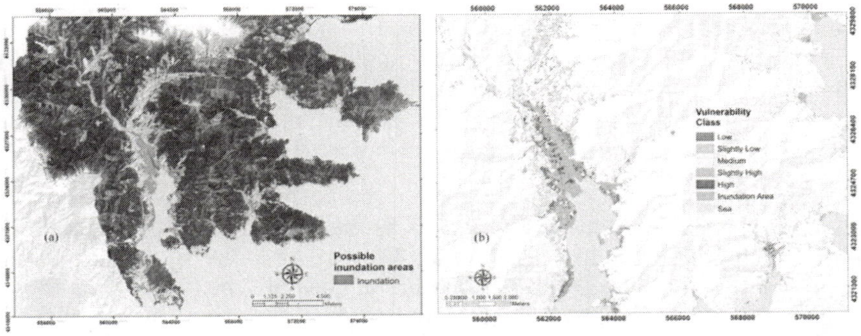

Figure 8. (a) Map of possible inundation area in ofunato; (b) Tsunami vulnerability class of building area in Ofunato.

In this research, five classes of vulnerability were used.

Vulnerability map describes that 73.63% of the area is low vulnerability, 22.30% is slightly low vulnerability, 1.75% is medium, 1.97% is slightly high vulnerability, and 0.34% is high vulnerability. The high vulnerability areas were mostly found in the coastal area with the sloping coast type. Inundation areas were predicted in areas that identified as high vulnerability and slightly high vulnerability area. In addition, we assumed that vegetation may play an important role as tsunami barriers to reduce the impact of the tsunami destruction. Based on the vegetation density that created from SAVI value in the study area, we found that 52.09% of the inundation area was in slightly low class of vegetation density and 34.05% of the inundation area was in medium class of vegetation density.

5. CONCLUSIONS

GIS application followed by satellite image processing and AHP approach is useful for tsunami vulnerability mapping and impact assessment. It can be used for the evacuation and reconstruction plan due to tsunami disaster. GIS indicate the vulnerability area due to tsunami and describe the possibility area that could be affected by tsunami wave. In this study we applied four parameters in order to create a tsunami vulnerability map and to predict the possibility of inundation area in the area of Ofunato. The vulnerability map showed that most of the coastal areas are vulnerable to tsunami. Most of buildings area and bare land along coastal area was in the high vulnerability of tsunami. The inundation pattern as the result of this study shown similar compared to inundation area of the 2011 Tohoku Earthquake in the area of Ofunato. ASTER GDEM version 2 is useful for tsunami vulnerability mapping for the area where the high resolution of DEM data is not available. Building and residence area in potentially affected areas should be provided with basic information on tsunami risk because awareness and preparedness are the most important factors to reduce potential losses due to a tsunami impact.

For the application in mitigation plan, we recommend to the user to be aware of the assumptions made, as well as the limitations within this study. Better mapping of tsunami vulnerability can be done by adding other appropriate parameters. The parameters of coastal type, relative direction of tsunami, and coastal bathymetry can be used. Moreover, the use of other DEM data, vulnerability mapping in other location and overlaying the vulnerability map with the current land use map will be a future works.

ACKNOWLEDGEMENTS

Abu Bakar Sambah, Fusanori Miura (JAXA) for the ALOS images and GSI for providing the base map of the study area. Authors are thankful to the 2011 Tohoku Earthquake Tsunami Joint Survey for the survey data in the study area. We also thank to Indonesia Directorate General for Higher Education for the scholarship and Yamaguchi University Japan for research support.

REFERENCES

1. M. Papathoma and D. Dominey-Howes, "Tsunami Vulnerability Assessment and Its Implications for Coastal Hazard Analysis and Disaster Management Planning, Gulf of Corinth, Greece," Natural Hazards and Earth System Sciences, Vol. 3, No. 6, 2003, pp. 733-747.
2. United Nations Disaster Relief Coordinator (UNDRO), "Mitigation Natural Disasters: Phenomena, Effects and Options, a Manual for Policy Makers and Planners," United Nations Publication, New York, 1991.
3. M. Pelling, "The Vulnerability of Cities: Natural Disasters and Social Resilience," Earthscan Publications, London, 2003.
4. C. Bollin, C. Cardenas, H. Hahn and K. S. Vatsa, "Natural Disaster Network; Disaster Risk Management by Communities and Local Governments," Inter-American Development Bank, Washington DC, 2003. http://www.iadb.org/sds/doc/GTZ%2DStudyFinal.pdf
5. P. Schmidt-Thome (Editor), "Natural and Technological Hazards and Risks Affecting the Spatial Development of European Regions," Special Paper 42, Geological Survey of Finland, 2006.
6. R. S. Mahendra, P. C. Mohanty, H. Bisoyi, K. T. Srinivasa and S. Nayak, "Assessment and Management of Coastal Multi-Hazard Vulnerability along the Cuddaloree

Villupuram, East Coast of India Using Geospatial Techniques," Ocean & Coastal Management, Vol. 54, No. 4, 2011, pp. 302-311.
7. S. J. Carver, "Integrating Multi-Criteria Evaluation with Geographical Information Systems," International Journal of Geo-Information System, Vol. 5, No. 3, 1991, pp. 321-339.
8. P. Jankowski, "Integrating Geographical Information Systems and Multiple Criteria Decision Making Methods," International Journal of Geo-Information System, Vol. 9, No. 3,1995, pp. 251-273.
9. T. P. T. Sinaga, N. Adhi, Y.-W. Lee and S. Yongcheol, "GIS Mapping of Tsunami Vulnerability; Case Study of the Jembrana Regency in Bali, Indonesia," KSCE Journal of Civil Engineering, Vol. 15, No. 3, 2011, pp. 537-543.
10. G. Strunz, J. Post, K. Zosseder, S. Wegscheider, M. Muck, T. Riedlinger, H. Mehl, S. Dech, J. Birkmann, N. Gebert, H. Harjono, H. Z. Anwar, Sumaryono, R. M. Khomarudin and A. Muhari, "Tsunami Risk Assessment in Indonesia," Natural Hazards and Earth System Sciences, Vol. 11, No. 1, 2011, pp. 67-82.
11. M. Papathoma, D. Dominey-Howes, Y. Zong and D. Smith, "Assessing Tsunami Vulnerability, an Example from Herakleio, Crete," Natural Hazards and Earth System Sciences, Vol. 3, No. 5, 2003, pp. 377-389.
12. T. Tachikawa, M. Kaku, A. Iwasaki, D. Gesch, M. Oimoen, Z. Zhang, J. Danielson, T. Krieger, B. Curtis, J. Haase, M. Abrams, R. Crippen and C. Carabajal, "ASTER Global Digital Elevation Model Version 2—Summary of Validation Results," The NASA Land Processes Distributed Active Archive Center and the Joint Japan-US ASTER Science Team, 2011.https://lpdaacaster.cr.usgs.gov/GDEM/Summary_GDEM2 _validation _report_final.pdf
13. H. Tsuyoshi and A. Iwamatsu, "Detailed Map of the Impacts of the 2011 Japan Tsunami. Vol. 1: Aomori, Iwate and Miyagi Prefecture," Kokon-shoin Publishers Ltd., Tokyo, 2012.
14. N. Mori and T. Takahashi, "Nationwide Post Event Survey and Analysis of the 2011 Tohoku Earthquake Tsunami," Coastal Engineering Journal, Vol. 54, No. 1, 2012, pp. 1-32.
15. S. Fraser, I. Matsuo, G. S. Leonard and H. Murakami, "Tsunami Evacuation: Lessons from the Great East Japan Earthquake and Tsunami of March 11th 2011," GNS Science Report 2012/17, Institute of Geological and Nuclear Sciences Limited, Avalon, 2012.
16. B. K. P. Horn, "Hill Shading and the Reflectance Map," Proceedings of IEEE, Vol. 69, No. 1, 1981, pp. 14-47.
17. K. Iida, "Magnitude, Energy and Generation Mechanisms of Tsunamis and a Catalogue of Earthquakes Associated with Tsunamis," Proceeding of Tsunami Meeting at the 10th Pacific Science Congress, Honolulu, July 1963, pp. 7-18.
18. R. A. Van Zuidam, "Guide to Geomorphologic-Aerial Photographic Interpretation and Mapping," International Institute for Geo-Information Science and Earth Observation, Enschede, 1983.
19. C. L. Bretschneider and P. G. Wybro, "Tsunami Inundation Prediction," Proceedings of the 15th ASCE Conference on Coastal Engineering, Honolulu, 11-17 July 1976, pp. 1006-1024.
20. A. R. Huete, "A Soil-Adjusted Vegetation Index (SAVI)," Remote Sensing of Environment, Vol. 25, No. 3, 1988, pp. 295-309.
21. L. S. Araujo, J. R. dos Santos and Y. E. Shimabukuro, "Relationship Between SAVI and Biomass Data of Forest and Savanna Contact Zone and the Brazilian Amazonia," International Archives of Photogrammetry and Remote Sensing, Vol. 33, 2000, pp. 77-81.
22. G. Chander and B. Markham, "Revised Landsat-5 TM Radiometric Calibration Procedures and Post Calibration Dynamic Ranges," IEEE Transactions on Geoscience and Remote Sensing, Vol. 41, No. 11, 2003, pp. 2674-2677.

23. M. Bouvet, G. Chander, P. Goryl and R. Santer, "Preliminary Radiometric Calibration Assessment of ALOS AVNIR-2," Geoscience and Remote Sensing Symposium, IGARSS, Barcelona, 2007, pp. 2673-2676.
24. National Aeronautics and Space Administration, "Data Users Handbook-Data Products, Chapter 11," 2011. http://landsathandbook.gsfc.nasa.gov/ pdfs/ Landsat7_Handbook.pdf
25. C. J. Tucker, "Red and Photographic Infrared Linear Combinations for Monitoring Vegetation," Remote Sensing of Environment, Vol. 8, No. 2, 1979, pp. 127-150.
26. R. D. Jackson, P. N. Slater and P. J. Pinter, "Discrimination of Growth and Water Stress in Wheat by Various Vegetation Indices Through Clear and Turbid Atmospheres," Remote Sensing of Environment, Vol. 13, No. 3, 1983, pp. 187-208.
27. C. J. Tucker, W. W. Newcomb, S. O. Los and S. D. Prince, "Mean and Inter-Year Variation of Growing-Season Normalized Difference Vegetation Index for the Sahel 1981-1989," International Journal of Remote Sensing, Vol. 12, No. 6, 1991, pp. 1133-1135.
28. M. C. Hansen, R. S. Defries, J. R. G. Townshend and R. Sohlberg, "Global Land Cover Classification at 1 km Spatial Resolution Using a Classification Tree Approach," International Journal of Remote Sensing, Vol. 21, No. 6-7, 2000, pp. 1331-1364.
29. P. Gong, R. Pu, G. S. Biging and M. R. Larrieu, "Estimation of Forest Leaf Area Index Using Vegetation Indices Derived from Hyperion Hyperspectral Data," IEEE Transactions on Geoscience and Remote Sensing, Vol. 41, No. 6, 2003, pp. 1355-1362.
30. A. R. Huete, G. Hua, J. Qi, A. Chehbouni and W. J. D. Van Leeuwen, "Normalization of Multidirectional Red and NIR Reflectances with the SAVI," Remote Sensing of Environment, Vol. 41, No. 2-3, 1992, pp. 143-154.
31. A. R. Huete, "A Soil-Adjusted Vegetation Index (SAVI)," Remote Sensing of Environment, Vol. 25, No. 3, 1988, pp. 295-309.
32. J. R. Jensen, "Remote Sensing of the Environmental: An Earth Resources Perspective," Prentice Hall, Upper Saddle River, 2000.
33. T. L. Saaty, "A Scaling Method for Priorities in Hierarchical Structures," Journal of Mathematical Psychology, Vol. 15, No. 3, 1977, pp. 234-281.
34. T. L. Saaty, "The Analytic Hierarchy Process, Planning, Priority Setting, Resource Allocation," McGraw-Hill, New York, 1980.
35. T. L. Saaty, "Decision Making with the AHP; Why Is the Principle Eigenvector Necessary," European Journal of Operational Research, Vol. 145, No. 1, 2003, pp. 85-91.
36. E. H. Forman and M. A. Selly, "Decision by Objectives," World Scientific Publishing Company, 2001.
37. T. L. Saaty, "Decision Making for Leaders: The Analytical Hierarchy Process for Decisions in a Complex World," The Analytical Hierarchy Process Series, RWS Publication, 1996.
38. Geospatial Information Authority of Japan (GSI), "Tsunami Inundation and Land Use Area of the 2011 Japan Earthquake," Geospatial Information Authority of Japan (GSI Japan), 2011.

CHAPTER 4

Tsunami Hazard Analysis of Future Megathrust Sumatra Earthquakes in Padang, Indonesia Using Stochastic Tsunami Simulation

Ario Muhammad[1,2]*, Katsuichiro Goda[1] and Nicholas Alexander[1]

[1]*Department of Civil Engineering, University of Bristol, Bristol, UK*

[2]*Department of Civil Engineering, University of Narotama, Surabaya, Indonesia*

ABSTRACT

This study assesses the tsunami hazard potential in Padang, Indonesia probabilistically using a novel stochastic tsunami simulation method. The stochastic tsunami simulation is conducted by generating multiple earthquake source models for a given earthquake scenario, which are used as input to run Monte Carlo tsunami simulation. Multiple earthquake source models for three magnitude scenarios, i.e., M_w 8.5, M_w 8.75, and M_w 9.0, are generated using new scaling relationships of earthquake source parameters developed from an extensive set of 226 finite-fault models. In the stochastic tsunami simulation, the effect of incorporating and neglecting the prediction errors of earthquake source parameters is investigated. In total, 600 source models are generated to assess the uncertainty of tsunami wave characteristics and maximum tsunami wave height profiles along coastal line of Padang. The results highlight the influence of the uncertainty of the scaling relationships on tsunami simulation results and provide a greater range of tsunamigenic scenarios produced from the stochastic tsunami simulation. Additionally, the results show that for the future major earthquakes in the Sunda megathrust, the maximum tsunami wave height in Padang areas can reach 20 m and, therefore, significant damage and loss may be anticipated in this region.

Keywords: stochastic tsunami simulation, earthquake source modeling, uncertainty and sensitivity of tsunami hazard, Sunda megathrust, West Sumatra

INTRODUCTION

Located among three major plates, namely the Indian-Australian, the Pacific, and the Eurasian, Indonesia archipelago is one of the most seismically active regions in the world. In the last 20 years (1994–2014), 528 earthquakes occurred in Indonesia, i.e., about 26 earthquakes per year (United States Geological Survey (USGS), 2015). Sumatra Island is the most seismically active region since it is located at the interface between the Indian-Australian and Eurasian Plates. Two major seismic sources are the 1,900-km long Sumatran fault located along the center of Sumatra Island and the Sunda megathrust zone traversing more than 2,000 km along the western coast of Sumatra (Sieh and Natawidjaja, 2000). In the past two decades, several large earthquakes occurred along the Sunda megathrust including the Aceh-Andaman earthquake in December 2004 (M_w 9.15), the Nias earthquake in March 2005 (M_w 8.6), two earthquakes of Bengkulu in September 2007 (M_w 8.4 and 7.9), the Mentawai tsunamigenic earthquake in October 2010 (M_w 7.7), and the Indian Ocean earthquake in April 2012 (M_w 8.6). Two of the most devastating earthquake events among those were the 2004 Aceh-Andaman earthquake triggering large tsunamis along the coastal line of Sumatra, Thailand, Sri Lanka, and India with the casualties of more than 250,000 people and the 2005 Nias earthquake which killed 2,000 people (Hsu et al., 2006; Banerjee et al., 2007).

The 2004 Aceh-Andaman earthquake was caused by the 1,600-km long rupture of the Sunda megathrust starting from North of Simeulue Island to North of Andaman Islands (Meltzner et al., 2006). This failure led to another 400 km rupture of the megathrust fault in the southern part of Simeulue Island (see Figure 1) causing the M_w 8.6 2005 Nias earthquake (Briggs et al., 2006). A paleotsunami study based on a 1,000-year long record of tsunami deposits in North-West of Sumatra suggests that the occurrence interval of tsunamigenic earthquakes (M_w 9.15) from the Sumatra-Andaman region is about 600 years (Monecke et al., 2008). Although the devastating tsunami event might not occur in the next few centuries in the Sumatra-Andaman segment, the ruptures of the megathrust fault have increased the failure probability of the Mentawai segment of the Sunda megathrust areas (see Figure 1) which is located in South of the fault rupture areas of the 2004 and 2005 events (Nalbant et al., 2005; Chlieh et al., 2008; Sieh et al., 2008; Collings et al., 2012, 2013). The past seismicity in the Mentawai segment indicates that there were two major tsunamigenic events occurred in 1797 and 1833 (M_w ~8.8) that affected the coastal areas of Padang and Bengkulu. Geodetic and paleogeodetic studies indicate that the slip deficit accumulated in the Mentawai segment has already exceeded the slip occurred during the 1797 and 1833 earthquakes (Collings et al., 2013). A large earthquake in 2007 (M_w 8.4) that ruptured the Sunda megathrust near the 1833 rupture area was significantly smaller than the accumulated slip since the twin events of the 1797 and 1833 earthquakes (McCloskey et al., 2005; Nalbant et al., 2005). Hence, the possibility of earthquake and tsunami hazards in West of Sumatra from the Mentawai segment which has a recurrence interval of 200 years according to paleoseismological studies remains large (Sieh et al., 2008). In addition, the slip deficit is sufficient to generate a M_w 8.8–9.0 earthquake (Zachariasen et al., 1999; Sieh et al., 2008).

Several earthquake source models have been developed with respect to the unruptured Mentawai segment (Borrero et al., 2006; Aydan, 2008; Griffin et al., 2016) and have been implemented to assess the earthquake and tsunami potential in several highly populated areas along the western coast of Sumatra, i.e., Padang, Painan, Bengkulu, and Pariaman (Borrero et al., 2006; McCloskey et al., 2008; Muhari et al., 2010, 2011). A wide range of rupture scenarios is essential for evaluating the earthquake and tsunami risk potential in coastal areas to capture worst (extreme) cases for emergency response preparedness and risk mitigation actions. However, except for the investigations by McCloskey et al. (2008) and Griffin et al. (2016), those studies implemented uniform slip models that oversimplify the earthquake source characteristics and considered a limited number of scenarios for future tsunamigenic events. On the other hand, McCloskey et al. (2008) considered the uncertainty of slip distribution by implementing the heterogeneous spatial distribution of slips based on a methodology proposed by Mai and Beroza (2002)and produced more than 100 scenarios to assess the tsunami hazards along the western coast of Sumatra. However, the events evaluated by Mai and Beroza (2002) were crustal earthquakes of magnitudes up to 8 and were not tsunamigenic. Recently, the Mai–Beroza method has been extended to apply to M_w 9.0 megathrust subduction earthquakes by adopting inverted source models from the 2011 Tohoku, Japan earthquake (Goda et al., 2014, 2015). Griffin et al. (2016) generated heterogeneous earthquake slips to assess tsunami hazard in Mentawai

Islands based on the random slip modeling proposed by Gallovič and Brokešová (2004). Up to 15 million random slip models were generated using existing scaling relationships that were based on only seven subduction earthquake events in Sumatra and eventually 1,000 tsunami simulations from those slip models were further performed to assess the tsunami hazard in Mentawai Islands.

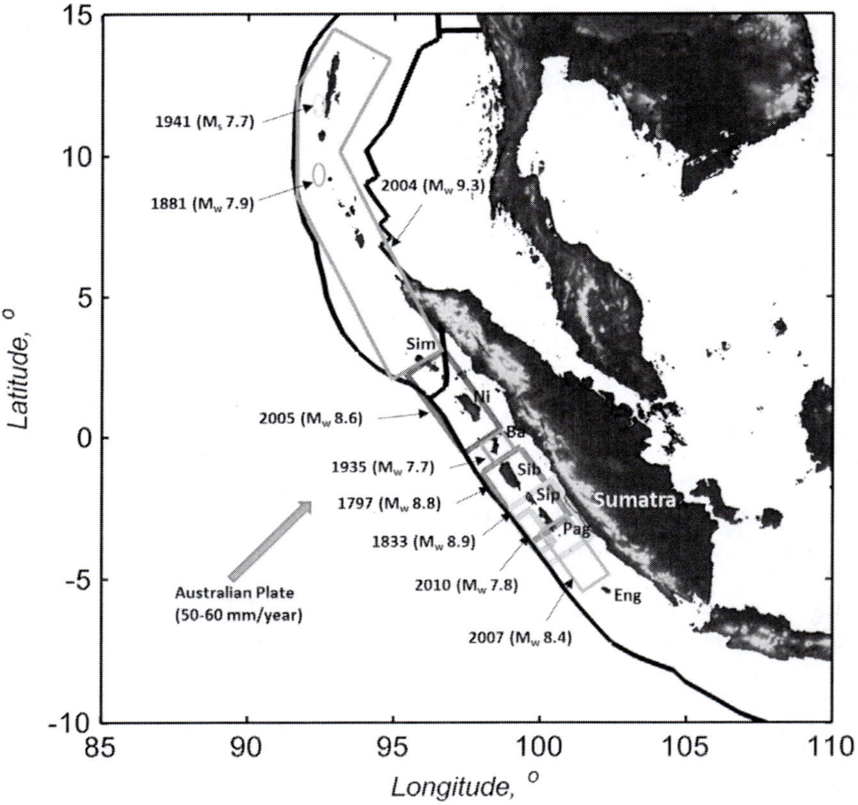

FIGURE 1. Historical recent seismic activities in the sumatra areas (sim: simeulue, ni: nias, ba: batu islands, sib: sibereut, sip: sipora, pag: pagai islands, and eng: enggano).

Moreover, those previous investigations for the Mentawai-Sunda subduction zone adopted the global empirical scaling relationships [e.g., Mai and Beroza (2002), Gallovič and Brokešová (2004), and Aydan (2008)] to generate only either deterministic fault geometry parameters (width and length) or slip distribution parameters without considering the uncertainty and relationships among earthquake source parameters. Recently, new probabilistic scaling relationships of fault geometry, slip statistics, and spatial slip heterogeneity parameters have been developed by Goda et al. (2016) using numerous inversion models (226 models) from the SRCMOD database (Mai and Thingbaijam, 2014) and can be used for tsunami hazard analysis. In those previous studies, the evaluation of regional earthquake source parameters from the finite-fault models of the past Sunda subduction earthquakes with respect to the global empirical relationships is also neglected. Hence, it is highly desirable to generate multiple earthquake source models by taking into account all relevant source parameters that are consistent with the regional source characteristics of the future tsunamigenic earthquakes in the Mentawai-Sunda zone.

Additionally, within the area of the Mentawai segment, Padang is one of the most anticipated areas to be affected by the tsunami compared to the other areas in the western coast of Sumatra. With the total population of 850,000 people, the social and economic impacts due to the future tsunamigenic earthquakes are high. The investigations considering multiple earthquake scenarios by McCloskey et al. (2008) and Griffin et al. (2016) only estimated the maximum tsunami height along the western coast of Sumatra and assessed the tsunami hazard in Mentawai Islands but excluded a rigorous evaluation in Padang areas. Moreover, the past tsunami hazard assessment studies in several important cities along the western coast of Sumatra, i.e., Padang, Painan, and Bengkulu, were performed using deterministic earthquake scenario approaches only (Borrero et al., 2006; Li et al., 2012). Therefore, it is important to apply the stochastic tsunami simulation method to assess the tsunami hazard in Padang probabilistically due to the future megathrust earthquakes in the Mentawai segment of the Sunda subduction zone.

The main objectives of this study are (1) to develop stochastic earthquake slip models for the future tsunamigenic earthquakes in the Mentawai segment of the Sunda subduction zone, (2) to evaluate the impact of stochastic earthquake slip on tsunami simulation results in terms of tsunami wave profiles and maximum tsunami height along the coastal line of Padang by considering the uncertainty and dependency of the earthquake source parameters, and (3) to assess the tsunami hazard in Padang using a wide range of earthquake scenarios generated from the novel stochastic tsunami simulation method. Extensive tsunami simulation for the future tsunamigenic earthquakes is conducted by developing a large number of stochastic earthquake slip models for different magnitude ranges. Three magnitudes, i.e., M_w 8.5, M_w 8.75, and M_w 9.0, are selected to develop stochastic source models. The earthquake source parameters from the finite-fault models of the past Sunda subduction earthquakes are first calculated and then compared with the global scaling relationships developed by Goda et al. (2016) to validate the applicability of the global models to the Sunda subduction zone. The verified scaling relationships are further used to generate the earthquake source models for tsunami simulation. Uncertainty and dependency of the earthquake source parameters are taken into account in producing earthquake source models stochastically which have not been implemented in the past studies of the tsunami hazard analysis in West of Sumatra. In total, 600 synthetic earthquake slip models are generated to obtain multiple realizations of maximum tsunami wave heights at various locations in Padang areas. For validation purposes, the simulated tsunami wave profiles for the M_w 9.0 scenario are used to compare with the results by Muhari et al. (2010) because they used M_w8.92 to define their earthquake source scenario. The tsunami hazard analysis in Padang areas are further performed by evaluating the tsunami wave height profiles and the maximum tsunami wave height along the coastal line of Padang.

In this study, the tsunamigenic earthquake potential of the Mentawai segment in the Sunda subduction zone is first discussed. The earthquake source parameters from the finite-fault models of the past Sunda subduction are further evaluated to determine the applicability of the global scaling relationships for the Mentawai-Sunda subduction zone. A summary of the stochastic tsunami simulation used in this study is then presented, and the stochastic source models for the Mentawai-Sunda megathrust are further developed. Subsequently, the main tsunami simulation results using different earthquake source models in Padang areas are discussed. To demonstrate the tsunami simulation results in comparison to the previous work, the results for the M_w 9.0 scenario are presented first. The tsunami simulation results for the other scenarios are then discussed to evaluate the tsunami hazard potential in Padang areas. Finally, the key conclusions of this work are drawn.

Tsunami Potential of the Mentawawi-Sunda Megathrust Zone

Extensive paleogeodetic, geodetic, and numerical modeling studies suggest that the potential of megathrust tsunamigenic earthquakes in the Mentawai segment of the Sunda subduction zone is high (Natawidjaja et al., 2006; McCloskey et al., 2008; Collings et al., 2012, 2013). The past seismicity in the Mentawai segment indicates that the most destructive historical event in this segment occurred in 1833. The shaking was reported from Bengkulu to Pariaman and near Pagai Islands. Tsunamis were observed along the western coast of Sumatra extending from Pariaman to Bengkulu due to this event. Bengkulu and Indrapura areas were greatly affected by the 1833 tsunamigenic event. The tsunami heights reaching 3–4 m were recorded near Padang. Another historical earthquake event occurred in 1797 produced a destructive tsunami at Padang and nearby. The shaking was the strongest in living memory in Padang, and the tsunami flow depth in Padang was about 5 m (Natawidjaja et al., 2006).

Rupture scenarios of the 1797 and 1833 earthquakes were developed based on the seismotectonic features, geodetic, and paleogeodetic measurements (see Figure 2). The seismotectonic study by Newcomb and McCan (1987) concluded that the rupture of the 1833 earthquake extended ~300 km from near Enggano Island in South to Batu Islands in North with the earthquake size of M_w 8.7–8.8. A paleogeodetic study by Natawidjaja et al. (2006) based on the measurements of the coral microatoll uplift confirmed that the uplift between 1 and 3 m occurred over a 170-km long stretch of the Sumatran outer arc ridge. Elastic dislocation modeling of those uplift data yielded the slip prediction of 9–18 m between 2°S and 5°S. The 220-km long rupture extension from the southern part of the uplift was defined as the south-eastern boundary of the 1833 fault rupture. In addition, the north-western limit of the rupture was likely to be at 2°S; the end point may be extended as much as 160 km farther North-West with much smaller amounts of slip. The North-West extension of 160 km beyond the rupture of the 1833 event was likely to stop at 0.5°S. The earthquake size was predicted in the range of M_w 8.7–8.9. On the other hand, the records of coral and microatoll uplift due to the 1797 earthquake showed that the 1797 event preceded the 1833 giant earthquake by 37 years. The south-eastern limit of the rupture was at about 3.2°S since the south-eastern limit of the uplift due to the 1797 event was on South Pagai Island. The slip was estimated to be in the range of 4–8 m with the depth from 34 to 50 km. In addition, the magnitude of the earthquake was estimated to be in the range of M_w 8.5–8.7 (Natawidjaja et al., 2006).

A recent significant earthquake occurred in the Mentawai segment was the M_w 8.4 12 September 2007 Bengkulu earthquake. Twelve hours later, a subsequent fault rupture produced another major earthquake of M_w 7.9. The geodetic and paleogeodetic modeling suggested that the rupture areas of these two events extended from North of Sipora Island at ~2°S to South of Pagai Islands at ~5°S. The maximum slip from this event is only a half of the maximum slip from the 1833 earthquake. Moreover, the total seismic moment released from the two earthquakes of Bengkulu in 2007 was significantly smaller than the 1833 rupture and the accumulated moment deficit since the last rupture in the Mentawai segment. Therefore, the potential for large tsunamigenic events in the Mentawai segment remains high (Konca et al., 2008). In addition, a 700-year sea-level change recorded

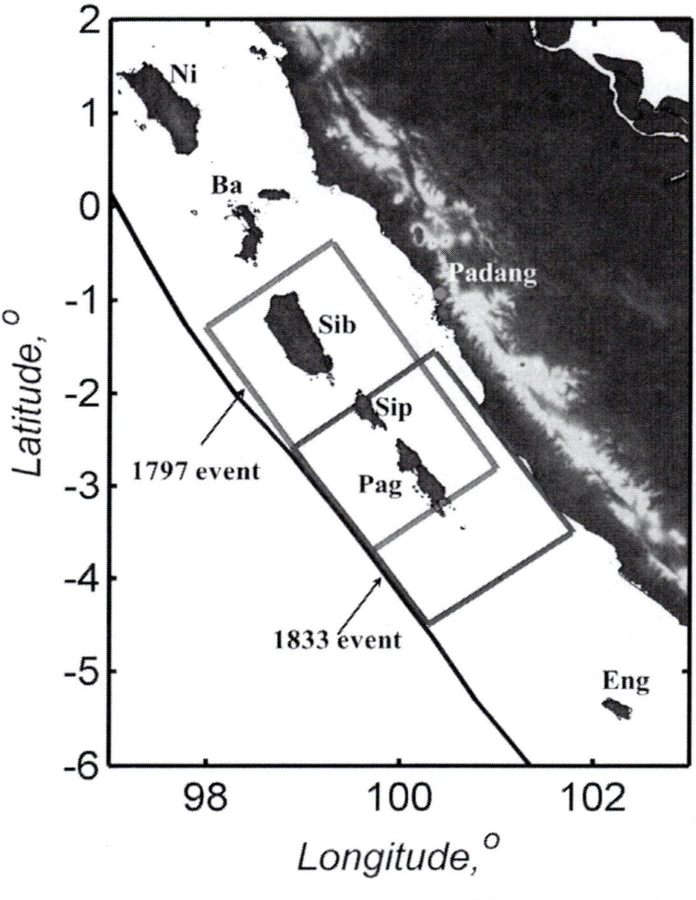

FIGURE 2. Geometry of earthquake source models of the 1797 and 1833 events (ni: nias, ba: batu islands, sib: sibereut, sip: siora, pag: pagai islands, and eng: enggano).

in the corals of the Mentawai segment implies that the recurrence time of the major earthquakes, i.e., the sequence of the 1797 and 1833 earthquakes, is ~200 years (Sieh et al., 2008). At least two of the three ancient sequences began with events that were smaller than the main events and in this context, the 2007 earthquakes may be considered to be only the beginning of an episode of the rupture of the Sunda megathrust in the Mentawai segment (Sieh et al., 2008). The failure of the Mentawai segment may significantly affect the western coast of Sumatra specifically in Padang areas. With the plain topographic features and high population density in urban areas, Padang will face significant economic and social losses due to the future tsunamigenic event in the Mentawai segment of the Sunda subduction zone.

Earthquake Source Parameters for Sumatra Earthquakes

This study implements the stochastic tsunami simulation to assess the tsunami hazard in Padang areas. To run the stochastic tsunami simulation, earthquake source models need to be generated stochastically. Predicting the earthquake source parameters, i.e., the geometry of the fault, slip statistics, and spatial slip distribution parameters, are needed to generate earthquake source models. In order to generate the earthquake source parameters, the fault length (L), fault width (W), mean slip (D_a), maximum slip (D_m), Box–Cox parameter (λ), correlation length along strike direction (A_x), correlation length along dip direction (A_z), and Hurst number (H) of 19 finite-fault models of the past Sunda subduction earthquakes are calculated. The width, length, strike, and dip angles define the geometry of the fault plane, while the mean slip, maximum slip, and Box–Cox parameter characterize the slip statistics values. In addition, the correlation lengths and Hurst number are used to model the spatial heterogeneity of the slip values. Subsequently, the calculated earthquake source parameters of the past Sunda subduction earthquake are evaluated against the global scaling relationships developed by Goda et al. (2016). The global scaling relationships developed by Goda et al. (2016) will be adopted only if the calculated earthquake source parameters from the 19 finite-fault models of the Sunda subduction earthquakes are consistent with these global relationships. Otherwise, the global scaling relationship should be adjusted to account for the regional differences of the source parameters based on the finite-fault models of the past Sunda subduction earthquakes. Calculations of the earthquake source parameters of the finite-fault models from the past Sunda subduction earthquakes are based on the effective dimension analysis (Mai and Beroza, 2000), Box–Cox analysis, and spectral analysis (Mai and Beroza, 2002; Goda et al., 2014). The effective dimension analysis is carried out to calculate the width, length, mean slip, and maximum slip, while the Box–Cox analysis is used to characterize the probability distribution of the slip values. In addition, the spectral analysis is conducted to define the correlation lengths along dip and strike and the Hurst number. Figure 3 illustrates the procedures of earthquake source parameter estimation using the Konca et al. (2007) model from the 2007 Bengkulu earthquake event.

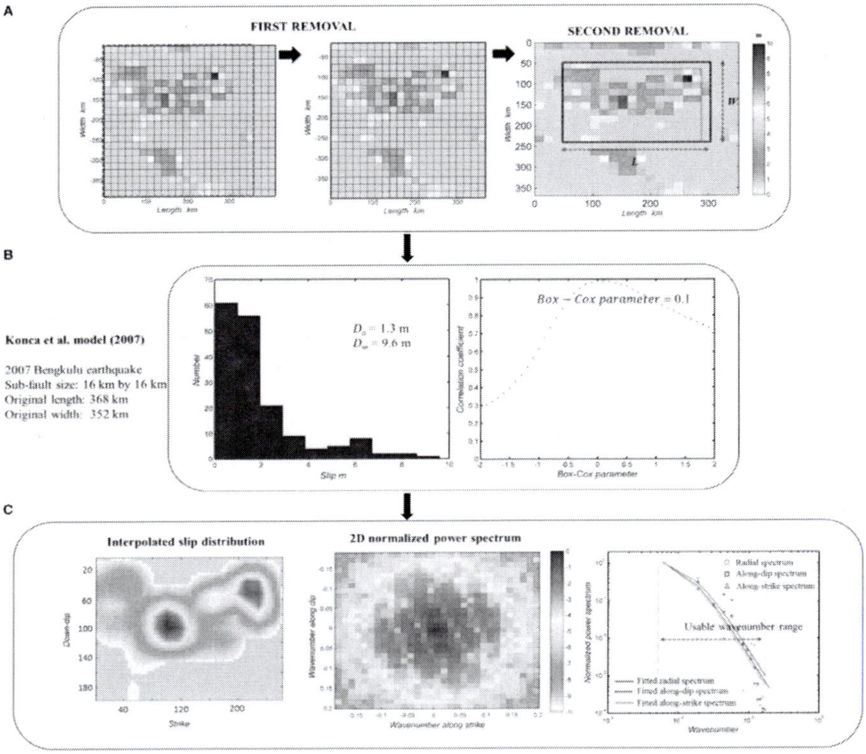

FIGURE 3. Procedures for source parameter estimation. **(A)** Effective dimension analysis. **(B)** Box–Cox analysis. **(C)** Spectral analysis.

First, the effective dimension analysis is carried out. The motivation to analyze the effective dimensions of the finite-fault models is because some of the finite-fault models have insignificant portions of slip located along the edges. The use of these insignificant portions may lead to overestimation of the rupture area and hence, it should be excluded from the source models. The insignificant portions of earthquake slip are removed in two steps. The first step is to simply trim the slip distribution when rows/columns having zero slip exist along the edges of the slip distribution. As shown in Figure 3A, three rows having zero slips are removed to produce a trimmed slip distribution. The second step is to determine the effective width and length by calculating the auto-correlation dimensions as defined by Mai and Beroza (2000). These dimensions, i.e., effective width and length, are then defined as W and L, respectively. Using the results from effective dimension analysis, the slip statistics parameters, D_a and D_m, are calculated. The mean slip may be changed from the original mean slip due to the effective dimension analysis.

Second, using the effective dimension, the Box–Cox analysis is conducted to characterize the probability distribution of slip values within the fault plane by identifying the best power parameter (λ) to transform a non-normal random variable (X) to a normal random variable (Y) as presented in Eq. 1.

$$Y = \frac{X^{\lambda} - 1}{\lambda} (\lambda \# 0) \tag{1}$$

The Box–Cox parameters corresponds to the lognormal transformation if the λ = 0. The best power parameter (Box–Cox parameter) can be determined by calculating the linear correlation coefficient of the standard normal variable and the transformed variable of the slip values (after standardization). The Box–Cox parameter (λ) is then obtained based on the value that achieves the maximum linear correlation coefficient (see Figure 3B).

Third, Fourier spectral analysis is carried out to calculate spatial heterogeneity parameters of the slip, i.e., A_x, A_z, and H. The Hurst number is used to characterize spectral decay as a function of wavenumber. Before carrying out spectral analysis, a cell-based grid of the finite-fault models is converted to a grid-based slip distribution and then the slip is interpolated using a selected grid spacing which is not smaller than one-fifth of the original grid resolution (left panel of Figure 3C). The interpolated grid-based slip distribution is then tapered using a Hanning window to control the edges of the rupture plane so that no significant slips occur along the rupture plane edges. Two-dimensional Fast Fourier Transform is calculated to obtain the 2D normalized power spectrum (middle panel of Figure 3C). The applicable wavenumber range for the spectral analysis is then defined by considering the original grid resolution and the characteristic size of the fault plane (right panel of Figure 3C). The circular average of the normalized wavenumber spectra is calculated, and the fractal dimension D_f is defined based on the least squares fitting. The Hurst number is then calculated after converting from the fractal dimension (i.e., $H = 3 - D_f$). Finally, A_z and A_x are obtained by minimizing the norm between the along-dip/along-strike wavenumber spectrum and the analytical von Kármán model (Mai and Beroza, 2002) calculated using Eq. 2 by varying the correlation length along-dip/along-strike.

$$P(k) \propto \frac{A_z A_x}{\left(1 + k^2\right)^{H+1}} \tag{2}$$

where $P(k)$ is the power spectrum of von Kármán model and k is the wavenumber defined as, $k = \left(A_z^{\,2} k_z^{\,2} + A_x^{\,2} k_x^{\,2} \right)^{0.5}$ (Goda et al., 2016).

The results of estimated source parameters for the nineteen source models are shown in Table 1 and Figure 4. Fourteen from the 19 source models shown by model number 1–14 in Table 1 are part of the 226 models used in Goda et al. (2015). In Figure 4, scaling relationships for tsunamigenic earthquakes by Goda et al. (2016) are also included. These equations are summarized in Table 2, noting that they are indeed probabilistic prediction models that allow characterizing the prediction errors. In the equations, epsilon terms represent the prediction errors of the equations, and furthermore, their correlation coefficients are given in Table 3. The results shown in Figure 4 indicate that the estimated source parameters are in agreement with the scaling relationships; for most cases, the estimated parameters fall within the 16th to 84th percentile confidence interval of the prediction equations. Therefore, the use of the scaling relationships by Goda et al. (2016) for generating stochastic source models for the future tsunamigenic earthquakes in the Mentawai segment of the Sunda subduction zone can be justified. In Figure 4, the source parameters for the model 19 by Muhari et al. (2010) are shown with the red circle (see also Table 1). This model is used as a benchmark to be compared with stochastic tsunami simulation results in Padang areas.

TABLE 1. Summary of earthquake source parameters for the 19 finite-fault models of the sumatra subduction earthquakes.

No	Date	M_w	W (km)	L (km)	Strike (°)	Dip (°)	Mean slip (m)	Maximum slip (m)	Box-Cox parameter	Correlation length along dip (km)	Correlation length along strike (km)	Hurst number	References
Model 1	26/12/2004	8.89	180	450	320	11	6.77	19.82	0.5	54.00	121.50	0.99	Ji (2005a)
Model 2	28/03/2005	8.68	220	340	326	8	3.14	12.50	0.4	64.60	98.80	0.99	Shao and Ji (2005)
Model 3	28/03/2005	8.5	192	304	325	10	2.13	15.20	0.0	58.24	99.84	0.83	Konca et al. (2007)
Model 4	28/03/2005	8.7	192	380	325	15	3.34	13.75	0.5	60.00	108.00	0.99	Ji (2005b)
Model 5	12/09/2007	7.9	90	225	319	19	0.68	3.89	0.3	31.30	84.51	0.99	Ji and Zeng (2007)
Model 6	12/09/2007	7.9	110	192	323	15	0.69	6.00	0.0	28.80	36.00	0.99	Konca et al. (2008)
Model 7	12/09/2007	7.9	100	168	319	19	0.83	7.40	–0.2	38.40	52.80	0.99	Sladen and Konca (2007)
Model 8	12/09/2007	8.5	225	300	328	9	1.55	5.44	0.1	64.00	80.00	0.99	Gusman et al. (2010)
Model 9	12/09/2007	8.4	159.5	460	323	12	1.08	3.61	0.7	56.00	100.80	0.99	Ji (2007)
Model 10	12/09/2007	8.4	192	256	324	15	1.30	9.60	0.1	44.00	68.00	0.46	Konca et al. (2008)
Model 11	20/02/2008	7.4	155	110	296	7.55	0.14	1.00	0.0	54.40	27.20	0.29	Hayes and Ji (2008)
Model 12	20/02/2008	7.4	112	152	302	7	0.15	1.08	–0.2	24.32	25.84	0.99	Sladen (2008)
Model 13	06/04/2010	7.8	144	144	313	12.04	0.43	3.17	–0.2	40.80	31.20	0.67	Hayes (2010a)
Model 14	09/05/2010	7.2	54	72	305	17.52	0.35	1.13	0.4	15.30	17.10	0.99	Hayes (2010b)
Model 15	25/10/2010	7.7	140	195	325	11.62	0.28	1.20	0.1	33.75	45.00	0.99	Hayes (2010c)
Model 16	25/10/2010	7.7	120	210	326	9.75	1.35	6.10	0.8	52.50	115.50	0.66	Satake et al. (2013)
Model 17	25/10/2010	7.7	72.5	180	324	7.5	4.10	23.41	0.2	25.20	41.40	0.99	Yue et al. (2014)
Model 18	25/10/2010	7.8	140	225	325	11.62	2.10	9.63	0.4	36.00	60.75	0.99	Newman et al. (2011)
Model 19	Future event	8.92	240	520	325	13.00	7.48	14.90	0.9	83.20	140.40	0.99	Muhari et al. (2010)

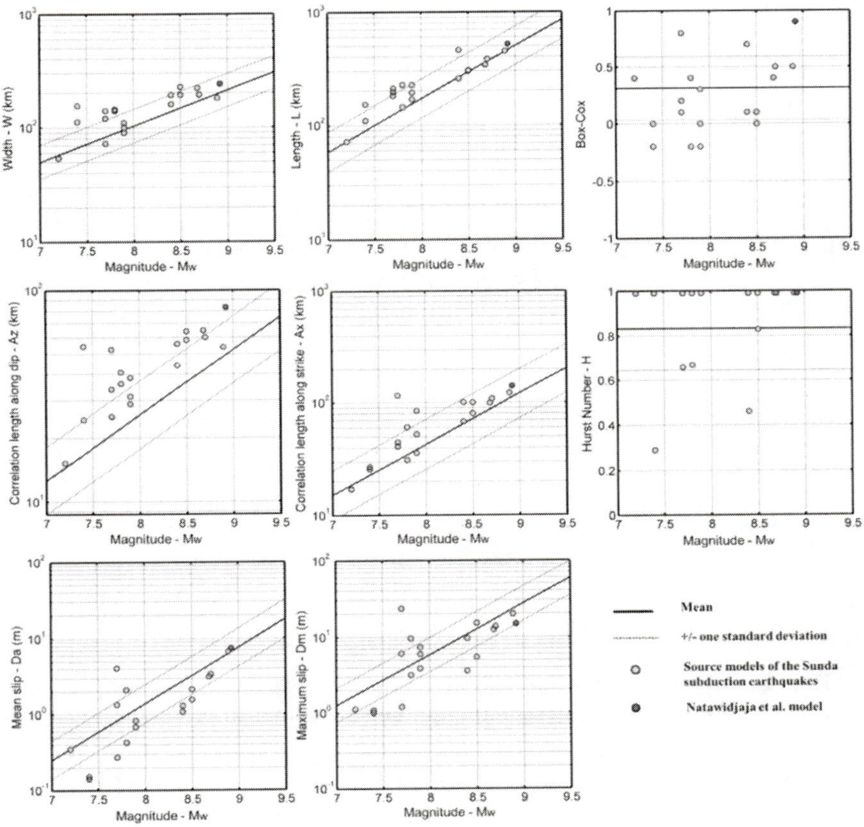

FIGURE 4. Comparison of the estimated source parameters for the 19 finite-fault models of the sunda subduction earthquakes with the corresponding scaling relationships.

Stochastic Tsunami Simulation

The stochastic tsunami simulation can be conducted by generating multiple source models for a given earthquake scenario and by performing tsunami forward modeling iteratively. Figure 5 presents a computational flowchart of the stochastic tsunami simulation. In the following, detailed procedures of the stochastic tsunami simulation for the future tsunamigenic earthquake in the Mentawai segment of the Sunda subduction zone are explained.

TABLE 2. Scaling relationships of the earthquake source parameters (**Goda et al., 2016**).

Parameter	Prediction models
W	$\log_{10} W = -0.4877 + 0.3125M_w + 0.1464\varepsilon_W$
L	$\log_{10} L = -1.5021 + 0.4669M_w + 0.1717\varepsilon_L$
D_a	$\log_{10} D_a = -5.7933 + 0.7420M_w + 0.2502\varepsilon_{D_a}$
D_m	$\log_{10} D_m = -4.5761 + 0.6681M_w + 0.2249\varepsilon_{D_m}$
A_z	$\log_{10} A_z = -1.0644 + 0.3093M_w + 0.1592\varepsilon_{A_z}$
A_x	$\log_{10} A_x = -1.9844 + 0.4520M_w + 0.2204\varepsilon_{A_x}$
Box–Cox power	A normal variable with mean = 0.312 and SD = 0.278
Hurst number	A value of 0.99 with probability of 0.43 and a normal variable with mean = 0.714 and SD = 0.172 with probability of 0.57

The prediction error terms of the scaling relationships are represented by epsilons, which are the standard normal variables (i.e., zero mean and unit SD).

TABLE 3. Linear correlation coefficients of regression residuals of the scaling relationships for the six earthquake source parameters.

Variables	ε_W	ε_L	ε_{D_a}	ε_{D_m}	ε_{A_z}	ε_{A_x}
ε_W	1	0.139	−0.68	−0.545	0.826	0.035
ε_L	0.139	1	−0.595	−0.516	0.249	0.734
ε_{D_a}	−0.68	−0.595	1	0.835	−0.62	−0.374
ε_{D_m}	−0.545	−0.516	0.835	1	−0.564	−0.337
ε_{A_z}	0.826	0.249	−0.62	−0.564	1	0.288
ε_{A_x}	0.035	0.734	−0.374	−0.337	0.288	1

Note that the Box–Cox parameter and the Hurst number are considered to be independent.

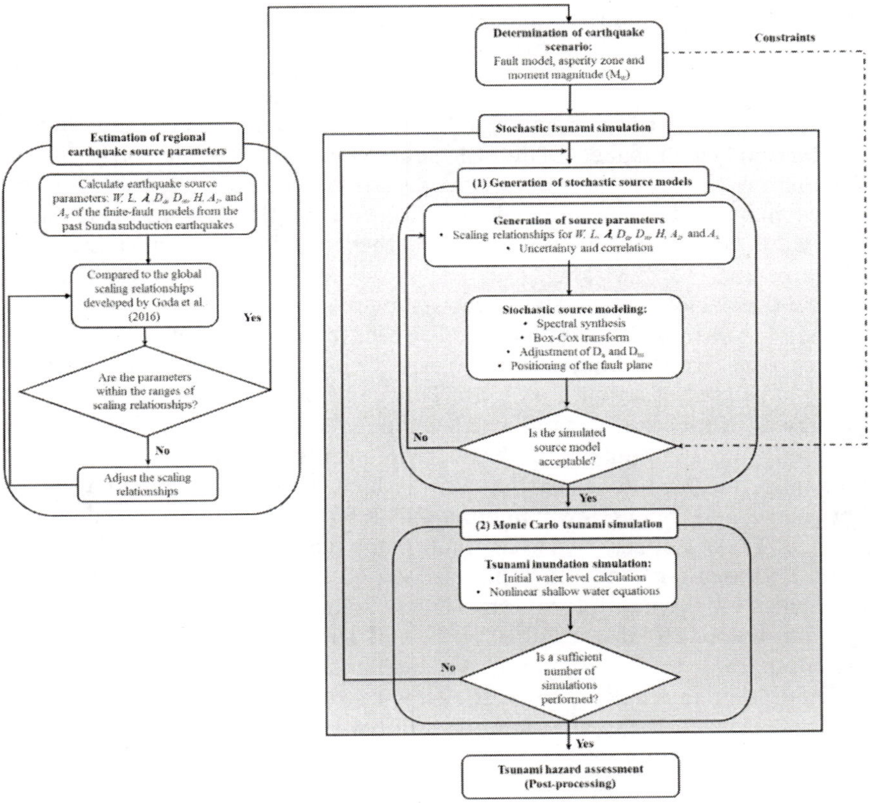

FIGURE 5. FLOWCHART OF STOCHASTIC TSUNAMI SIMULATION.

Earthquake Scenarios and Fault Zone Model

Before running the stochastic tsunami simulation, earthquake scenarios (e.g., magnitude and source zone) need to be selected and a suitable fault rupture zone model (e.g., geometry and asperity zone) needs to be defined. The fault plane is used to model the source zone of the earthquake, while within the fault plane the so-called asperity zone is set up. When generating stochastic earthquake source models, the source zone of the earthquake is needed to define the area of earthquake source and the asperity zone is essential as the concentration region for certain amount of slip. In addition, the selected magnitude should be determined based on the purpose of the analysis. Since this work aims at assessing the tsunami hazards in Padang due to megathrust earthquakes from the Mentawai segment of the Sunda subduction zone, three magnitudes are considered: M_w 8.5, M_w 8.75, and M_w 9.0.

First, a possible source zone of the future tsunamigenic earthquakes in the Mentawai segment is defined based on source models for the past Sunda subduction earthquakes. The 19 source models of the past Sunda subduction earthquakes are used to determine the rupture areas for the future megathrust tsunamigenic earthquakes in the Mentawai segment (Table 1). The strike and dip angles of these models are typically in the range of 296° to 326° and 7° to 19°, respectively. A generic fault model covers the region of the Mentawai segment starting from North of Batu Islands to South of Enggano Island. The length and width of the Mentawai source zone are 920 and 250 km, respectively. The top-edge of the fault plane is located at a depth of 3 km. This depth is consistent with the past Mentawai finite-fault models developed for the 2010 Mentawai tsunamigenic earthquakes and the twin events of the 1797 and 1833 tsunamigenic Mentawai earthquakes which have the top-edge depth between 2 and 5 km (Newman et al., 2011; Satake et al., 2013; Philibosian et al., 2014; Yue et al., 2014). The fault plane has a constant strike angle of 325°. On the other hand, dip angles are varied depending on the depth. The dip angles for the depth from 3 to 10 km, from 10 to 17 km, from 17 to 29 km, and below 29 km are 8°, 10°, 12°, and 16°, respectively. These values are comparable to the slab models for the Sunda subduction zone produced by the USGS (Hayes et al., 2009, 2012). For stochastic source modeling and Monte Carlo simulation, the Mentawai source zone is discretized into 10 km by 10 km sub-faults. In stochastic source modeling, slip values that are consistent with considered spatial slip distribution features are generated.

Second, an asperity zone is set up within the fault plane of the source zone along with the required slip concentration range. The seismological knowledge of earthquake rupture in the target region must be reflected in the asperity zone. Basically, the asperity zone serves as crude constraints of the generated source model regarding the spatial distribution of earthquake slip within the fault plane. In generating the stochastic source models, a certain amount of slip must be concentrated within the target region. Generally, the determination of asperities of the future tsunamigenic earthquake in the Mentawai segment of the Sunda subduction zone is complex and involves large uncertainty (McCloskey et al., 2008; Philibosian et al., 2014). The interseismic coupling and coseismic slip modeling based on the paleogeodetic study confirm that the asperities in the Mentawai segment must be multiple. A future megathrust earthquake in the Mentawai segment may rupture similar to the scenarios of either the 1797 event or the 1833 event. This assumption is based on the paleogeodetic study that finds the potential of giant earthquakes is high in the Mentawai segment as the source earthquake region for the 1797 and 1833 events (Chlieh et al.,

2008; Sieh et al., 2008). The asperity areas of the 1797 event are closer to Padang in comparison with those of the 1833 event (Philibosian et al., 2014). In addition, effects of the 1797 event in Padang in terms of ground shaking and flow depth were greater than the 1833 event. Therefore, the asperity zone of the future megathrust earthquake for tsunami hazard assessment in Padang is assumed to follow the asperity areas of the 1797 event (see Figure 6A).

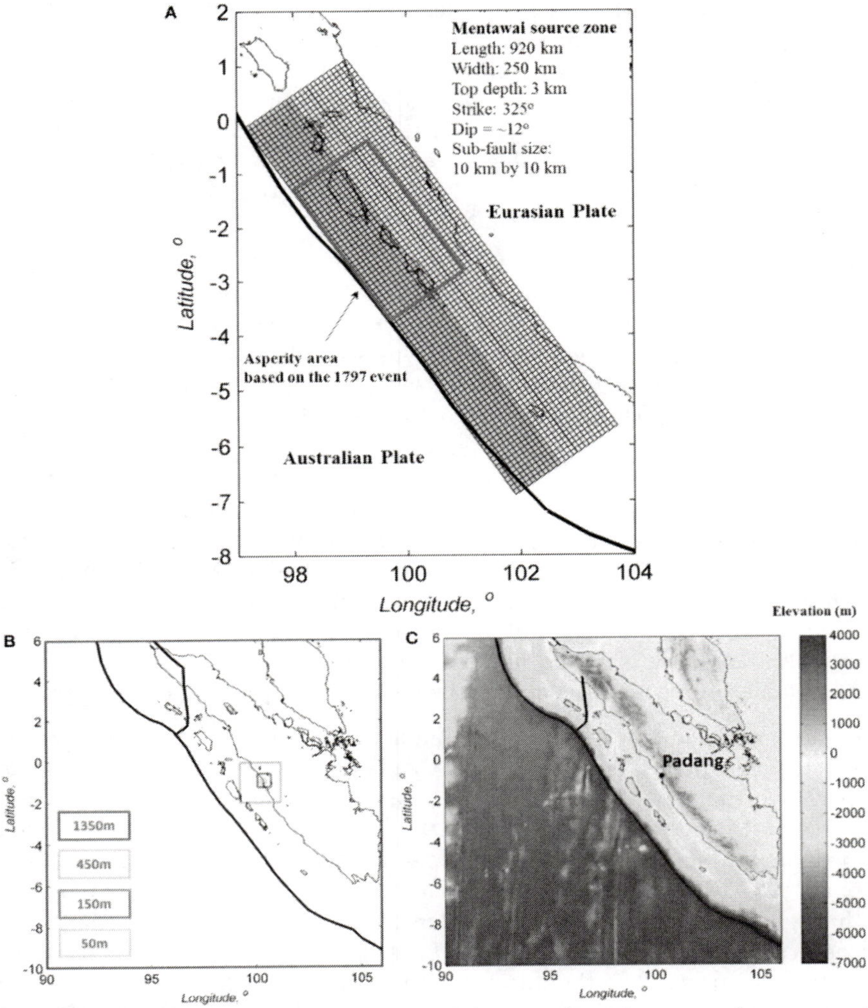

FIGURE 6. (A) Tsunami source zone. **(B)** Nested grid system for tsunami simulation in Padang. **(C)**GEBCO2014 bathymetry for the Sumatra region.

Stochastic Tsunami Simulation

Essentially, the stochastic tsunami simulation involves two main calculations, i.e., generation of stochastic source models and Monte Carlo tsunami simulation (Figure 5). First, earthquake source parameters, i.e., W, L, D_a, D_m, λ, A_z, A_x, and H, are generated using the prediction models (Tables 2 and 3). The uncertainty and correlation associated with the prediction models are taken into account in sampling the values of the earthquake source parameters from the multivariate normal distribution. To control the consistency of the generated earthquake source parameters, the simulated seismic moment ($M_o = \mu WLD_a$, where μ is the rock rigidity which is set to 40 GPa) is compared with the target seismic moment defined by the scenario magnitude. The combination of W, L, and D_a are calculated iteratively until the seismic moment criterion is satisfied (note: a tolerance of ±0.05 magnitude units is permitted).

Subsequently, a random slip field is generated using a Fourier integral method (Pardo-Iguzquiza and Chica-Olmo, 1993). The synthesized slip distribution is converted *via* Box–Cox transformation to achieve slip distribution with realistic positive skewness (Goda et al., 2014). To achieve the target mean slip D_a and to avoid very large slip values exceeding the target maximum slip D_m, the transformed slip distribution is further adjusted. Next, the synthesized fault plane position is randomly located within the source region. The final synthesized earthquake slip model should be consistent with the seismotectonic features of the target region. For this purpose, two criteria for acceptance of the candidate slip model are implemented, i.e., the asperity area ratio of the candidate slip distribution is within the range of 0.2 and 0.3 and the simulated earthquake slip is more concentrated in the considered asperity region with the percentage range of 50–80%. Multiple slip distributions are generated iteratively until an acceptable source model is obtained which has all the expected features.

Once a realistic stochastic source model is generated, the initial water surface elevation is calculated using Okada (1985) and Tanioka and Satake (1996) formulae which consider the deformation due to both vertical and horizontal displacements of seafloor. Tsunami wave propagation is then evaluated by solving non-linear shallow water equations with run-up (Goto et al., 1997). The effects of surface roughness on tsunami flows are modeled through the Manning's bottom friction formula with a uniform roughness coefficient of 0.025 $m^{-1/3}s$. The fault rupture is assumed to occur instantaneously, while the duration of the simulation is set to 2 h and the time step for the simulation is 0.5 s, which satisfies the C.F.L. criterion for the bathymetry and elevation data for the Mentawi region.

For tsunami forward modeling, digital elevation model (DEM) and bathymetry data are needed. For the western coast of Sumatra, the bathymetry and elevation data are constructed from the publicly available data, i.e., GEBCO2014 (http://www.gebco.net/data_and _products /gridded_bathymetry_data/) and GDEM2 (https://asterweb.jpl .nasa.gov/gdem.asp). The nested grid systems of bathymetry and DEM having four resolutions are developed to assess the tsunami hazard in Padang due to the future tsunamigenic earthquakes in the Mentawai segment. The crudest grid is 1,350 m, while the finest grid is 50 m. To connect grid systems with different resolutions, two grids having different grid resolutions by a factor of 3 are considered. Therefore, the nested grid resolutions to carry out tsunami simulation in Padang are 1,350, 450, 150, and 50 m (Figure 6B). The 1,350 m region covers the entire region of West Sumatra.

The GEBCO2014 dataset is adopted for bathymetry data with the resolution of 30 arc-sec (~900 m). The bathymetry plot of the Sumatra region from GEBCO2014 is presented in Figure 6C. For DEM, the GDEM2 dataset having the resolution of 1 arc-sec (~30 m) is used. In this study, bathymetry data with a 50-m resolution are adopted to run the tsunami simulation in the shallow water and land regions. To develop such bathymetry datasets, GEBCO2014, GDEM2, and SRTM Water Body Data (SWBD), are merged by considering the resolution of 1 arc-sec (same as GDEM2). The merging of the datasets is conducted by simply substituting NaN (Not a Number) values (i.e., ocean areas) in GDEM2 with the GEBCO2014 data at the same coordinate, while all land elevations from the GEBCO2014 data are neglected. In addition, the coastal line data from SWBD are defined as "zero" in the merged dataset. Subsequently, linear interpolation is performed to produce a 1 arc-sec of the merged data. The merged and interpolated data are used to produce a 50 m resolution dataset along the coastal line of Padang areas. The use of the linear interpolation scheme is deemed as appropriate over other more complex schemes, such as spline interpolation, because at the near coastal line areas (where drastic changes of the spatial density of the data points are inevitable), complex interpolation methods may over-interpolate the topographical features.

Finally, the above simulation procedure is run iteratively until a sufficient number of acceptable source models are generated and their tsunami inundation heights at locations of interest are evaluated. The results from the Monte Carlo tsunami simulation are useful for evaluating variability of tsunami simulation results at different locations and for developing stochastic tsunami hazard maps (Goda et al., 2014).

RESULTS AND DISCUSSION

The tsunami simulation results including the simulated tsunami wave height profiles and maximum tsunami heights along the coastal line are presented in this section by considering three magnitude scenarios (M_w 8.5, 8.75, and 9.0) and two cases ignoring and incorporating the uncertainty of the scaling relationships. The motivation to involve the uncertainty of the scaling relationships in the analysis is to assess the effect of incorporating the uncertainty of the scaling relationships to the tsunami simulation results. In addition, the height presenting in this section corresponds to the height of water flow above sea level. For each combination of magnitude scenario and uncertainty case, 100 stochastic source models are generated and used in Monte Carlo tsunami simulation. The stochastic tsunami simulation results from the M_w 9.0 scenario are used to investigate the sensitivity of the tsunami simulated wave profile and are compared with the reference results based on the Muhari et al. source model.

Tsunami Simulation Results: M_w 9.0 Scenario

Using the stochastic tsunami simulation results for the M_w 9.0 scenario, sensitivity analysis of the tsunami simulated wave profiles is carried out and presented in this section. The simulated tsunami wave profiles produced from Muhari et al. (2010) are used as a benchmark to demonstrate the tsunami simulation results. The earthquake source model considered by Muhari et al. (2010) is shown in Figure 7A. The model was developed based on the slip accumulation of the current locked zone in the Mentawai segment with the total sub-fault number of 348 and the sub-fault size of 20 km by 20 km. Its moment magnitude was M_w 8.92. The comparison results show that the Muhari et al. model is in agreement with the global scaling relationships (Figure 4). For each tsunami simulation run, tsunami waveforms are recorded at three points with the water depth of 5 m: Tabing (0.85°S and 100.34°E), Purus (0.88°S and 100.345°E), and Teluk Bayur (1°S and 100.38°E) as shown in Figure 9A. These points are selected because they were also considered by Muhari et al. (2010). In addition, the maximum tsunami wave height contours are recorded to investigate the inundated area in Padang.

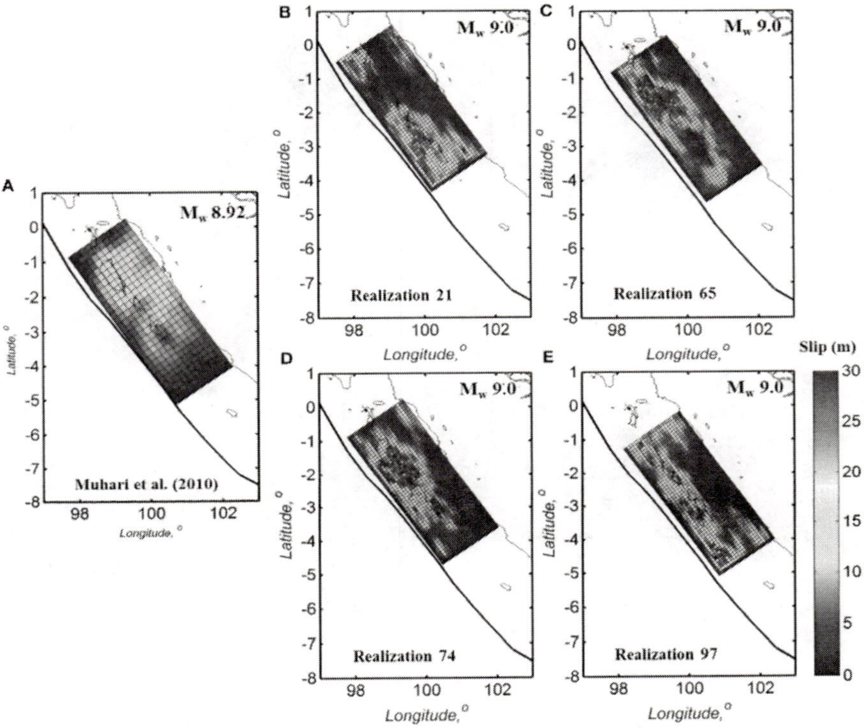

FIGURE 7. (A) Muhari et al. source model. **(B–E)** Stochastic source models for the future tsunamigenic earthquake scenario without considering the uncertainty of the scaling relationships.

Two sets of 100 stochastic source models are generated for the M_w 9.0 earthquake scenario to carry out the Monte Carlo tsunami simulation. The first set takes into account the uncertainty of the scaling relationships of the source models, while the second set does not. Figures 7 and 8 illustrate four realizations of the stochastic source models for the cases of including and excluding the uncertainty of the scaling relationships, respectively. The figures show that neglecting the uncertainty of the scaling relationships leads to identical dimensions (L and W) of the generated earthquake source models and the same slip statistics values for different realizations. For example, the generated values of D_a and D_m for the M_w 9.0 scenario are 8 and 30 m, respectively, while the spatial slip distribution and location of the fault rupture within the overall source zone are varied. By contrast, incorporating the uncertainty of the scaling relationships results in variability of dimensions and slip statistics.

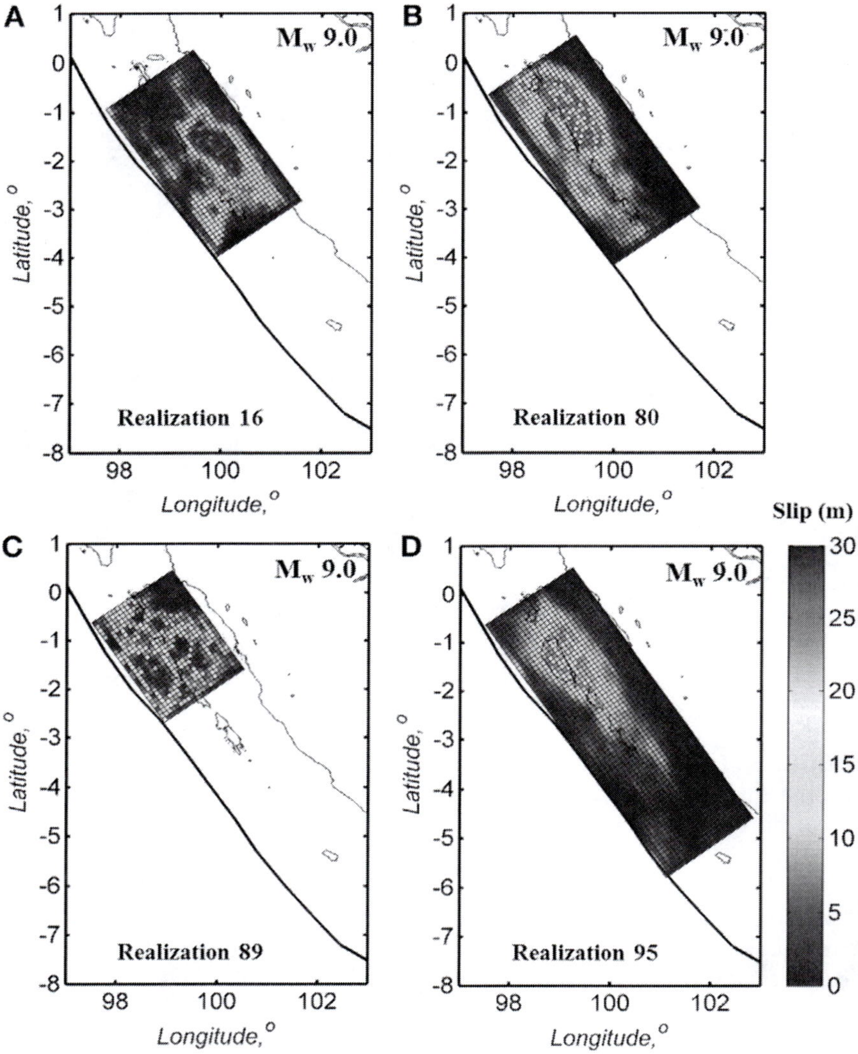

FIGURE 8. (A–D) Stochastic source models for the future tsunamigenic earthquake scenario by considering the uncertainty of the scaling relationships.

The tsunami wave profiles at three recording points produced from the 100 stochastic source models for the M_w 9.0 scenario generated without considering the uncertainty of the scaling relationships (Figure 7) are shown in the middle panels of Figures 9B–D, whereas similar results obtained based on the Muhari et al. source model are shown in the top panels of Figures 9B–D. The raw simulated data are shown with gray color and the median and 10th/90th percentiles of the simulated tsunami waveforms are illustrated with red line and blue line, respectively. Large variations in the temporal tsunami wave profiles are observed at the recording points 1 to 3 (P1 to P3). For instance, at the Teluk Bayur station (P3), the variations in the simulated results range from −5 to 15 m, while the range between the 10th and 90th percentiles varies from −5 to 12.5 m. These trends are also observed at the other two stations. The medians of the simulated tsunami wave profiles recorded at three stations demonstrate results that are comparable to the Muhari et al. model with the maximum tsunami height of about 5 m. In addition, the tsunami arrival time to the coastal areas of Padang based on the median simulation results and Muhari et al. results are also consistent, i.e., 20–25 min after the event. Furthermore, the tsunami simulation results based on the 100 stochastic source models generated by incorporating the uncertainty of the scaling relationships for the M_w 9.0 (Figure 8) are presented. The simulated tsunami wave profiles at three recording points are shown in the bottom panels of Figures 9B–D. In general, the medians of tsunami wave profiles are similar to those produced without considering the uncertainty of the scaling relationships. Large variations in tsunami wave heights are observed at those three points with the maximum tsunami height of 15 m. From the medians of the tsunami waveforms at three recording points, consistent tsunami waveforms compared to the Muhari et al. results are also demonstrated.

To evaluate the differences of tsunami simulation results between excluding and including the uncertainty of the scaling relationships, 150 points are selected to record the maximum tsunami wave height along the coastal line of Padang starting from Tabing to Teluk Bayur (see Figure 10A). The median and the 10th/90th percentiles of maximum tsunami wave height profiles along the coastal line of Padang are shown in Figure 10B. In addition, the maximum tsunami wave height profiles from the 200 stochastic models excluding the uncertainty and including the uncertainty are presented in Figures 10C,D, respectively. The ranges between the 10th/90th percentiles of the maximum tsunami heights from those two cases show that the models with the uncertainty have greater variability in comparison to the models without the uncertainty. The range of the maximum tsunami wave height of the models considering the uncertainty is between 2.5 and 20 m, while the corresponding range for the models excluding the uncertainty is between 5 and 17.5 m. However, as observed from the tsunami wave profiles at the three recording stations, i.e., Tabing, Purus, and Teluk Bayur stations, the median of maximum tsunami wave profiles produced from these two calculation cases are similar (see black and green lines in Figure 10A).

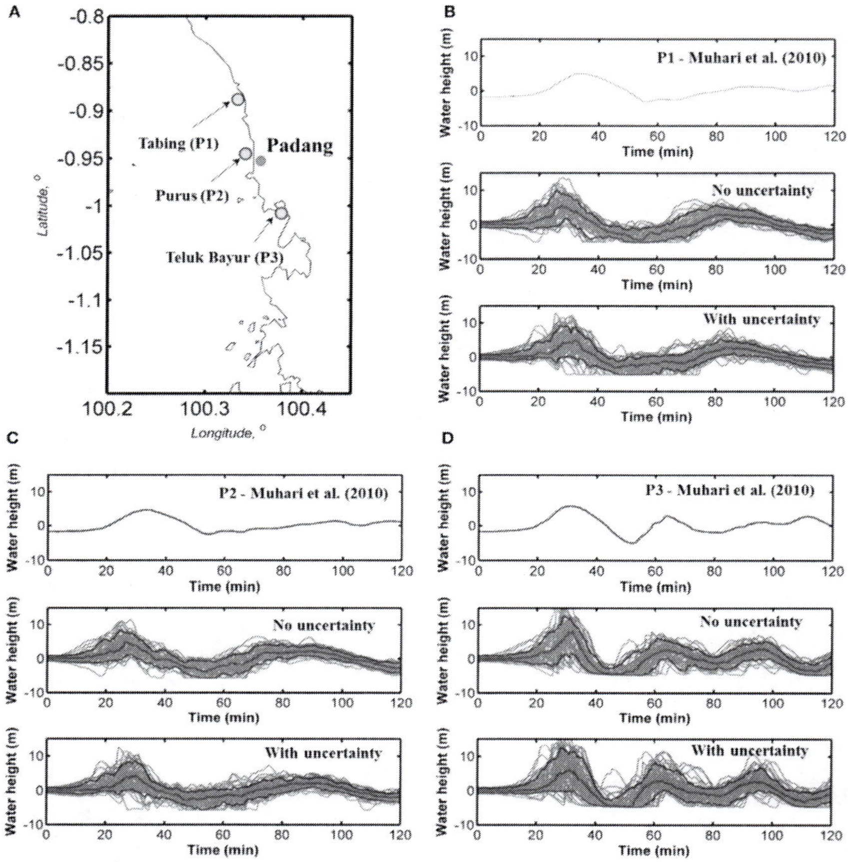

FIGURE 9. (A) Site location. **(B)** Tsunami wave profiles at the Tabing station (P1). **(C)** Tsunami wave profiles at the Purus station (P2). **(D)** Tsunami wave profiles at the Teluk Bayur station (P3).

Tsunami Hazard Assessment in Padang: All Magnitude Scenarios

The tsunami simulation results produced from different magnitude scenarios are presented in this section. Figures 11–13 show the tsunami waveform results based on the stochastic source models for three scenario magnitudes, i.e., M_w 8.5, M_w 8.75, and M_w 9.0, both including and excluding the uncertainty of the scaling relationships recorded in Tabing, Purus, and Teluk Bayur stations, respectively. One hundred stochastic source models

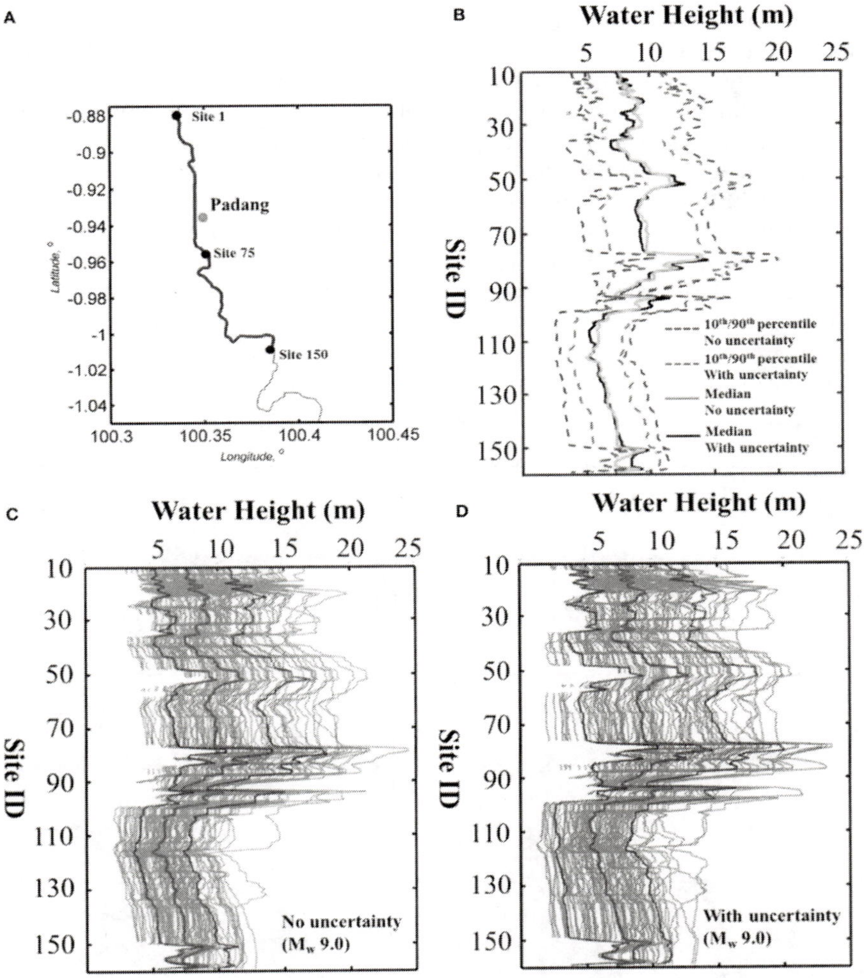

FIGURE 10. (A) Site location. **(B)** 10th/90th percentile of maximum tsunami wave height along the coastal line. **(C)** Maximum tsunami wave height profile along the coastal line without incorporating uncertainty. **(D)** Maximum tsunami wave height profile along the coastal line by incorporating uncertainty.

are used to run the tsunami simulation for each magnitude and uncertainty consideration (600 cases in total). In general, the large variation of wave amplitudes from the 10th and 90th percentile curves suggests that the earthquake slip model is an important source of uncertainty for the tsunami prediction. In addition, the tsunami waveforms from all three locations exhibit that the variation of tsunami wave heights increases with the magnitude. It can be seen that the 10th and 90th percentile curves vary significantly from the M_w 8.5 scenario to the M_w 9.0 scenario. For instance, at the Tabing station (P1), the percentiles range from −2 to 2 m for the M_w 8.5 scenario and the range increases to −5 to 10 m for the M_w 9.0 scenario. The median of the tsunami waveforms shows that the tsunami wave height increases by a factor of 2 for the increasing magnitude by 0.25 U. Additionally, the tsunami hazard in Padang areas can also be assessed from the maximum tsunami wave height along the coastal line of Padang areas (see Figure 10). These data show that the maximum tsunami wave height in Padang can reach 20 m in urban areas (Tabing–Purus) where many important public facilities exist (e.g., school, hospital, and gas station). Therefore, the economic and social losses can be significant.

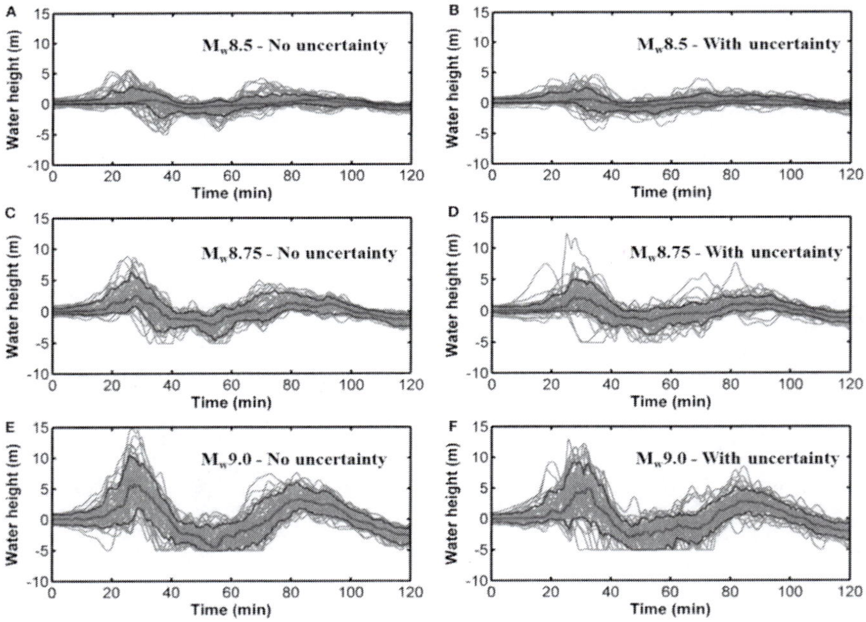

FIGURE 11. TSUNAMI WAVE PROFILES AT THE TABING STATION (P1). **(A)** M_w 8.5 scenario without considering uncertainty. **(B)** M_w 8.5 scenario considering uncertainty. **(C)** M_w 8.75 scenario without considering uncertainty. **(D)** M_w 8.75 scenario considering uncertainty. **(E)** M_w 9.0 scenario without considering uncertainty. **(F)** M_w 9.0 scenario considering uncertainty.

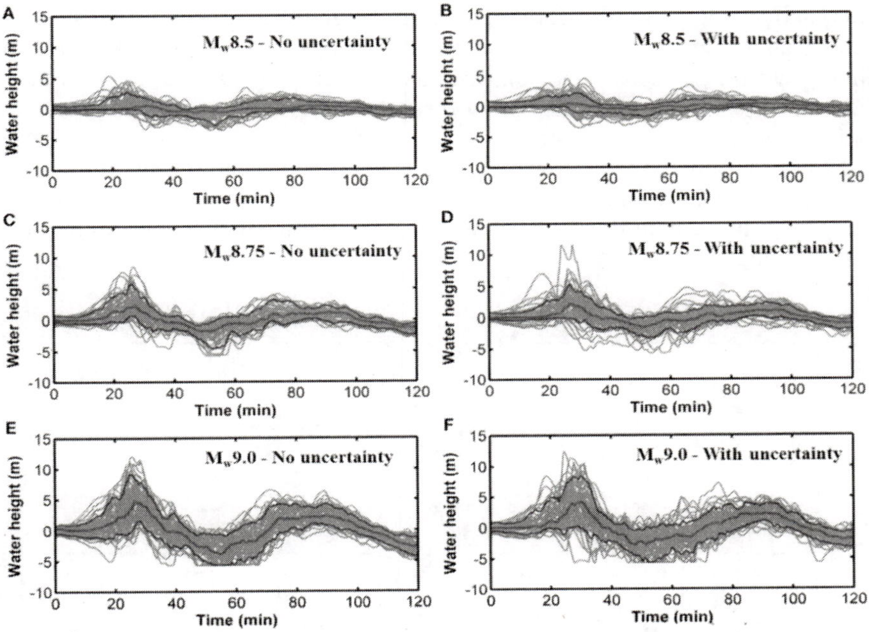

FIGURE 12. TSUNAMI WAVE PROFILES AT THE PURUS STATION (P2). **(A)** M_w 8.5 scenario without considering uncertainty. **(B)** M_w 8.5 scenario considering uncertainty. **(C)** M_w 8.75 scenario without considering uncertainty. **(D)** M_w 8.75 scenario considering uncertainty. **(E)** M_w 9.0 scenario without considering uncertainty. **(F)** M_w 9.0 scenario considering uncertainty.

Another important characteristic illustrated from the tsunami wave profiles for all scenarios is the feature of secondary waves (second and third waves) which are important to design an evacuation plan for the areas of interest. As presented at P1 and P3, the secondary wave heights for the scenarios of M_w 8.5 and M_w 8.75 are insignificant with the heights of below 1 m. However, the heights increase significantly when the M_w 9.0 scenario are considered. The medians of the second waves in Tabing and Purus areas (P1 and P2) reach ~3 m for both uncertainty considerations with the maximum height of 5 m for the 90th percentile. In the Teluk Bayur region (P3), the second and third waves increase drastically with the increasing magnitude. For instance, the maximum height from the 90th percentile of the second and third waves for the M_w 9.0 scenario without considering the uncertainty is ~5 m, while it is ~7.5 m for the case considering the uncertainty of the scaling relationships. The striking times of the second and third waves in those three locations are in the range of 60–90 min, respectively. With the expected maximum height of the secondary waves as much as 7.5 m within the period of 90 min after the earthquake, people living within the coastal region of Padang must evacuate earlier and stay out of the inundated areas otherwise the secondary waves may cause additional human loses. Moreover, by neglecting the uncertainty of the scaling relationships in generating the earthquake scenarios for the worst case (M_w 9.0), the tsunami hazard in terms of the maximum secondary wave heights is underestimated and hence, may oversimplify the evacuation plan in the target region.

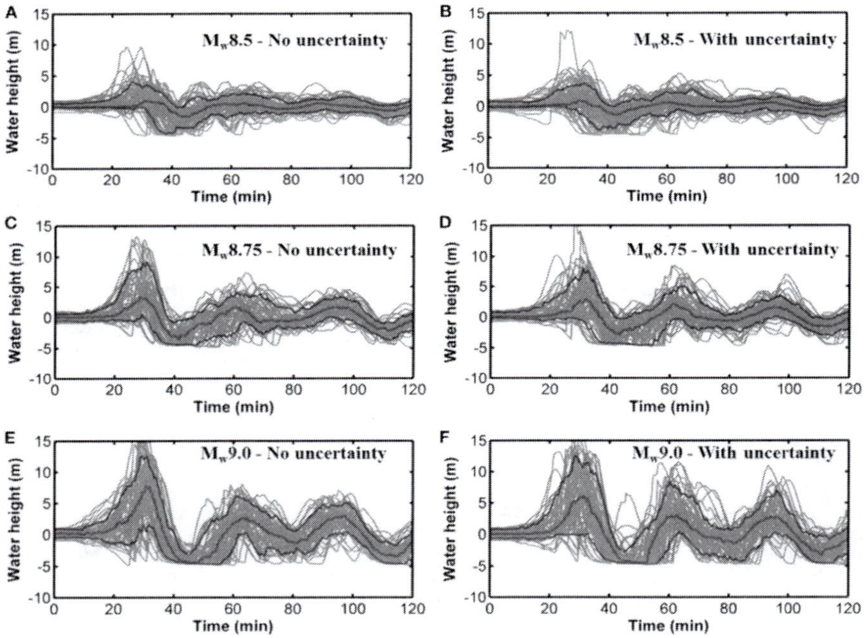

FIGURE 13. TSUNAMI WAVE PROFILES AT THE TELUK BAYUR STATION (P3). **(A)** M_w 8.5 scenario without considering uncertainty. **(B)** M_w 8.5 scenario considering uncertainty. **(C)** M_w 8.75 scenario without considering uncertainty. **(D)** M_w 8.75 scenario considering uncertainty. **(E)** M_w 9.0 scenario without considering uncertainty. **(F)** M_w 9.0 scenario considering uncertainty.

The tsunami hazard evaluations using multiple scenarios show that Padang may face a significant risk due to the future tsunamigenic event from the Mentawai segment. In comparison to the multiple-earthquake scenario approach, a deterministic scenario is not complex and requires a less computational effort. Although the single scenario approach is straightforward to communicate with the hazard results with emergency officers and relevant stakeholders, the multiple-scenario approach can produce a greater range of tsunami scenarios and therefore, more informed decisions regarding evacuation and mitigation actions can be made. In addition, for the risk assessment, the worst scenarios may be more relevant for critical facilities, such as public evacuation facilities. It is critically important to capture the most devastating effect that may occur in the target region. Using the stochastic tsunami simulation approach, the worst scenario ($M_w > 9.0$ at different percentile levels) can be considered to predict the future tsunamigenic earthquake impact, and therefore, a probabilistic approach is recommended to be implemented for preparing a better tsunami mitigation system for the future event.

CONCLUSION

The main objective of this study was to assess the tsunami hazard in Padang due to the future tsunamigenic event from the Mentawai source region in terms of near-shore tsunami wave profiles and maximum tsunami wave height data along the coastal of Padang using a novel method namely stochastic tsunami simulation. Extensive tsunami simulation for the future tsunamigenic earthquakes was conducted by developing a large number of stochastic earthquake slip models for different magnitude ranges. The earthquake source parameters from the finite-fault models of the past Sunda subduction earthquakes were firstly calculated and then compared with the corresponding scaling relationships for global tsunamigenic earthquakes developed by Goda et al. (2016). The verified scaling relationships were further used to build the earthquake source models for tsunami simulation. Uncertainty and dependency of earthquake source parameters were taken into account in producing earthquake source models stochastically. In total, 600 synthetic earthquake slip models were generated to obtain multiple realizations of maximum tsunami wave heights at various locations in Padang areas. Three scenarios magnitudes, i.e., M_w 8.5, M_w 8.75, and M_w 9.0, were considered to generate the stochastic earthquake source models, while the asperity zone was based on the significant slip areas from the 1797 tsunamigenic event. The simulated tsunami wave profiles from the M_w 9.0 scenario were compared with the results from Muhari et al. (2010) that predicted the tsunami hazard in Padang areas using the earthquake source models developed from the slip accumulation. The tsunami hazard in Padang was further evaluated using the tsunami wave profiles and maximum tsunami wave height data based on 600 stochastic tsunami simulations.

The stochastic earthquake source models for the future tsunamigenic earthquake in the Mentawai segment of the Sunda subduction zone have been successfully developed and further used in stochastic tsunami simulation. The estimated median of the simulated tsunami wave profiles produced from stochastic tsunami simulation is acceptable in comparison to the results from Muhari et al. (2010). Incorporating the uncertainty of the scaling relationships results in a larger variability of tsunami hazard parameters in comparison to excluding the uncertainty of the scaling relationships. The magnitude of earthquake scenarios has significant influence on the hazard assessment. In particular, the tsunami hazard assessment in Padang indicated that this region may experience a significant tsunami event with the maximum inundation height of 20 m in main urban areas (Tabing-Purus). Therefore, it can be concluded that the tsunami risk potential in Padang is high due to the future tsunamigenic event from the Mentawai segment of the Sunda subduction zone. Importantly, multiple scenarios of tsunami simulation using the stochastic methodology can produce a greater range of tsunami scenarios and hence, can inform emergency officers and stakeholders of the tsunami risk for improving the better tsunami mitigation system in the target region.

AUTHOR CONTRIBUTIONS

The three co-authors contributed equally to this work.

CONFLICT OF INTEREST STATEMENT

The authors declare that the research was conducted in the absence of any commercial or financial relationships that could be construed as a potential conflict of interest.

ACKNOWLEDGMENTS

The first author is grateful to the Directorate General of Resources for Science, Technology and Higher Education, Ministry of Research, Technology and Higher Education of Indonesia which sponsor his PhD study. This work is also funded by the Engineering and Physical Sciences Research Council (EP/M001067/1). The bathymetry and elevation data for the Sumatra region were obtained from the GEBCO2014 database (http://www.gebco.net/data_and_products/gridded_bathymetry_data/)

and the GDEM2 database (https://asterweb.jpl.nasa.gov/gdem.asp), respectively. The authors are grateful to Abdul Muhari, Hilman Natawidjaja, and Widjo Kongko who provide an earthquake source model for the future tsunamigenic event in the Mentawai segment and the bathymetry and elevation data for Padang.

REFERENCES

1. Aydan, O. (2008). Seismic and tsunami hazard potentials in Indonesia with a special emphasis on Sumatra Island. *J. Sch. Mar. Sci. Tech.*6, 19–38.
2. Banerjee, P., Nagarajan, B., Burgmann, R., and Pollitz, F. (2007). Coseismic slip distributions of the 26 December 2004 Sumatra-Andaman and 28 March 2005 Nias earthquakes from GPS static offsets. *Bull. Seismol. Soc. Am.* 97, S86–S102.
3. Borrero, J. C., Sieh, K., Chlieh, M., and Synolakis, C. E. (2006). Tsunami inundation modeling for western Sumatra. *Proc. Natl. Acad. Sci. U.S.A.* 103, 19673–19677.
4. Briggs, R. W., Sieh, K., Meltzner, A. J., Natawidjaja, D., Galetzka, J., Suwargadi, B., et al. (2006). Deformation and slip along the Sunda megathrust in the great 2005 Nias-Simeulue earthquake. *Science* 311, 1897–1901.
5. Chlieh, M., Avouac, J. P., Sieh, K., Natawidjaja, D. H., and Galetzka, J. (2008). Heterogeneous coupling of the Sumatran megathrust constrained by geodetic and paleogeodetic measurements. *J. Geophys. Res.* 113, 1–31.
6. Collings, R., Lange, D., Rietbrock, A., Tilmann, F., Natawidjaja, D., Suwargadi, B., et al. (2012). Structure and seismogenic properties of the Mentawai segment of the Sumatra subduction zone revealed by local earthquake traveltime tomography. *J. Geophys. Res.* 117, B01312.
7. Collings, R., Rietbrock, A., Lange, D., Tilmann, F., Nippress, S., and Natawidjaja, D. (2013). Seismic anisotropy in the Sumatra subduction zone. *J. Geophys. Res. Solid Earth.* 118, 5372–5390.
8. Gallovič, F., and Brokešová, J. (2004). On strong ground motion synthesis with k^{-2} slip distributions. *J. Seismol.* 8, 211–224.
9. Goda, K., Mai, P. M., Yasuda, T., and Mori, N. (2014). Sensitivity of tsunami wave profiles and inundation simulations to earthquake slip and fault geometry for the 2011 Tohoku earthquake. *Earth Planet Space* 66, 105.
10. Goda, K., Yasuda, T., Mori, N., and Mai, P. M. (2015). Variability of tsunami inundation footprints considering stochastic scenarios based on a 2 single rupture model: application to the 2011 Tohoku earthquake. *J. Geophys. Res. Oceans* 120, 4552–4575.
11. Goda, K., Yasuda, T., Mori, N., and Maruyama, T. (2016). New scaling relationships of earthquake source parameters for stochastic tsunami simulation. *Coastal Eng. J.* 58, 40
12. Goto, C., Ogawa, Y., Shuto, N., and Imamura, F. (1997). *Numerical Method of Tsunami Simulation with the Leap-Frog Scheme. IOC Manual*. Paris, France: UNESCO. UNESCO, No. 35.
13. Griffin, J. D., Pranantyo, I. R., Kongko, W., Haunan, A., Robiana, R., Miller, V., et al. (2016). "Assessing tsunami hazard using heterogeneous slip models in the Mentawai Islands, Indonesia," in *Geohazards in Indonesia: Earth Science for Disaster Risk Reduction*, eds P. R. Cummins and I. Meilano (London: Geological Society). Special Publications, 441.
14. Gusman, A. R., Tanioka, Y., Kobayashi, T., Latief, H., and Pandoe, W. (2010). Slip distribution of the 2007 Bengkulu earthquake inferred from tsunami waveforms and InSAR data. *J. Geophys. Res.* 115, B12316.

15. Hayes, G. P. (2010a). *Updated Result of the Apr 6, 2010 Northern Sumatra Earthquake*. Available at: http://equake-rc.info/SRCMOD/
16. Hayes, G. P. (2010b). *Preliminary Result of the May 9, 2010 Mw 7.2 Northern Sumatra Earthquake*. Available at: http://equake-rc.info/SRCMOD/
17. Hayes, G. P. (2010c). *Result of the Oct 25, 2010 Mw 7.7 Southern Sumatra Earthquake*. Available at: http://equake-rc.info/SRCMOD/
18. Hayes, G. P., and Ji, C. (2008). *Preliminary Result of the Feb 20, 2008 Mw 7.4 Simeulue Earthquake*. Available at: http://equake-rc.info/SRCMOD/
19. Hayes, G. P., Wald, D. J., and Johnson, R. L. (2012). Slab1.0: a three-dimensional model of global subduction zone geometries. *J. Geophys. Res.* 117, B01302.
20. Hayes, G. P., Wald, D. J., and Keranen, K. (2009). Advancing techniques to constrain the geometry of the seismic rupture plane on subduction interfaces a priori – higher order functional fits. *Geochem. Geophys. Geosyst.* 10, Q09006.
21. Hsu, Y. J., Simons, M., Avouac, J. P., Galetzka, J., Sieh, K., Chlieh, M., et al. (2006). Frictional afterslip following the 2005 Nias-Simeulue earthquake. *Sumatra. Sci.* 312, 1921–1926.
22. Ji, C. (2005a). *Preliminary Rupture Model for the December 26, 2004 Earthquake, off the West Coast of Northern Sumatra, Magnitude 9.1*. Available at: http://neic.usgs.gov/neis/eq_depot/2004/eq_041226/neic_slav_ff.html
23. Ji, C. (2005b). *Preliminary Result 05/03/28 (Mw 8.7), Sumatra Earthquake*. Available at: http://www.tectonics.caltech.edu/slip_history/2005_sumatra/sumatra.html
24. Ji, C. (2007). *Rupture Process of the Sep 12, 2007 Mw 8.4 Sumatra Earthquake: Phase II*. Available at: http://www.geol.ucsb.edu/faculty/ji/big_earthquakes/2007/09/sumatra_seismic.html
25. Ji, C., and Zeng, Y. (2007). *Preliminary Result of the Sep 12, 2007 Mw 7.9 Kepulauan Earthquake*. Available at: http://equake-rc.info/
26. Konca, A. O., Avouac, J. P., Sladen, A., Meltzner, A. J., Sieh, K., Fang, P., et al. (2008). Partial rupture of a locked patch of the Sumatra megathrust during the 2007 earthquake sequence. *Nature* 456, 631–635.
27. Konca, A. O., Hjorleifsdottir, V., Song, T. A., Avouac, J. P., Helmberger, D. V., Ji, C., et al. (2007). Rupture kinematics of the 2005, Mw 8.6, Nias-Simeulue earthquake from the joint inversion of seismic and geodetic data. *Bull. Seismol. Soc. Am.* 97, S307–S322.
28. Li, L., Huang, Z., Qiu, Q., Natawidjaja, D. H., and Sieh, K. (2012). Tsunami-induced coastal change: scenario studies for Painan, West Sumatra, Indonesia. *Earth Planet Sp.* 64, 2.
29. Mai, P. M., and Beroza, G. C. (2000). Source scaling properties from finite-fault-rupture models. *Bull. Seismol. Res. Lett.* 85, 1348–1357.
30. Mai, P. M., and Beroza, G. C. (2002). A spatial random field model to characterize complexity in earthquake slip. *J. Geophys. Res. Solid. Earth* 107, ESE 10-1–ESE 10-21.
31. Mai, P. M., and Thingbaijam, K. K. S. (2014). SRCMOD: an online database of finite source rupture models. *Seismol. Res. Lett.* 85, 1348–1357.
32. McCloskey, J., Antonioli, A., Piatanesi, A., Sieh, K., Steacy, S., Nalbant, S., et al. (2008). Tsunami threat in the Indian Ocean from a future megathrust earthquake west of Sumatra. *Earth Planet. Sci. Lett.* 265, 61–81.
33. McCloskey, J., Nalbant, S. S., and Steacy, S. (2005). Earthquake risk from co-seismic stress. *Nature* 434, 291.
34. Meltzner, A. J., Sieh, K., Abrams, M., Agnew, D. C., Hudnut, K. W., Avouac, J. P., et al. (2006). Uplift and subsidence associated with the great Aceh-Andaman earthquake of 2004. *J. Geophys. Res.* 111, B02407.
35. Monecke, K., Finger, W., Klarer, D., Kongko, W., McAdoo, B. G., Moore, A. L., et al. (2008). A 1,000-year sediment record of tsunami recurrence in northern Sumatra. *Nature* 455, 1232–1234.
36. Muhari, A., Imamura, F., Koshimura, S., and Post, J. (2011). Examination of three practical run-up models for assessing tsunami impact on highly populated areas. *Nat. Haz. Earth Syst. Sci.* 11, 3107–3123.

37. Muhari, A., Imamura, F., Natawidjaja, D. H., Diposaptono, S., Latief, H., Post, J., et al. (2010). Tsunami mitigation efforts with pTA in West Sumatra Province, Indonesia. *J. Earthq. Tsunami* 4, 341–368.

38. Nalbant, S. S., Steacy, S., Sieh, K., Natawidjaja, D., and McCloskey, J. (2005). Seismology: earthquake risk on the Sunda trench. *Nature*435, 756–757.

39. Natawidjaja, D. H., Sieh, K., Chlieh, M., Galetzka, J., Suwargadi, B. W., Cheng, H., et al. (2006). Source parameters of the great Sumatran megathrust earthquakes of 1797 and 1833 inferred from coral microatolls. *J. Geophys. Res.* 111, B06403.

40. Newcomb, K. R., and McCan, W. R. (1987). Seismic history and seismotectonics of the Sunda arc. *J. Geophys. Res.* 92, 421–439.

41. Newman, A. V., Hayes, G., Wei, Y., and Convers, J. (2011). The 25 October 2010 Mentawai tsunami earthquake, from real-time discriminants, finite-fault rupture, and tsunami excitation. *Geophys. Res. Lett.* 38, 1–7.

42. Okada, Y. (1985). Surface deformation due to shear and tensile faults in a half-space. *Bull. Seismol. Soc. Am.* 75, 1135–1154.

43. Pardo-Iguzquiza, E., and Chica-Olmo, M. (1993). The Fourier integral method: an efficient spectral method for simulation of random fields. *Math. Geol.* 25, 177–217.

44. Philibosian, B., Sieh, K., Avouac, J. P., Natawidjaja, D. H., Chiang, H., Wu, C., et al. (2014). Rupture and variable coupling behavior of the Mentawai segment of the Sunda megathrust during the super cycle culmination of 1797 to 1833. *J. Geophys. Res. Solid Earth* 119, 7258–7287.

45. Satake, K., Nishimura, Y., Putra, P. S., Gusman, A. R., Sunendar, H., Fujii, Y., et al. (2013). Tsunami source of the 2010 Mentawai, Indonesia earthquake inferred from tsunami field survey and waveform modeling. *Pure Appl. Geophys.* 170, 1567–1582.

46. Shao, G., and Ji, C. (2005). *Preliminary Result of the Mar 28, 2005 Mw 8.68 Nias Earthquake*. Available from: http://www.geol.ucsb.edu/faculty/ji/big_earthquakes/2005/03/smooth/nias.html

47. Sieh, K., and Natawidjaja, D. H. (2000). Neotectonics of the Sumatran fault. *Indonesia. J. Geophys. Res. Solid Earth.* 105, 28295–28326.

48. Sieh, K., Natawidjaja, D. H., Meltzner, A. J., Shen, C. C., Cheng, H., Li, K. S., et al. (2008). Earthquake super cycles inferred from sea-level changes recorded in the corals of West Sumatra. *Science* 322, 1674–1678.

49. Sladen, A. (2008). *Preliminary Result 02/20/2008 (Mw 7.4), Simeulue Earthquake, Indonesia*. Available at: http://www.tectonics.caltech.edu/slip_history/2008_n_sumatra /simeulue.html

50. Sladen, A., and Konca, A. O. (2007). *Preliminary Result 09/12/2007 (Mw 7.9), Central Sumatra Earthquake*. Available at: http://www.tectonics. caltech.edu/slip_history/2007_c_sumatra/c-sumatra.html

51. Tanioka, Y., and Satake, K. (1996). Tsunami generation by horizontal displacement of ocean bottom. *Geophys. Res. Lett.* 23, 861–864.

52. United States Geological Survey (USGS). (2015). *Latest Earthquakes*. Available at: http://earthquake.usgs.gov/earthquakes/map/

53. Yue, H., Lay, T., Rivera, L., Bai, Y., Yamazaki, Y., Cheung, K. F., et al. (2014). Rupture process of the 2010 Mw7.8 Mentawai tsunami earthquake from joint inversion of near-field hr-GPS and teleseismic body wave recordings constrained by tsunami observations. *J. Geophys. Res. Solid Earth.* 119, 5574–5593.

54. Zachariasen, J., Sieh, K., Taylor, F. W., Edwards, R. L., and Hantoro, W. S. (1999). Submergence and uplift associated with the giant 1833 Sumatran subduction earthquake: evidence from coral microatolls. *J. Geophys. Res.* 104, 895–919.

CHAPTER 5

Earthquake and Tsunami Knowledge, Risk Perception and Preparedness in the SE Bangladesh

Alam E*

University of Chittagong, Chittagong, Bangladesh

ABSTRACT

Frequent tropical cyclones and floods, and less frequent earthquakes and tsunamis affect Bangladesh. In absence of recent large earthquake and tsunami experience, the country's disaster risk reduction strategies have been developed from the learning dealing with frequent tropical cyclones and floods. However, an analysis of historical earthquake and tsunami suggests that the country is also vulnerable to earthquakes and tsunamis. For example, the 1762 earthquake originating within the Arakan Subduction Zone generated a tsunami in the northern Bay of Bengal that struck SE Bangladesh. This research aims to understand how local residents perceive and prepare for earthquake and tsunami in SE Bangladesh. In order to achieve this objective, this research used both quantitative (i.e. questionnaire survey) and qualitative (i.e. focus group discussions and informal interviews) data collection techniques in SE Bangladesh. The results of data analysis suggest that the local residents have lesser level of risk perception and preparedness in absence of their direct experience of earthquakes and tsunamis. As such, this research recommends further assessment of community vulnerability to earthquakes and tsunamis and implementing risk reduction strategies.

Keywords: Earthquake; Knowledge; Perception; Preparedness; Risk; Tsunami

INTRODUCTION

Bangladesh (**FIGURE 1**) is currently ranked as one of the world's most disaster prone countries. Approximately ninety seven percent of the total land area and all inhabitants are at risk of multiple hazards including tropical cyclones, floods, droughts, riverbank and coastal erosion. Bangladeshi communities have significant experience of frequent tropical cyclones and floods [1-3]. However, the recurrence interval of great earthquakes and tsunamis is very long in the northern Bay of Bengal region [4-8]. As such, people may not perceive earthquakes and tsunamis as major hazards, but the main hazard lies in potentials for rare but catastrophic high-magnitude earthquake and tsunami events. In this regard, this research is an attempt to understand that local residents' risk perception and preparedness about earthquake and tsunami in SE Bangladesh that has recently been identified prone to these types of hazards [4].

The Himalayan Mountains are located 200 km from the northern border of Bangladesh. Between the Himalayan Mountains and Bangladesh, the Dauki Fault and the Assam Seismic Fault are located (**FIGURE 1**). In addition, the Bogra Fault System (BGF), the Jamuna Fault (JF), the Madhupur Fault (MF) and the Sylhet Fault (SF) are located within Bangladesh (**FIGURE 1**). To the south, the country is bordered by the Bay of Bengal. The Arakan Subduction Zone (ASZ) along the northern end of the Bay of Bengal extends up to the SE Bangladesh. It is a tropical low-lying country having a long, funnel-shaped coast (**FIGURE 1**). Bangladesh has a coastline over 710 km long, joining the vast 47,201 km^2 areas of coastal plain with the Bay of Bengal [9]. The country has 64 administrative districts and 19 districts are considered as coastal [10]. The coastal zone accounts for 32% of the total area of the country and 28% of the people lives in this zone. The triangular shape of the Bangladesh coast at the northern Bay of Bengal and high astronomical tidal ranges increase exposure to tropical cyclone [1,11].

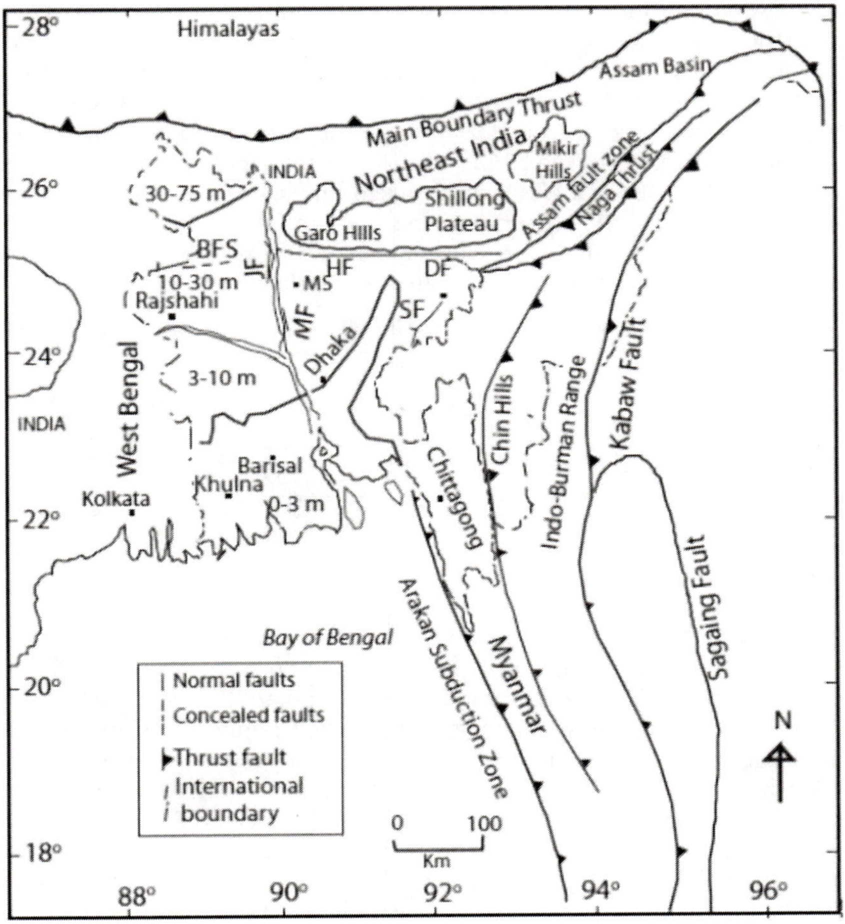

FIGURE 1: Location of Bangladesh in relation to regional active fault sources. ASZ: Arakan Subduction Zone, BFS: Bogra Fault System, DF: Dauki Fault, HF: Haluaghat Fault, JF: Jamuna Fault, MF: Madhupur Fault, SF: Sylhet Fault and MS: Mymensingh. The dark solid contour lines show the demarcation of different zones from the mean sea level (Source: map prepared adapting tectonic elements from Alam et al. [51], Ali et al. [21], Khan [7], Mukherjee [52].

The first record of an earthquake dates back to an event occurring in 1548 that affected both Bangladesh and India [12]. The 1548 earthquake may have ruptured large areas within Bangladesh and NE India. Few destructive earthquakes (**TABLE 1**) can be identified from the published and unpublished literature. However, in historical and recent times, the major cities, Dhaka and Chittagong, (**FIGURE 1**) have experienced ground shaking with intensities between III and VIII on the Modified Mercalli Intensity (MMI) scale from both distant and local severe earthquakes [12]. Earthquakes that occurred in 1897, 1934 and 1950 ruptured large areas adjacent to Bangladesh. The 1762 and 1897 earthquakes severely shook the floodplain of the Bengal Basin and caused landslides and liquefaction in Bangladesh [13,14]. Thus, earthquakes in Bangladesh are low-frequent catastrophic events and the vast majority of people have lack of experiencing significant earthquakes in their lifetimes.

TABLE 1: The occurrence of large earthquakes in and around Bangladesh [12,53].

Date	Name of earthquakes	Latitude /Longitude	Affected areas	Magnitude	Death/ Injuries
1548	-	26/ 92	India and Bangladesh	-	-
02-04-1762	Chittagong Earthquake	22/ 92	India, Bangladesh and Myanmar	7.8?	200
10-01-1869	Cachar Earthquake	24.75/ 93.25	India and Bangladesh	7.5	-
14-07-1885	Bengal Earthquake	24.8/ 89.5	Bangladesh and India	7.0	75
12-06-1897	Great Indian Earthquake	26/ 91	India and Bangladesh	8.7	1626
18-07-1918	Srimangal Earthquake	24.3/ 91.7	Bangladesh and India	7.6	9
02-07-1930	Dhubri Earthquake	25.8/ 90.2	India and Bangladesh	7.1	1
15-01-1934	Bihar-Nepal Earthquake	26.5/ 86.5	Nepal, India and Bangladesh	8.3	13772

The 2004 Indian Ocean Tsunami has already challenged the level of regional risk, rupturing around 1300 km of seafloor from Sumatra to the Andaman Island [15,16]. During the 2004 Indian Ocean Tsunami, the tsunami reached the coast of Bangladesh within 2.2 hours. The largest wave amplitude was 31 cm inland of the most southern coast of Bangladesh [17]. Officially, two human deaths occurred in Bangladesh [18]. The 2004 Sunda Subduction Zone earthquake was strongly felt [12]. Old buildings in Dhaka and Chittagong (two major cities in Bangladesh) developed cracking and some people were frightened following the 2004 Indian Ocean Tsunami [19]. People who were residing in these old buildings suddenly became panic and came out at street. The effects of the 2004 SSZ earthquake ranged between V-VI on the MMI scale from the coast to inland of Bangladesh [12].

Recently debate has arisen regarding the 1762 earthquake that originated in the Arakan Subduction Zone (ASZ) in the northern Bay of Bengal and whether or not it generated a 'mega-tsunami' similar to the one that devastated several Indian Ocean countries in December 2004 [5]. Cummins [5] refers to old 'historic' accounts of the effects of the earthquake in the region of Chittagong, Bangladesh. Later, Alam et al. [20] and Alam and Dominey-Howes [4] provide a detailed analysis of earthquakes and tsunamis in the northern Bay of Bengal. They reveal evidence of major historical earthquake and tsunami occurrence in SE Bangladesh. For example, they suggest that the 1762 earthquake generated several local tsunamis which flooded the Chittagong coast [4].

In the following sections, at first, I discuss earthquake and tsunami preparedness activities in Bangladesh. I then provide an overview of theoretical framework of earthquake and tsunami risk perception and preparedness, and a review of earthquake and tsunami literature in Bangladesh. Followed by this, I outline methods used for this research. In the subsequent sections, I present results and discussion of this paper. Finally, I make a series of recommendations for future research and implementing risk reduction strategies.

Earthquake and Tsunami Preparedness Activities in Bangladesh

The government of Bangladesh has a well-planned disaster response and mitigation mechanism for frequently occurring floods and tropical cyclones. However, it's programs and policies to mitigate the effects of less frequently occurring earthquakes and tsunamis are still in progress. In the late 1970s and early 1980s, many bridges, buildings and industrial structures were constructed. Following this, the assessment of seismic risk in different regions in the country was of great concern of engineers and scientists [21]. Research was undertaken to identify areas liable to severe seismic effects in Bangladesh [22,23]. The seismic zoning maps of Bangladesh have been revised ever since. The revised standing orders on disasters published in 2010, emphasised the need for mapping areas liable to earthquake damage.

The 12 May 1977 earthquake reaffirmed the need for improving seismic zoning in Bangladesh. Consequently, in June 1977, the Government formed a committee of experts to undertake the seismic zoning of Bangladesh and formulate policy options. The committee reviewed all the available information, revised seismic risk maps and outlined the building codes for earthquake resistant designs. In 1992, this exercise was repeated. Both seismic risk maps and building codes were revised to provide better guidelines of earthquake resistant design. For this mapping exercise, Bangladesh was divided into three zones (**FIGURE 2**): zone 3 (liable to severe damage); zone 2 (liable to moderate damage); and zone 1 (liable to slight damage).

The Disaster Management Bureau, a technical division of the government of Bangladesh, has attempted to increase public awareness in order to enhance preparedness for the pre- and post- earthquake periods in Bangladesh. Awareness activities included the publication of preparedness manuals and launching of programs in national broadcasting media, such as TV and radio [24]. In 2009, the Disaster Management Bureau published a manual that outlined pre-disaster measures and recommended activities that should be carried out during earthquake events [24] (**TABLE 2**).

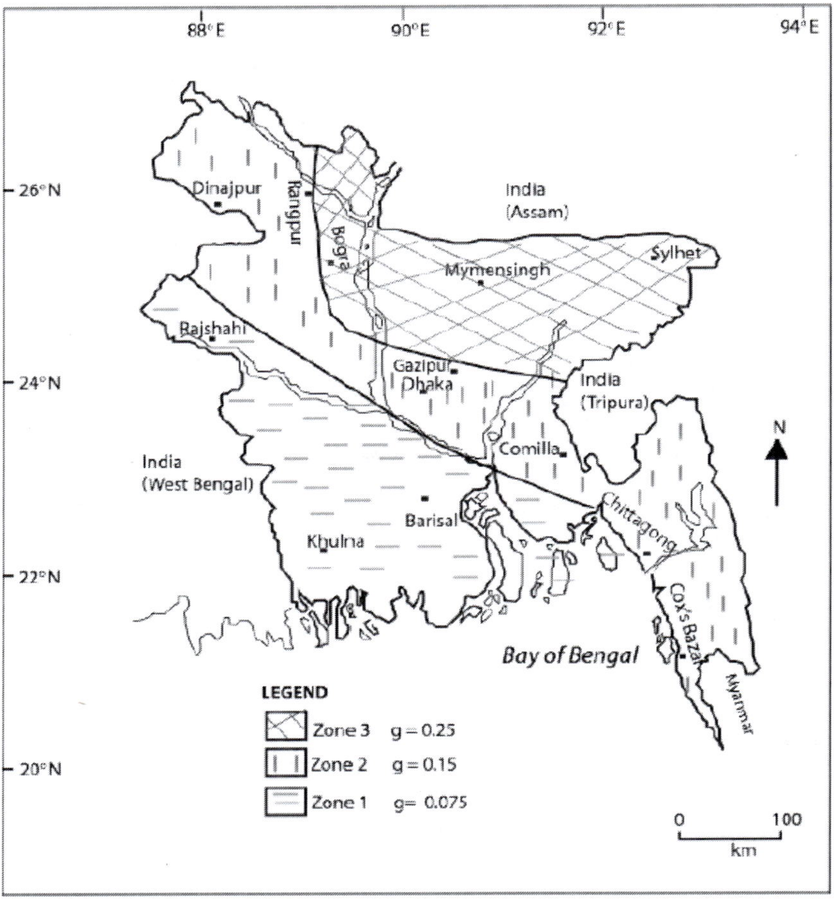

FIGURE 2: The current seismic zoning map of Bangladesh with probable earthquake zone coefficients. The g value equals maximum ground acceleration that can be expected based upon the associated from historic earthquakes of last 200 years. Zone 1 includes NE of Bangladesh and Sylhet City lies within this zone. Zone 2 includes Dhaka and Chittagong cities. Zone 3 includes SW Bangladesh [21].

TABLE 2: Recommended earthquake preparedness measures and activities to perform immediately after an earthquake, for individuals and households [24].

Pre earthquake preparedness measures
Having information about morphological characteristics and active fault before building house
Identification of strong structures inside the house to take shelter during the earthquake
Discussing preparedness plan (evacuation route) with family members
Having always first aid kit at home
Keep a battery-operated working torch at home
Having a battery-operated working radio at home
Having a bed and table made of hard materials at home
Manage to have a helmet for each family member
Taking part in the preparedness activities launched by the Government of Bangladesh or NGOs
Activities to perform immediately after an earthquake
Leave the house/apartment as quickly as possible after first tremor and stay in open space
Put helmet on and ask everyone else to do so
While exiting ask the neighbours to leave house/apartment
Switch off electric and gas lines and do not make any fire
Do not try to carry any belongings
Keep safe distance to buildings and electric poles
In case unable to leave house, take refuge in one corner of the house; if the house is built with concrete columns, take refuge at the bottom of a column
If the resided house is made of tin, take refuge under bed
Provide first aid for the injured and manage to go to hospital as soon as possible

Although some measures have been adopted (**TABLE 2**), there is no effective earthquake preparedness strategic plan. The policies and practices, which are based on those applying to tropical cyclones and floods, and they may not be adequate to mitigate the consequences of earthquakes. The review of the work of the Ministry of Food and Disaster Management and Disaster Management Bureau suggests that the government of Bangladesh is still working towards indentifying earthquake prone areas and improving disaster risk reduction for earthquake-related hazards.

Being a member of the Indian Ocean Tsunami Warning System (IOTWS), Bangladesh receives tsunami-related information for the Indian Ocean. Although the effects of the 2004 Indian Ocean Tsunami along the Bangladesh coast were minimal, this event raised considerable concern about the risk of tsunami hazard in Bangladesh [20]. Following the 2004 Indian Ocean Tsunami, the government of Bangladesh has incorporated tsunami hazard into its disaster risk reduction program, specifically in the second phase of the comprehensive disaster management program. The Disaster Management Bureau developed guidelines for tsunami preparedness (**TABLE 3**) that complement the cyclone preparedness strategies [4]. Because the tsunami is a rapid onset hazard and damage associated with its occurrence is different to that associated with tropical cyclones, these activities (**TABLE 3**) cannot adequately deal with the consequences of tsunamis. This research may help reducing community vulnerability to earthquakes and tsunamis in Bangladesh.

TABLE 3: Suggested tsunami preparedness activities to perform in pre, during and post tsunami period [24].

Pre tsunami preparedness measures
Organise meetings, seminars, symposium and workshops to create awareness about tsunami
Arrange improved training for rescue team
Organise grassroot level voluntary team
Providing of primary treatment training in household levels
Strengthening *Upazila* and *Union* level disaster management committee
Recommended activities during tsunami
Take shelter in nearest distant cyclone shelter after receiving tsunami warning
Measures should be taken to evacuate people at risk before tsunami arrived in the coast
Prioritise evacuating women and children in tsunami period
Activities to perform following tsunami
Launch rescue operation following the tsunami involving all public and private organisations
In emergency rescue operations, providing food, clothing, medicine, and safe water. Additionally strengthening rehabilitation program following the tsunami
Disseminating information to the grassroot levels by mobile phone

Theoretical Framework of Disaster Risk Perception and Preparedness

In order to mitigate the impacts of earthquake and tsunami hazards, it is essential to understand the many factors that influence people's ability to effectively respond to future events. A review of literature indicates that the following three issues are particularly important in relation to this study.

- Hazard knowledge
- Risk perception
- Disaster preparedness

In order to mitigate the effects of natural hazards, it is essential to understand how people living in places at risk perceive hazards, risk and their knowledge and preparedness in relation to particular hazards [25,26]. Enhancement of earthquake hazard knowledge may result in an increase in risk perception and preparedness actions [27] to reduce future vulnerability. In contrast, Mullis et al. [28] revealed that even when individuals have a better understanding of hazards and perceived risk, they may not undertake preparedness activities. In relation to volcanic hazards. Johnston et al. [29] suggest that only direct experiences may increase hazard knowledge and risk perception. When people perceive hazards [30], they are supposed to take more protective measures [31]. Again, this may not essentially lead to better preparedness [29]. Despite this, hazard knowledge and risk perception are important components of risk mitigation approaches [32]. The reason is that it is unlikely that an individual will be prompted to take part in mitigation measures for natural hazards without any appreciation of the hazard and risk [33].

The adoption of personal preparedness measures to reduce the risk of earthquake and tsunami hazard consequences may include but not limited to: education, age of residents, income level, ownership of residential unit in which currently residing, previous property loss due to an earthquake, knowledge of local alert systems and emergency response plans [34,35]. However, many factors influence and complicate people's decisions to adopt preparedness measures [36].

The review of the literature on the earthquake hazards in Bangladesh shows that much research was conducted to understand the morphology and seismic risk in Dhaka [12,37,38] and Chittagong [22] and Mymensingh cities (**FIGURE 1**) [39]. Paul et al. [34] conducted a quantitative survey of 444 residents of Dhaka to understand risk perceptions and preparedness about earthquakes. The findings of their research suggest that 83% of the residents were not prepared for a future major earthquake. Alam [4] suggested that the 1762 earthquake caused liquefaction, compaction, landslides and co-seismic subsidence, deaths and injuries in SE Bangladesh. However, in the past, limited research was conducted in order to understand risk perception of and preparedness to less frequent earthquakes and tsunamis in SE Bangladesh. As such, this research aims to fill this gap in knowledge by exploring communities' risk perceptions and their preparedness to mitigate risks of earthquake and tsunami hazards.

METHODS AND MATERIALS

In order to gain a better understanding of community perception of hazard and risk, and their preparedness [36,40] in SE Bangladesh, both quantitative and qualitative methods were utilized in this study. Both approaches have limitations [41]. However, mixed methods may bring in more robust evidence than either qualitative or quantitative approaches provide when they are used separately [42]. As such, household survey questionnaires, focus group discussions (FGDs) and informal interviews were used to gain a deeper understanding of hazard perception and preparedness.

RATIONALE OF RESEARCH AREA SELECTION

The SE Bangladesh was selected as the research area for mainly three reasons. First, as outlined in section two, limited research was conducted in order to understand earthquake and tsunami risk perception and preparedness in SE Bangladesh. Although, Alam [4] provided significant evidence of earthquake and tsunami occurrence in SE Bangladesh, the government of Bangladesh considers this region as a moderate seismic risk zone (**FIGURE 2**) due to lack of knowledge about major earthquake occurrences [21].

Secondly, with over six million people, the second largest Chittagong City in Bangladesh is situated in this region. Due to agglomeration of industries and business and the fact that the main seaport is located in this region, Chittagong City is called the commercial capital of Bangladesh. Over 80% of the export-import of the country has been operated by this city. Like unplanned development in Dhaka City, in the last three decades, residential, commercial and industrial buildings in Chittagong City have also been developed without following proper planning and building codes. Such unplanned development may increase susceptibility to earthquakes and tsunamis.

Thirdly, the effects of the Indian Ocean Tsunami 2004 in the SE Bangladesh were moderate. After this tsunami occurrence, the significant concern arose from scientists, policy planners, the government of Bangladesh and international development organizations about future risk along the Bangladesh coast.

SURVEY DESIGN AND IMPLEMENTATION STRATEGIES

The fieldwork of this research can be divided into two phases. In the first phase, the primary knowledge of earthquake and tsunami risk perception and preparedness were acquired whilst conducting field visits and investigations during November 2010 and January 2011 for the research work of Alam [4]. The lead author of that research went for field visits to identify the locations of effects of the 1762 earthquake and tsunami in SE Bangladesh. While ground-truthing the locations of the effects of the 1762 earthquake described in historical documents, "impromptu" opportunities arose to consult 30 local residents at Bajalia, Dohazari, Baharchhara (BH1 in **FIGURE 3**), Burumchhara and Akilpoor in Chittagong (**FIGURE 3**). They were asked about history of earthquakes and tsunamis and knowledge, awareness and risk perceptions they had of future earthquake and tsunami occurrence. This type of knowledge development about the research area and local residents can be coined as 'scoping exercises' (conducted to identify existing knowledge and develop guideline for further research [43]. Synthesis and analysis of the primary informal discussion and interviews guided to the development of detailed household survey questionnaires, FGDs and informal interviews [44].

In the second phase of the research (December 2012 to February 2013), detailed questionnaire surveys, FGDs and informal interviews were conducted. Before this, a list of households from the selected study areas was prepared. Selection of these households was based on the provision that household heads would be available for interview. The information from in-depth questionnaires were collected from the 25 heads of households (i.e. 15 male and 10 female headed households) as they play the main role in economic activities as well as disaster response processes [36]. Five in-depth household questionnaire surveys were conducted in each location. Five FGDs were conducted, which include two in each location (**FIGURE 3**), consisting of 10 people in each group (equal number of male and female in each group) at a convenient time and place for the participants. The respondents in the household questionnaire survey and FGDs were randomly selected for interview. The duration of the each FGD was approximately 1 hour. The ages of the participants in the FGDs were over twenty-five. The occupations of the FGDs participants included labourer, local vehicle drivers, farmers, fishermen, housewives and small shopkeepers.

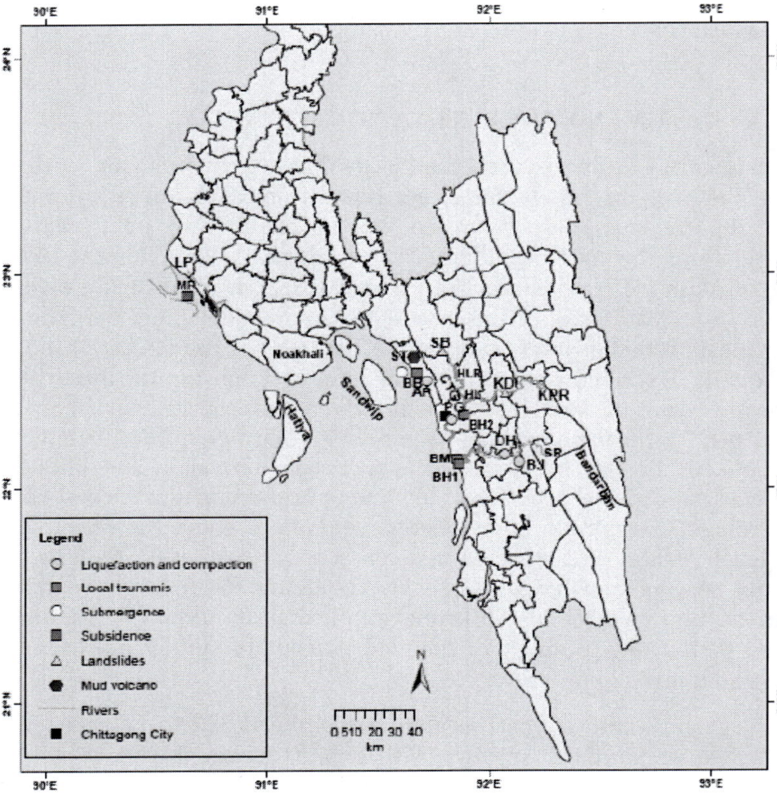

FIGURE 3: Location of field visits and the effects of the 1762 earthquake along the SE Bangladesh coast [4]. AP: Akilpoor, BB: Bansbaria, BM: Burumchhara, BH1: Baharchhara 1, BH2: Baharchhara 2, BJ: Bajalia, CTG: Chittagong City, DH: Dohazari, HL: Howla, KD: Kodala, LP: Lakshmipur, PG: Pathorgatha, ST: Sitakund, SB: Suabil, KPR: Karnafuli River, MR: Meghna River, SR: Sangu River.

Informal interviews were conducted with 20 local educated and influential people, comprising of four individuals each from Bajalia, Dohazari, Baharchhara, Burumchhara and Akilpoor (**FIGURE 3**). The respondents were selected on the basis of general recommendation by the local community. It was believed that these respondents possessed better knowledge of earthquakes and tsunamis. The participants were over forty years of age, male, and had completed at least 12 years of schooling. Professionally, they were local councillors, journalists, community leaders and school teachers.

Discussion in the questionnaire survey, FGDs and informal interviews was concentrated on the following questions:

- What people knew about the term earthquake and tsunami?
- Perceptions of earthquake and tsunami risk,
- Personal experience of earthquake and tsunami particularly about the 2004 IOT;
- Whether people and their communities have any oral stories available about the 1762 earthquake;
- Whether people thought that their area could be affected either by an earthquake or a tsunami in the future;
- What is the most perceived natural hazard in the area?
- Preparedness for the most perceived hazard;
- Awareness and preparedness activities for earthquake and tsunami; and
- Knowledge about disaster preparedness program by Governmental Organizations and Non-Governmental Organizations (NGOs) and their involvement with any past or current disaster risk reduction programs.

RESULTS

In the following sections, key findings are provided that ought to be of interest to organizations responsible for developing earthquake and tsunami risk mitigation strategies and for assisting the government of Bangladesh to determine the appropriateness of its preparedness strategy. **TABLE 4** provides a 'quick-look reference' to the key findings of this research. Sequentially, this research explores the following:

- Knowledge about earthquake and tsunami;
- **KNOWLEDGE** about historical and contemporary earthquakes and tsunamis;
- **RISK** perception and preparedness about earthquakes and tsunamis; and
- Knowledge about disaster responding organizations

TABLE 4: The SE Bangladeshi communities' earthquake and tsunami knowledge, risk perception and preparedness.

Earthquake and tsunami knowledge, risk perception and preparedness	Sources of information
Earthquake and tsunami knowledge	
Have general understanding of the natural processes of earthquake and tsunami generation	Interviews and FGDs
Have heard the IOT 2004 and aware about its devastative effects	Interviews
Lack of knowledge about the occurrence of historical earthquakes and tsunamis	Interviews
Lack of knowledge about damage associated with historical earthquakes and tsunamis	Interviews
Low risk perception and subsequent lack of preparation	
Lack of direct experiences of earthquakes and tsunamis	Informal interviews and FGDs
Apathetic belief about possible occurrence of earthquakes and tsunamis in the region. Dependency on *Allah's* wish about future occurrence of earthquakes and tsunamis	FGDs
Lack of preparation for the occurrence of future earthquakes and tsunamis	Interviews
Lack of government and NGOs disaster risk reduction preparation for earthquakes and tsunamis	Interviews
Local residents have preparation for the occurrence of frequent hazards (i.e. tropical cyclones). However, they were apathy towards less frequent hazards (i.e. earthquakes and tsunamis)	Interviews and FGDs

KNOWLEDGE ABOUT EARTHQUAKE AND TSUNAMI

In order to gain a better understanding of the local residents' hazard knowledge and risk perception, the research participants were asked if they knew about earthquakes or tsunamis. All the people I spoke to, they knew the term Bhumikompo. Bhumikompo is the Bengali term for earthquake, which consists of two words Bhumi (land) and kompo (shaking). Along with its scientific name, the local residents refer to tsunami as Jalhosti, which consists of two Bengali words Jal (water) and Hosti (Elephant). The reason for the use of this term is that, the participants suggested that the ground shaking of an earthquake caused the turbulence of fresh pond water resulting in viscosity. The informal interviews and group discussions revealed that the participants also knew about certain observable aspects of earthquakes and tsunamis. They were aware of the sudden onset nature of these hazards that leave people with limited time to prepare and evacuate.

KNOWLEDGE ABOUT HISTORICAL AND CONTEMPORARY EARTHQUAKES AND TSUNAMIS

All participants in interviews and discussions suggest that they did not have any knowledge about historical earthquake (**TABLE 1**) and tsunami occurrence in their region. None of them had ever heard about the 1762 event. Even none of the participants were aware of the tsunami events that had occurred before the 2004 Indian Ocean **TSUNAMI**. During the 2004 Sumatra-Andaman earthquake the participants experienced severe shaking, observed agitation of water in ponds and witnessed the rise of sea water adjacent to the Bay of Bengal coast. Experiencing the 2004 Indian Ocean Tsunami, the local residents are now familiar with the possible destructive effects (i.e. deaths, injuries and damage to houses and other infrastructures) of tsunami and the generation earthquake-induced tsunami.

RISK PERCEPTION AND PREPAREDNESS ABOUT EARTHQUAKES AND TSUNAMIS

Having directly experienced severe ground shaking during the 2004 Sumatra-Andaman earthquake, 75% of the respondents thought that Bangladesh was at risk of earthquake and tsunami. However, in response to possible attack either by an earthquake or a tsunami in the selected study area, 100% respondents replied that they had no idea what will happen in the future. For the same question, all the participants in the group discussions replied that this depends on Allah's (God) wish. Apparently, none of the participants in the interviews and group discussions had any direct experience of facing damage, injury and deaths as a result of any known earthquake and tsunami. When they were asked about their preparedness to earthquakes and tsunamis, 100% replied that they had not taken any preparedness actions to reduce the future risks that may occur due to these events. This means that Bangladesh resident's perceptions of and preparedness for earthquake and tsunami hazards are significantly influenced by their lack of experience of these hazards (perhaps during their life time) or by the low frequencies of these hazards.

Talking about natural hazards, the participants ranked the tropical cyclones as the damaging events in their areas. In relation to this, they recalled their own experiences and oral stories passed down from past generations' experiences of tropical cyclones (i.e. 1970, 1991 and 2007 that affected the

Bangladesh coast). They also predicted that they would face tropical cyclones in the future and, as such, had their personal preparedness plans for that. As such, the findings of this research coincide with seminal hypothesis of the White [26] that hazard perception and preparation depend on frequency and magnitude of past hazard experiences.

KNOWLEDGE ABOUT DISASTER RESPONDING ORGANIZATIONS

In Bangladesh, the Ministry of Food and Disaster Management (MoFDM) (https://www.dmrd.gov.bd/) is the main organisation responsible for dealing with disaster risk reduction. This ministry develops the legislative framework (i.e. the Disaster Management Act), and the policies and plans for disaster risk reduction in Bangladesh. In the field level, the local governmental organisations (i.e. Deputy Commissioner Offices, City Councils, Upazila Nirbahi Offices and Union Parishad Councils) and NGOs (i.e. CARE, ActionAid, Plan International and Oxfam etc.) worked both collaboratively and independently for implementing disaster risk reduction programs. All the respondents were aware about disaster responding governmental organisations and NGOs' preparedness activities for tropical cyclones and they were also involved at least one time in their lifetime in such programs. However, the findings revealed that not any risk reduction programs were in operation for earthquake and tsunami. None of the participants were involved in disaster risk reduction programs on earthquake and tsunami.

CONCLUSIONS AND RECOMMENDATIONS

Interviews and discussions with people living in SE Bangladesh reveal that they have knowledge about less frequently occurring earthquake and tsunami. However, they perceived the hazards as a low risk because they believe that hazard occurrence depends on Allah's wish. In addition, the majority have not considered the hazards as a personal risk because they did not experience severe earthquakes and tsunamis in the past. People intended to take preparedness for tropical cyclones because they have past experiences, but they showed apathy towards less frequently occurring earthquake and tsunami due to lack of personal experiences. Thus, vulnerability to earthquakes and tsunamis along the Bangladesh coast will heighten because they have very low risk perception, leading to non-preparedness to hazards. Here it should be noted that prior to the 1991 tropical cyclone, the most local residents on the Bangladesh coast had strong beliefs that the occurrence of tropical cyclones was a matter of Allah's (God) wish. Such a belief led them to perceive low hazard risk that resulting in obstruction to long-term preparedness to hazard [35,45,46]. Consequently, tropical cyclones have taken numerous lives in Bangladesh.

This research has shown that the resident populations of SE Bangladesh are at high risk from future earthquakes and tsunamis because they do not have any preparation for such less frequently occurring hazards. As such, this research recommend taking proactive measures in relation to earthquakes and tsunamis in Bangladesh by launching of appropriate awareness programmes and implementing risk management strategies. For example, the following preparations should be made for earthquake and tsunami hazard mitigation in SE Bangladesh:

- Enhancing community knowledge by informing local communities about local and distant tsunami risk by citing examples of the 1762 tsunami and the 2004 IOT;
- Launching a continuous education program in order to reduce the vulnerability of local communities to earthquakes and tsunamis by informing them of losses associated with past events, for example;
    - Land level changes (i.e. liquefaction, compaction, landslide, subsidence, submergence);
    - Human casualties and injuries;
    - Damage to houses;
    - Damage to lifelines (i.e. electricity and water); and
    - River seiches;
- Enhancing community knowledge about the occurrence of historical earthquakes in Bangladesh (i.e. 1762 and 1885) [47];
- enhancing community knowledge about potential damage in Bangladesh associated with earthquakes from India and Myanmar [48];
- Launching an awareness program and risk management strategies;
- Whilst this research was conducted for assessing community vulnerability to earthquakes and tsunamis in SE Bangladesh, such findings should be further validated by conducting in depth community surveys on the other areas particularly in major urban population centers across Bangladesh; and
- This research recognizes that the quantification of hazard and risk can be more extensively conducted with rigorous analysis of the physical and social (i.e. local community) context [40,49]. A comprehensive analysis of the local context and the community's living knowledge would need to be undertaken to achieve this [50-52].

REFERENCES

1. Salek JA(1998) Coastal trapping and funnelling effects on storm surges in the Meghna Estuary in relation to cyclones hitting Noakhali-Cox's Bazar coast of Bangladesh. Americal Meteorological Society 28: 227-249.
2. Haque U, Hashizume M, Kolivras KN, Overgaard HJ, Das B,et al. (2012) Reduced death rates from cyclones in Bangladesh: what more needs to be done? Bulletin of World Health Organisation 90: 150-156.
3. Paul BK(2009)Why relatively fewer people died? The case of Bangladesh's Cyclone Sidr. Natural Hazards 50: 289-304.
4. Alam E,Howes DD(2014)An analysis of the AD1762 earthquake and tsunami in SE Bangladesh. Natural Hazards 70: 903-933.
5. Cummins PR(2007)The potential for giant tsunamigenic earthquakes in the northern Bay of Bengal. Nature 449: 75-78.
6. Gupta H, Gahalaut V(2009)Is the northern Bay of Bengal tsunamigenic? Bulletin of the Seismological Society of America 99: 3496-3501.
7. Khan AA(2012)Seismogenic sources in the Bay of Bengal vis-a`-vis potential for tsunami generation and its impact in the northern Bay of Bengal coast. Natural Hazards 61: 1127-1141.
8. Sukhtankar RK, Pandian RS, Guha SK(1993)Seismotectonic studies of the coastal areas of India, Pakistan, Bangladesh, and Burma. Natural Hazards 7: 201-210.
9. BBS(2011) Bangladesh Population Census-2011. Bangladesh Bureau of Statistics (BBS), Government of Bangladesh, Dhaka.
10. Karim MF, Mimura N(2008) Impacts of climate change and sea-level rise on cyclonic storm surge floods in Bangladesh. Global Environmental Change 18: 490-500.
11. Talukder J, Roy GD, Ahmad M(1992)Living with Cyclone: study on storm surge prediction and disaster preparedness. Community Development Library Dhaka.
12. Akhter SH(2010) Earthquakes of Dhaka. Environment of Capital Dhaka-Plants Wildlife Gardens Parks Air Water and Earthquake. Asiatic Society of Bangladesh pp: 401-426.
13. Ambraseys N, Bilham R(2003)Reevaluated Intensities for the Great Assam Earthquake of 12 June 1897, Shillong, India. Bulletin of the Seismological Society of America 93: 655-673.
14. Verelst(1763) An account of the earthquakes that have been felt in the province of Islamabad, with the damages attending them, from 2nd to the 19th of April, 1762: translated from the Persian, and communicated to Henry Vansittart, Esq; President and Governor of Fort William in Bengal, by Mr.Verelst, Chief of the Hon. East India Company's Affairs at Islamabad. Philosophical Transactions 53: 265-269.
15. Satake K, Atwater B(2007) Long-term perspectives on giant earthquakes and tsunamis at subduction zones. Annu Rev Earth Planet Sci 35: 349-74.
16. Satake K, Aung TT, Sawai Y, Okamura Y, Win KS, et al. (2006) Tsunami heights and damage along the Myanmar coast from the December 2004 Sumatra-Andaman earthquake. Earth, Planets and Space, 58: 243-252.
17. Sarker NCD(2008) Tsunami simulation and hazard assessment on the Bangladesh Coast. Bulletin of the International Institute of Seismology and Earthquake Engineering 42: 121-126.
18. Ioualalen M, Pelinovsky E, Asavanant J, Lipikorn R, Deschamps A(2007) On the weak impact of the 26 December Indian Ocean Tsunami on the Bangladesh coast. Natural Hazards and Earth System Sciences 7: 141-147.
19. Anon(2004)The death of earthquake exceeded 25000: Mass graveyard: the devastation of tsunamis extended from Sumatra to Somalia. The Daily Azadi, 29th December, Chittagong, Bangladesh.

20. Alam E, HowesDD, Goff J,Goff C(2012) Tsunamis of the northeast Indian Ocean with a particular focus on the Bay of Bengal region – a synthesis and review. Earth-Science Reviews 114: 175-193.
21. Ali MH, Choudhury JR(2001) Assessment of seismic hazard in Bangladesh. Disaster Research Training and Management Centre, University of Dhaka, Bangladesh.
22. Kamal ASMM(2008) Seismic hazard assessment for Chittagong City Corporation area, Bangladesh. International Geological Congress Oslo 2008, August 6-14.
23. Sarker JK, Ansary MA, Rahman MS,Safiullah AMM(2010) Seismic hazard assessment for Mymensingh, Bangladesh. Environmental Earth Sciences 60: 643-653.
24. DMB(2009)New warning signal for cyclone and weather and mass disaster report: actions to be taken by people (in Bengali). Disaster Management Bureau, Ministry of Food and Disaster Management, Government of Bangladesh.
25. Burton I, Kates RW, White GF(1993) The Environment as Hazard. Annals of the Association of American Geographers 70: 306-311.
26. White GF (1974) Natural hazards: Local, national, global. Oxford University Press, New York.
27. Hurnen F, McClure J (1997)The effect of increased earthquake knowledge on perceived preventability of earthquake damage. Australasian Journal of Disaster Trauma Studies.
28. Mulilis JP, Duval TS(1995) Negative threat appeals and earthquake preparedness: a Person-Relative-to-Event (PrE) model of coping with threat. Journal of Applied Social Psychology 25: 319-39.
29. Johnston DM, Bebbington MS, Lai CD, Houghton BF, Paton D(1999)Volcanic hazard perceptions: comparative shifts in knowledge and risk. Disaster Prevention and Management 8: 118-126.
30. Lindell MK(1994) Perceived characteristics of environmental hazards. International Journal of Mass Emergencies and Disasters 12: 303-326.
31. Mileti DS, Fitzpatrick C(1992)The casual sequence of risk communication in the Parkfield earthquake prediction experiment. Risk Analysis 9: 20-8.
32. Gaillard JC,Dibben CJL(2008) Volcanic risk perception and beyond. Journal of Volcanology and Geothermal Research 172: 163-169.
33. Wegscheider S, Post J, Zosseder K, Muck M, Strunz G, et al. (2011) Generating tsunami risk knowledge at community level as a base for planning and implementation of risk reduction strategies. Natural Hazards and Earth System Sciences 11: 249-258.
34. Paul BK, Bhuiyan RH(2010)Urban earthquake hazard: perceived seismic risk and preparedness in Dhaka City, Bangladesh. Disasters 34: 337-359.
35. Tekeli Y, Dedeoglu, Fahrlaender B, Tanner (2011) Earthquake awareness and perception of risk among the residents of Istanbul. Natural Hazards 59: 427-446.
36. Alam E, Collins AE(2010) Cyclone disaster vulnerability and response experiences in coastal Bangladesh. Disasters 34: 931-954.
37. Islam MS, Hossain MT(2010) Earthquake induced liquefaction potential of reclaimed areas of Dhaka City. GeoShanghai 2010 International Conference, Shanghai, China.
38. Khan AA,Hossain MM(2005) Recurrence of 1885 Bengal earthquake and hazard vulnerability status of Dhaka Metropoliton City, Bangladesh. Oriental Geographer 49: 205-216.
39. Morino M, Kamal ASMM, Muslim D, Ali RME, Kamal MA, et al. (2011) Seismic event of the Dauki Fault in 16th century confirmed by trench investigation at Gabrakhari Village, Haluaghat, Mymensingh, Bangladesh. Journal of Asian Earth Sciences 42: 492-498.
40. Howes D,Minopoulos DM(2004) Perceptions of hazard and risk on Santorini. Journal of Volcanology and Geothermal Research 137: 285-310.
41. Gregory D, Johnston R, Pratt G, Watts MJ, Whatmore S(2009)The Dictionary of Human Geography. Wiley-Blackwell, West Sussex.

42. Creswell JW(2009) Research Design: Qualitative, Quantitative, and Mixed Methods Approaches. London: Sage.
43. Ehrich K, Freeman GK, Richards SC, Robinson IC, Shepperd S(2002) How to do a scoping exercise. Research Policy and Planning 20: 25-29.
44. Bryman A(2008) Social Research Methods. Oxford: Oxford Univeristy Press.
45. Alam E(2003)Coping with cyclone: an occupational group perspective. The Journal of Asiatic Society of Bangladesh (Humanities) 48: 59-74.
46. Alam E(2003) Post cyclone adjustment process: basic needs perspective. Oriental Geographer 47: 47-60.
47. Chen WP, Hsu L(2013) Historic seismicity near the source zone of the great 2008 Wenchuan earthquake: implications for seismic hazards. Tectnophysics 584: 114-118.
48. Gaillard JC, Mercer J(2013)From knowledge to action: bridging gaps in disaster risk reduction. Progress in Human Geography 37: 93-114.
49. HowesDD, Humphreys GS, Hesse PP(2006) Tsunami and palaeotsunami depositional signatures and their potential value in understanding the late-Holocene tsunami record. The Holocene 16: 1095-1107.
50. Miles M,Huberman A(1994)An expanded sourcebook: qualitative data analysis. London: Sage.
51. Alam M, Alam MM, Curray JR, ChowdhuryMLR, Gani MR(2003) An overview of the sedimentary geology of the Bengal Basin in relation to the regional tectonic framework and basin-fill history. Sedimentary Geology 155: 179-208.
52. Mukherjee M, Fryar AE, Thomas WA(2009) Geologic, geomorphic and hydrologic framework and evolution of the Bengal basin, India and Bangladesh. Journal of Asian Earth Sciences 34: 227-244.
53. Sharfuddin M(2010) Earthquake hazard analysis for Bangladesh. MSc Thesis in the Department of Civil Engineering, Bangladesh University of Engineering and Technology, Dhaka, Bangladesh.

CHAPTER 6

Estimation of Tsunami Hazard Vulnerability Factors by Integrating Remote Sensing, GIS and AHP Based Assessment

Mohammad Reza Poursaber[1,2*], Yasuo Ariki[1]

[1]*Graduate School of System Informatics, Kobe University, Kobe, Japan*

ABSTRACT

Geographic Information Systems (GIS), image processing in remote sensing and analytical hierarchy process (AHP) were used to estimate and classify vulnerability and inundation areas under the Tohoku tsunami event 2011 in the Ishinomaki, Miyagi prefecture, Japan. Acceptable data were obtained from Geoeye-1 satellite image, GSI DEM and field survey. Five factors of elevation, slope, shoreline distance, river distance and vegetation were used to classify the vulnerability and be weighted via AHP. By assessing the estimated and classified vulnerability map and comparing it with the inundation map of the study area, we found that a 13.44 km^2 area came under the tsunami vulnerability zone. Inundation areas were located in high and slightly high vulnerability classifications. Kitakami river and the Unga water canal played the role of flooding strips by transporting tsunami waves into the hinterland. This research is important to understand the roles of main topographical factors in a tsunami disaster.

Keywords: GIS, Remote Sensing, AHP and Tsunami Vulnerability Map

SUBJECT AREAS: Environmental Sciences

1. INTRODUCTION

Tsunami is a hazardous event as a natural phenomenon when happening in a populated area causes unacceptably large numbers of fatalities or property damage. By means of a vulnerable area map and performing a primary assessment, minimizing the impact of tsunami event is possible. Vulnerability is defined as an element to estimate the disaster risk within the context of hazard and risk. Since vulnerability is a result of capacity lack, increasing in capacity means reducing vulnerability, and high vulnerability means low capacity [1] [2] .

The evolution of remote sensing technology and its applications have enabled us to use satellite imagery data for mapping damage areas due to a natural disaster and also for assessing the vulnerable areas and damage distribution. As a further matter, introduction of Geographic Information System (GIS) as a database system has allowed us to analyze and display spatial data by using digitized maps for planning, mitigation and decision making [3] .

GIS software with its powerful tools is widely employed by decision-makers and planners for analyzing the spatial information through spatial multi-criteria analysis. Spatial multi-criteria analysis of GIS requires standard values and geographical location features, weighted topographical factors, according to users' preferences [4] - [6] . GIS is also applicable to evaluate the strategies required for creation of coastal vegetation belts against tsunami risk and to analyze tsunami risk using a multi-scenario approach [7] [8] .

Previous studies have developed and analyzed integrative remote sensing techniques in assessing building vulnerability to tsunami hazard. GIS approaches and a set of building vulnerability classification rules have been developed and successfully applied to categorize building vulnerability classes. Another study developed the Papathoma Tsunami Vulnerability Assessment (PTVA) to provide first-order assessments of building vulnerability to tsunami. In another study to undertake assessments of coastal vulnerability, analysis data are gathered from available ASTER satellite images (Advanced Space born Thermal Emission and Reflection Radiometer) and 3s-SRTM-v3 (3 arc-seconds Shuttle Radar Topography Mission, version 3) digital elevation models which the vulnerability assessment can be displayed via GIS as a series of thematic maps. The uses of multiple geospatial variables of topographic elevation, relation to tsunami direction, coastal proximity, and coastal shape are incorporated by the Analytic Hierarchy Process (AHP) to construct a weighting scheme for the geospatial variables and assessing tsunami vulnerability [9] - [12] .

2. SPATIAL DATA GATHERING AND ANALYSIS

In this study, remote sensing data captured by very high resolution (VHR) optical satellites with a pixel size about 0.5 meter is used to identify critical geographical elements such as buildings, transport infrastructure and possible inundation areas due to a tsunami event in order to estimate the vulnerability and risk in coastal areas. Specifically in this study, the area of Ishinomaki City in Miyagi Prefecture, (in the Tohoku region of northern Japan) was studied (Figure 1). Latitude and Longitude for Ishinomaki city in decimal degrees are: 38.4345°N, 141.3029°E respectively. As of September 2015, the city has an estimated population of 145,805 and a population density of 263 persons per km^2. The total area was 554.50 square kilometers. Ishinomaki has a humid subtropical climate with warm summers and cold winters. These climates usually occur on the eastern coasts and eastern sides of continents like east side of Japan, from 26 to 45 latitude and it is called cfa within Köppen climate classification [13] .

We extracted the factors of elevation and slope from a digital elevation model (DEM) obtained from the Geospatial Information Authority of Japan (here after referred to as GSI DEM), while the NDVI for vegetation density is extracted from Geoeye-1 image and Image Analysis toolbar in ArcGIS 10.2.1 The shoreline distance and river distance were measured from the vector maps of the study area.

The vulnerability by tsunami is then estimated by applying AHP to factors of elevation, slope, shoreline distance, river distance and vegetation. The vulnerability assessment can be displayed via GIS in terms of spatial multi-criteria analysis on a map of the tsunami vulnerability area. The general steps used in this study are shown in Figure 2.

2.1. Elevation

A digital elevation model was created from elevation data obtained from GSI (Geospatial information authority of Japan). GSI DEM was downloaded. The mesh elevation data created by interpolating to the elevation point at center point in 0.2 seconds (about 5 meter) mesh that is from the value of elevation (ground data) measured by airborne laser. The height accuracy of the 5m mesh elevation from the airborne laser is less than 0.3 m and the standard deviation of the altitude acquisition position (from the photogrammetry) is less than 1.0 meter. The data was in JPGIS (format) converted to shape file in point format using base map viewer converter software version 4.00 (FGDV) provided by GSI.

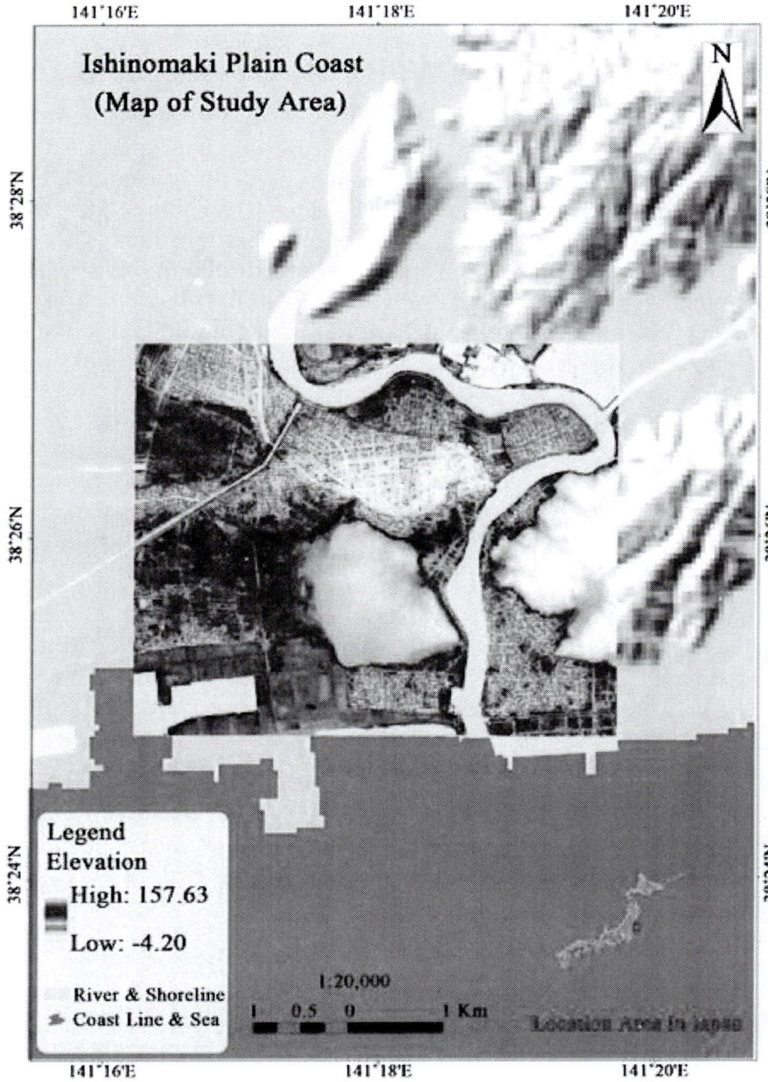

Figure 1. Case study site in Ishinomaki, Japan.

Finally, this point format was converted to raster for creating the digital elevation model via Arc GIS 10.2.1 software and Raster Interpolation toolset function.

2.2. Slope

The slope was determined as the rate of maximum change in z value from each cell of the satellite image. The use of a z-factor is essential for correct slope calculations when the surface z units are expressed in units different from the ground x, y units. The range of values in the output depends on the type of measurement units. The range of slope values is 0 to 90 for degrees and 0 to essentially infinity for percent rise. We created a slope map using the surface creation and analysis tools of the ArcGIS 10.2.1 software to use a third-order finite-difference method for calculating the slope [14] .

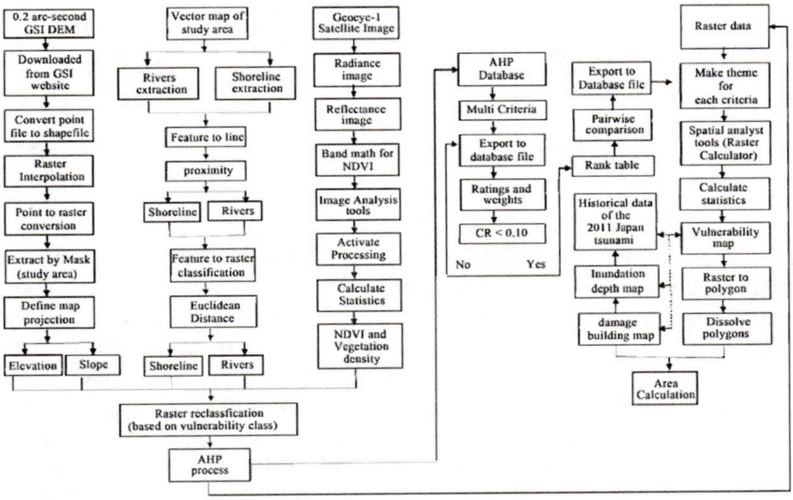

Figure 2. Framework for the study. GSI: Geospatial information authority of Japan, DEM: Digital Elevation Model, Geoeye-1: Very high resolution optical satellite image, NDVI: Normalized Difference Vegetation Index, AHP: Analytical Hierarchy Process, CR: Consistency Ratio.

2.3. Distance from Shoreline

The distance from the shoreline was created in a polyline file for buffering the distance from the shoreline to the land. We computed the distance using the proximity and the Euclidean distance analyst tool in the ArcGIS 10.2.1 software. The distance is based on the historical report of the maximum run-up in the area of study. We used Equation (1) to classify coastal proximity and shoreline distance buffering from Bretschneider and Wybro (1976) [12] :

$$\log X_{max} = \log 1400 + \frac{4}{3}\log\left(\frac{Y_0}{10}\right) \tag{1}$$

where X_{max} is the maximum reach of the tsunami over land, and Y_0 is the tsunami height at the shore line.

The maximum run-up of the tsunami in the study area was 8.6 m according to the 2011 Tohoku Earthquake Tsunami Joint Survey Group, 2011 at Ayukawa in Ishinomaki City, Miyagi Prefecture. We classified distance buffers in five classes based on Equation (1) in order to create a tsunami vulnerability map. It explains that 4.55 m to 7.09 m of run-up can reach a distance of 489.94m from the shoreline, 7.09 m to 9.64 m of run-up can reach 885.76 m, 9.64 m to12.18 m of run-up can reach 1332.84, 12.18 m to 14.73 m of run-up can reach 1821.46 m and 14.73 m to 17.27 m of run-up can reach more than 2345.53 m.

2.4. Distance from the (Kitakami) River and (Unga) Water Canal

We know the tsunami propagation in the rivers has higher speed than the tsunami propagation over land and it may maintain and propagate the wave energy further upstream and may cause damages in area far from the shoreline. The Kitakami River, the fourth largest river in Japan and the downstream branches into two water canal, Old-Kitakami river and Kitakami river. Old-Kitakami river flows on a fertile plain which has been highly developed for agriculture and industries and this water canal passes through the study area, while Kitakami river flows through a narrow valley into a small bay facing the Pacific Ocean. The Old-Kitakami river mouth had parallel jetties. The bed slope was 0.00017, which is the average bed slope from the river mouth up to the intrusion limit. On the contrary, the Kitakami river mouth was mostly sandy soil with 0.0001 bed slope and not as deep as the Old-Kitakami river mouth. It can be seen that the tsunami had reached 8 to 10 km or more off the coastline towards inland around Kitakami River and Old Kitakami River and the elevation in this region was extremely low, ranging between 0 and 2 m. while the devastated damages to the building were seen till the 3 km of the river area while the inundation area was about 8 km in the study area [15] [16] .

Kitakami-unga or Kitakami Canalis an artificial watercourse (canal) with an average elevation of-9 meter below sea level along 15.8 km. The soil in the area is high in andosols (an), soils composed of volcanic materials, usually dark colored. Kitakami Canal ran across the city and then inundated the inland area. Similarly to the river study, based on the historical report of the inundation area and inundation depth maps, we evaluated the inundated area along the water canal. Accordingly the influence of water canal width on the inundation flows was classified into three classes which are shown in Figure 7.

The elevation, river distance and shoreline distance were classified into five classes of vulnerability using the Jenks natural breaks method. This classification method indicates by picking the class breaks in best group similar values, maximize the differences between classes and minimize value differences between data within the same class and emphasize the differences between the reclassified classes [17] . The tsunami vulnerability map based on the elevation, slope, shoreline distance, river distance and Unga canal distance are shown in Figure 3-7.

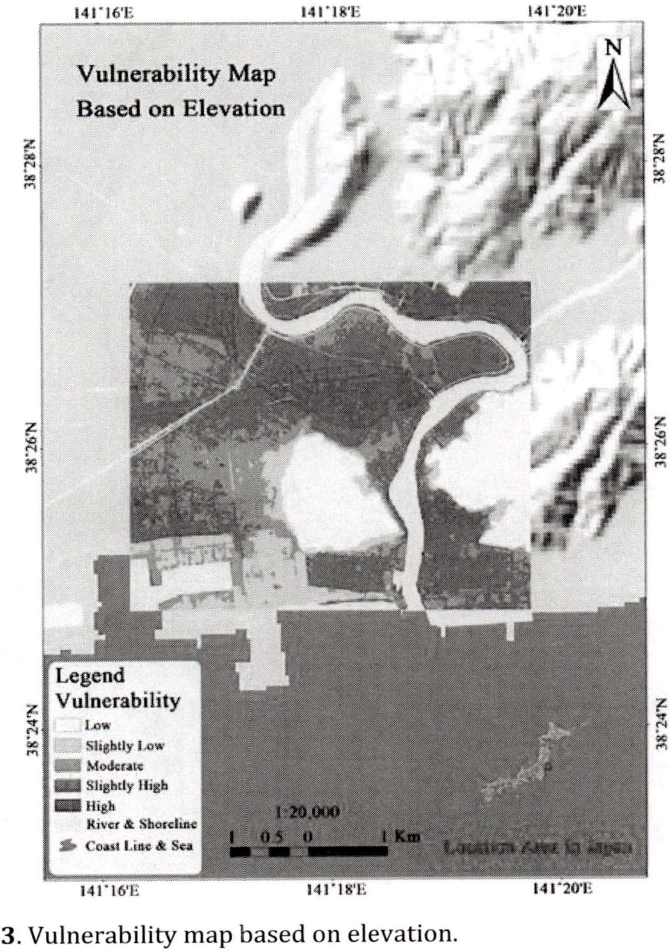

Figure 3. Vulnerability map based on elevation.

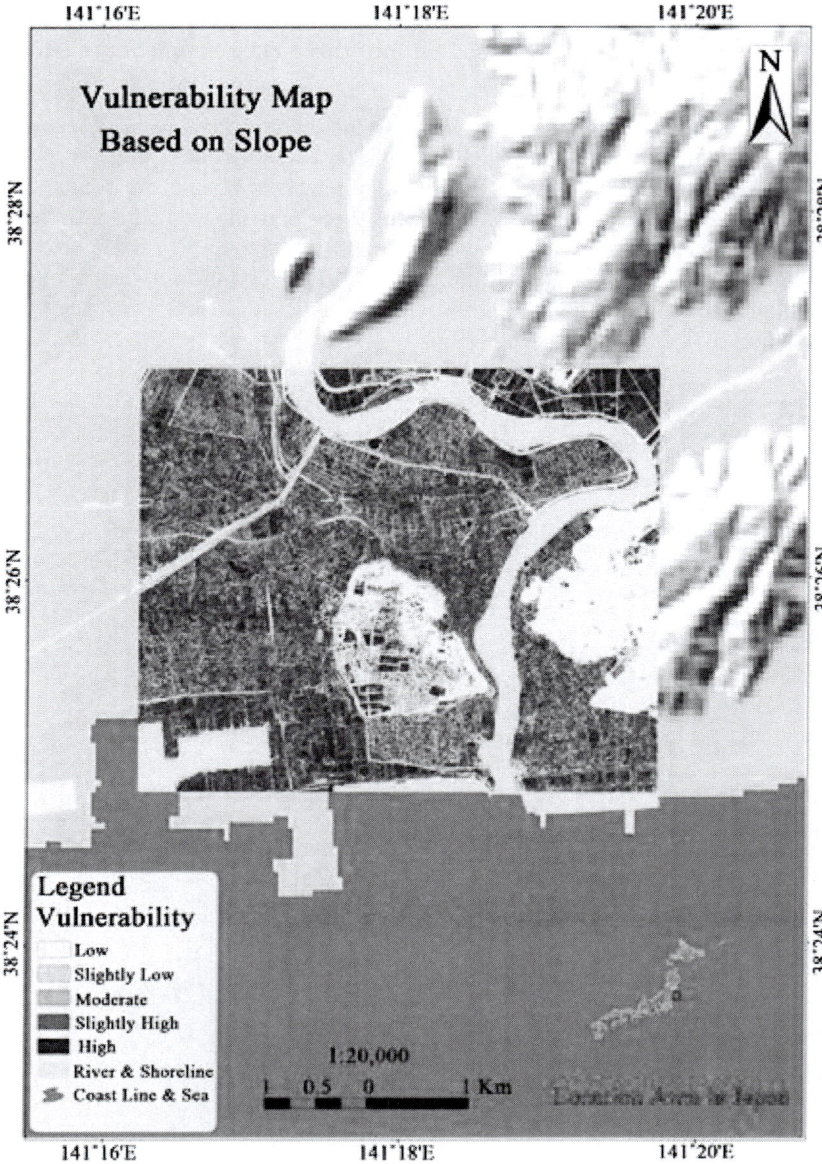

Figure 4. Vulnerability map based on slope.

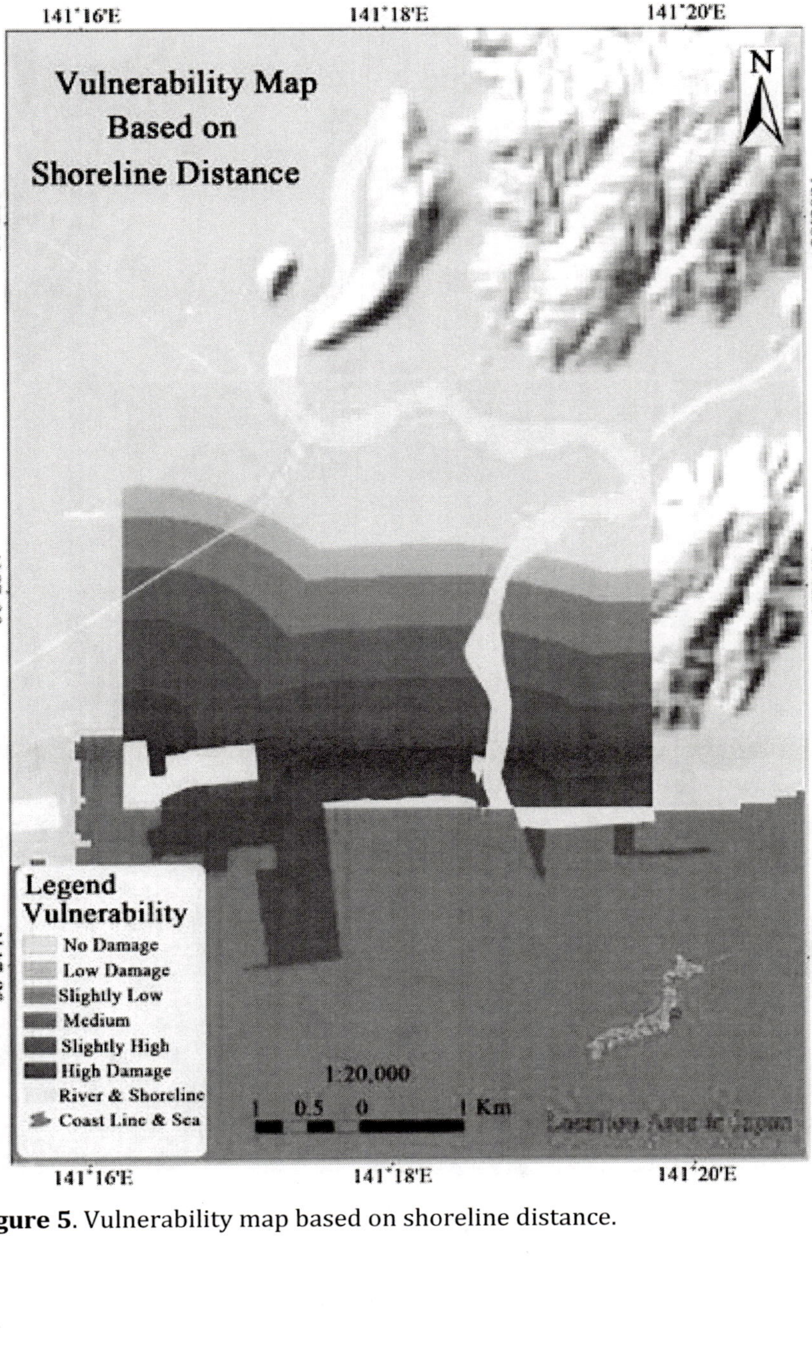

Figure 5. Vulnerability map based on shoreline distance.

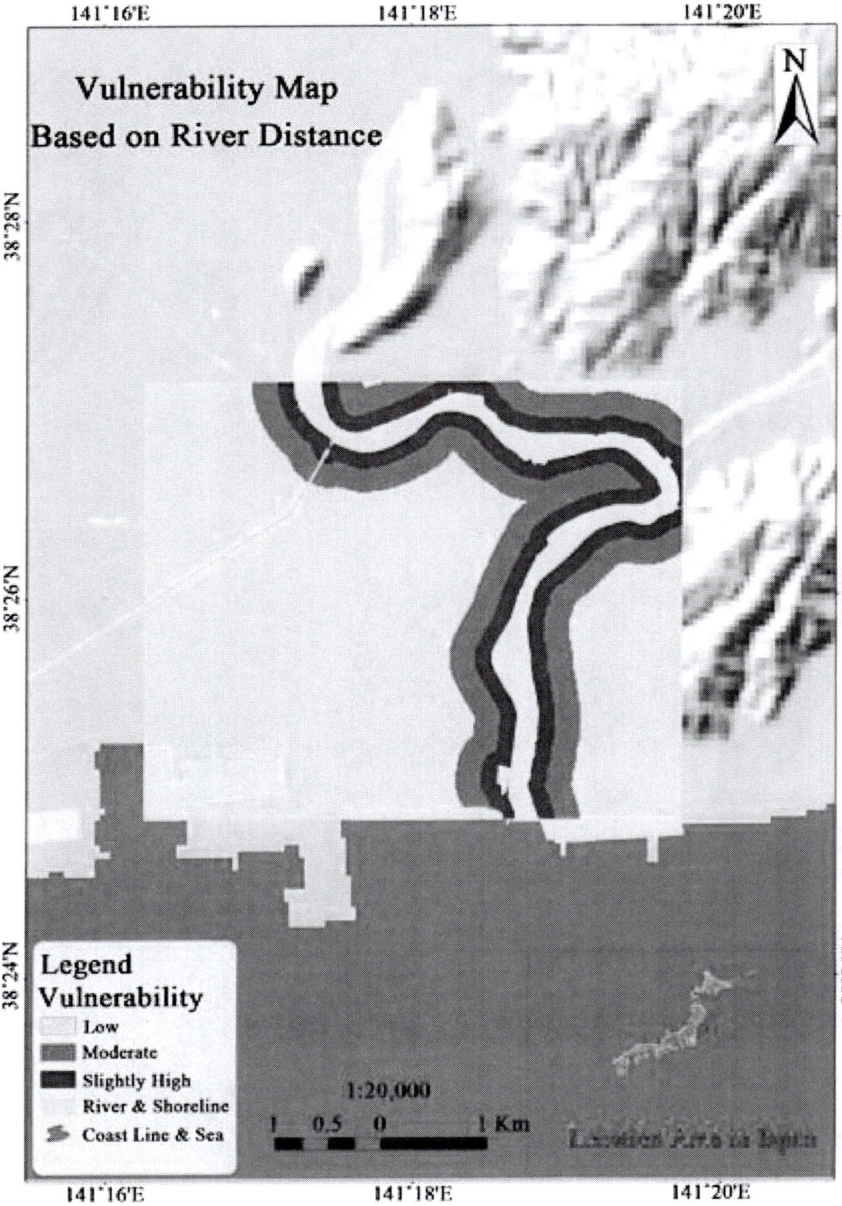

Figure 6. Vulnerability map based on river distance.

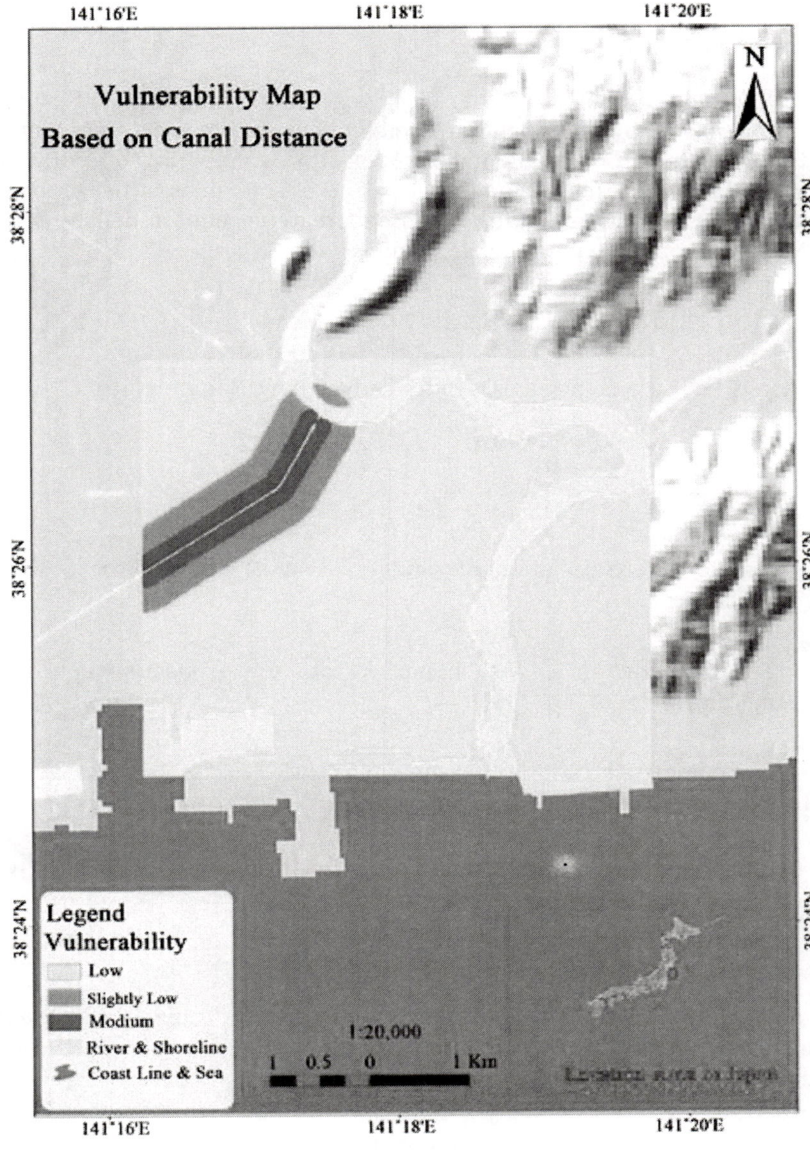

Figure 7. Vulnerability map based on the Canaldistance.

3. GEOEYE-1 PROCESSING

We calculated normalized difference vegetation index (NDVI), and vegetation density by using GeoEye-1 satellite image. The spectral radiance can be calculated from the digital number (DN) values in the GeoEye-1 image product using the radiometric gain and offset values in the product metadata by Equation (2) and the steps to create vegetation density map are described in the following subsection:

3.1. Conversion of DN to Radiance

Equation (2) describes the algorithm for converting DN to radiance:

$$L_\lambda = \mathrm{Gain}_\lambda \cdot DN_\lambda + \mathrm{Offset}_\lambda \tag{2}$$

In which,

λ = Specific spectral band of image: Near-IR, Red, Green, Blue orPanchromatic;

L_λ = Spectral radiance for band λ at the sensor's aperture ($mW/cm^2/\mu m/str$);

Gain_λ = Radiometric calibration gain ($mW/cm^2/\mu m/str/DN$) for band λ from product metadata;

DN_λ = Digital number values for band λ of image product;

Offset_λ = Radiometric calibration offset ($mW/cm^2/\mu m/str$) for band λ from product metadata [18] .

The bandwidths for the GeoEye-1 bands are given in Table 1, which are calculated by integrating over the relative spectral response curve of each band filter.

3.2. Conversion Radiance to Reflectance

Equation (3) describes the converting from radiance to reflectance [18] .

Table 1. GeoEye-1 band-dependent factors.

GeoEye-1 Band (λ)	spectral response for each band	
	Bandwidth (μm)	Esunλ (mW/cm²/μm)
Blue	0.0584	196.0
Green	0.0646	185.3
Red	0.0316	150.5
Near IR	0.1012	103.9

$$\rho_P = \frac{\pi \cdot L_\lambda \cdot d^2}{Esun_\lambda \cdot \cos\theta_S} \tag{3}$$

θ_S = 90°-Solar Elevation Angle (from metadata).

where,

ρ_P = Unitless planetary reflectance;

d = Earth-Sun distance (astronomical units) from nautical handbook;

$Esun_\lambda$ = Mean solar exoatmospheric spectral irradiances (mW/cm²/μm), at an Earth-Sun distance of one astronomical unit (A.U.);

θ_S = Solar zenith angle in degrees, (The solar zenith angle is calculated from the solar elevation angle;

L_λ = Spectral radiance for band λ at the sensor's aperture (mW/cm²/μm/str).

3.3. NDVI and Vegetation

The Normalized Difference Vegetation Index (NDVI) is a common tool for identifying and characterizing vegetation and a measure of the difference in reflectance between these wavelength ranges that takes values between −1 and 1, while vegetated areas produce values starting around 0.4 and approaching 1.0 and values <0 indicating no vegetation [19] .

Equation (4) was used for NDVI calculation and specification:

$$\text{NDVI} = \frac{(\text{NIR} - \text{RED})}{(\text{NIR} + \text{RED})} \tag{4}$$

In which for GeoEye-1 Band 4 is NIR (near infrared band) and Band 3 is red.

We classified elevation and the distance for river, canal and shoreline based on the values described in Table 2.

As shown in Figure 8, a vulnerability map based on vegetation density, explained that most of high-vulnera- bility areas were located in the coastal areas where the vegetation is less.

4. GIS AND AHP FOR VULNERABILITY MAPPING

Cell-based modeling in spatial analysis was used to specify the vulnerability area due to tsunami hazard. Cells are classified into five classes of vulnerability in the numbers of 1, 2, 3, 4 and 5, which represent low, slightly low, medium, slightly high, and high vulnerability classes.

AHP is a multi-criteria decision-making analysis (MCDA) approach introduced by Saaty (1977, 1980), which is developed by a series of pair-wise comparisons between each factor relative to other factors to make a scaled set of preferences. By the AHP measurement theory, pair wise comparisons and relies on the judgments of experts to conclude (derive) priority scales that measure intangibles in relative terms. The AHP is an Eigen value technique to the pair-wise comparisons approach and a numerical fundamental scale, which ranges from 1 to 9 to calibrate the quantitative and qualitative performances of priorities to score the importance of each factor (Saaty, 2008). Table 3 describes the fundamental scale of absolute numbers which is named the Saaty nine- point comparison scale. As we mentioned each number explains (depends on) the relative importance of each factor [20] .

Table 2. Tsunami vulnerability classes based on elevation, Shoreline distance, river distance and vegetation.

Vulnerability class	Factors name						
	Elevation (m)	Slope (%)	Shoreline distance (m)	River distance (m)		Vegetation	
				Kitakami	Unga canal	Index	Density
High (5)	−4 - −3	0 - 1	0 - 489.94	-	-	−0.998 - 0.151	Low
Slightly high (4)	3 - 4	1 - 3	489.94 - 1198.53	-	-	0.151 - 0.164	Slightly low
Medium (3)	4 - 5	3 - 6	1198.53 - 1700.48	0 - 160	1 - 150	0.164 - 0.188	Medium
Slightly low (2)	5 - 6	6 - 8	1700.48 - 2184.28	160 - 400	150 - 400	0.188 - 0.218	Slightly high
Low (1)	>6	>8	2184.28 - 2491.91	>680	>400	0.218 - 0.556	High

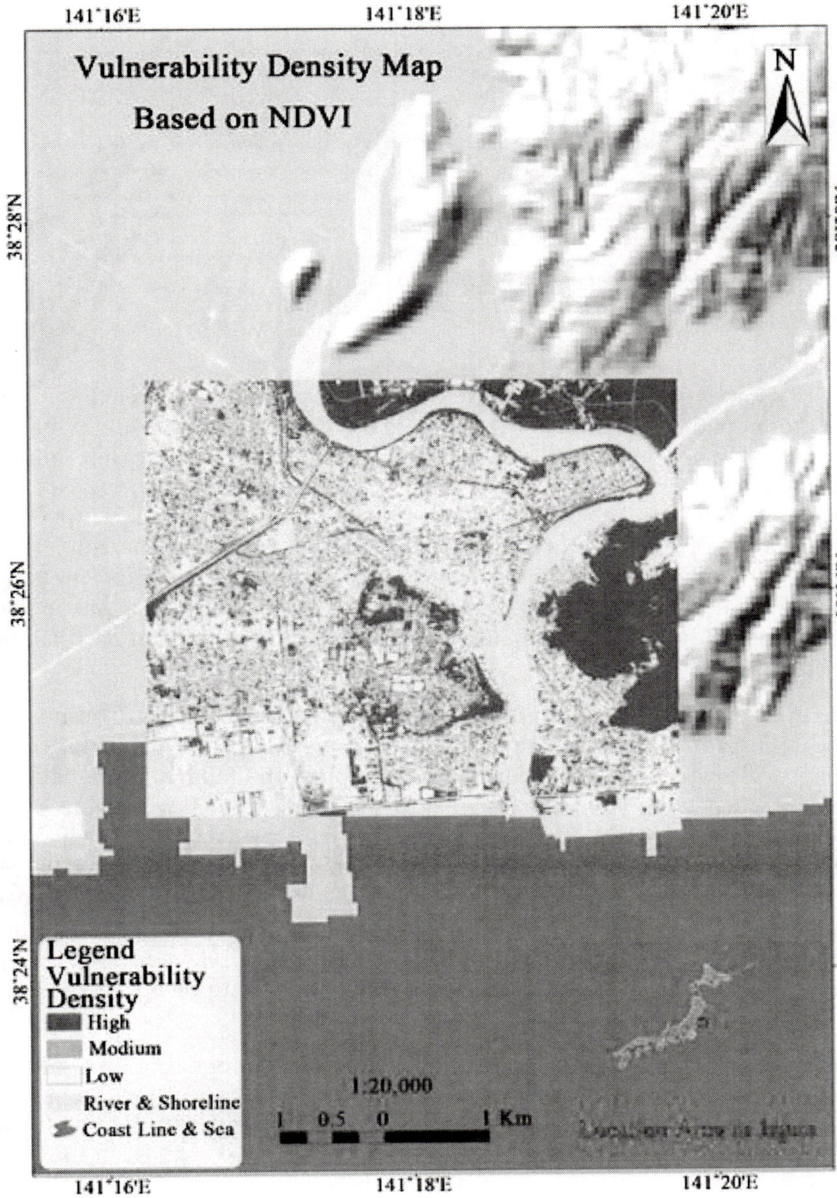

Figure 8. Vulnerability map based on vegetation density.

Table 3. The saaty nine-point comparison scale [21] .

Intensity of importance (Score)	Definition and Explanation	
	Definition	Explanation
1	Equal importance	Two factors contribute equally to the objective
3	Weak importance of one over another	The judgment is slightly favor one factor over another
5	Essential or strong importance	The judgment is to strongly favor one factor over another
7	Demonstrated importance	A factor is strongly favored and its dominance is demonstrated in practice
9	Absolute importance	The evidence favoring one factor over another is of the highest possible order of affirmation
2, 4, 6, 8	Intermediate values between the two adjacent judgments	When compromise is needed

All of tsunami vulnerability factors are overlaid and weighted based on their dominant influences in determining the class of tsunami vulnerability. The relative importance of each factor within the hierarchy is determined by their weights (Saaty, 1977 & 1980) with a pair wise comparison as shown in Table 4. The hierarchical interactions based on the respective importance of each factor were computed by estimating the numerical score. When there are evaluation criteria/objectives, decision makers must carry out a pairwise comparison. The scores are made by the subjective definition of the investigator in determining the importance of each factor [22] .

The first eigenvector computed based on the pairwise comparison matrix shown in Table 4. an approximation of Eigen vector (and Eigen value) of a reciprocal matrix can be obtained by or through the following method: 1) sum of each column of the reciprocal matrix; 2) Then we divide each element of the matrix with the sum of its column, we have normalized relative weight where the sum of each column is 1 as shown in Figure 9. The normalized principal Eigen vector can be obtained by averaging across the rows. The normalized principal eigenvector explains that shoreline distance has the highest weight (42.16%), followed by elevation (14.92%), slope (8.57%), river distance (26.60%) and vegetation density (7.74%) as shown in Figure 9.

AHP is subjective and tolerates inconsistency through the amount of redundancy by providing a measure of inconsistency assessment, which is shown by consistency ratio (CR). If the value of Consistency Ratio is smaller or equal to 10%, the inconsistency is acceptable. CR indicates the probability that the matrix judgments were randomly generated and it is defined as the ratio of the consistency index (CI), which is the degree of logical consistency among pair-wise comparisons, to the random consistency index (RI) which is the average CI value of randomly-generated comparison matrices.

Equations (5) and Equation (6) describe the algorithms for CR and CI calculation:

$$CR = \frac{CI}{RI} \tag{5}$$

$$CI = \frac{\lambda_{max} - n}{n - 1} \tag{6}$$

In which:

λ_{max} is the maximum eigenvalue of the judgment matrix and calculated from the sum of all factors and is multiplied by its eigenvector which is 42.16%;

n is the size of the comparison matrix. In this study, n = 5.

The RI is based on the random consistency index as shown in Table 5. The RI of 1.11 was used for five factors of normalized matrix in Figure 10.

So the consistency index, CI, is calculated 0.093 and CR is 8.37%.

$$\begin{matrix} \text{Eevation} \\ \text{Slope} \\ \text{Shoreline distance} \\ \text{River distance} \\ \text{Vegetation} \end{matrix} \begin{bmatrix} 0.1429 & 0.1818 & 0.1460 & 0.1224 & 0.1538 \\ 0.0714 & 0.0909 & 0.1095 & 0.0816 & 0.0769 \\ 0.4286 & 0.3636 & 0.4380 & 0.4898 & 0.3846 \\ 0.2857 & 0.2727 & 0.2190 & 0.2449 & 0.3077 \\ 0.0714 & 0.0909 & 0.0876 & 0.0612 & 0.0769 \end{bmatrix} \begin{bmatrix} 14.92\% \\ 8.57\% \\ 42.16\% \\ 26.60\% \\ 7.74\% \end{bmatrix}$$

Figure 9. Normalized matrix and eigenvector calculation.

Table 4. Pairwise comparison.

Pairwise Comparison	Factors Name				
	Elevation	Slope	Shoreline Distance	River Distance	Vegetation
Elevation	1.00	2.00	0.33	0.50	2.00
Slope	0.50	1.00	0.25	0.33	1.00
Shore-Line Distance	3.00	4.00	1.00	2.00	5.00
River Distance	2.00	3.00	0.50	1.00	4.00
Vegetation	0.50	1.00	0.20	0.25	1.00

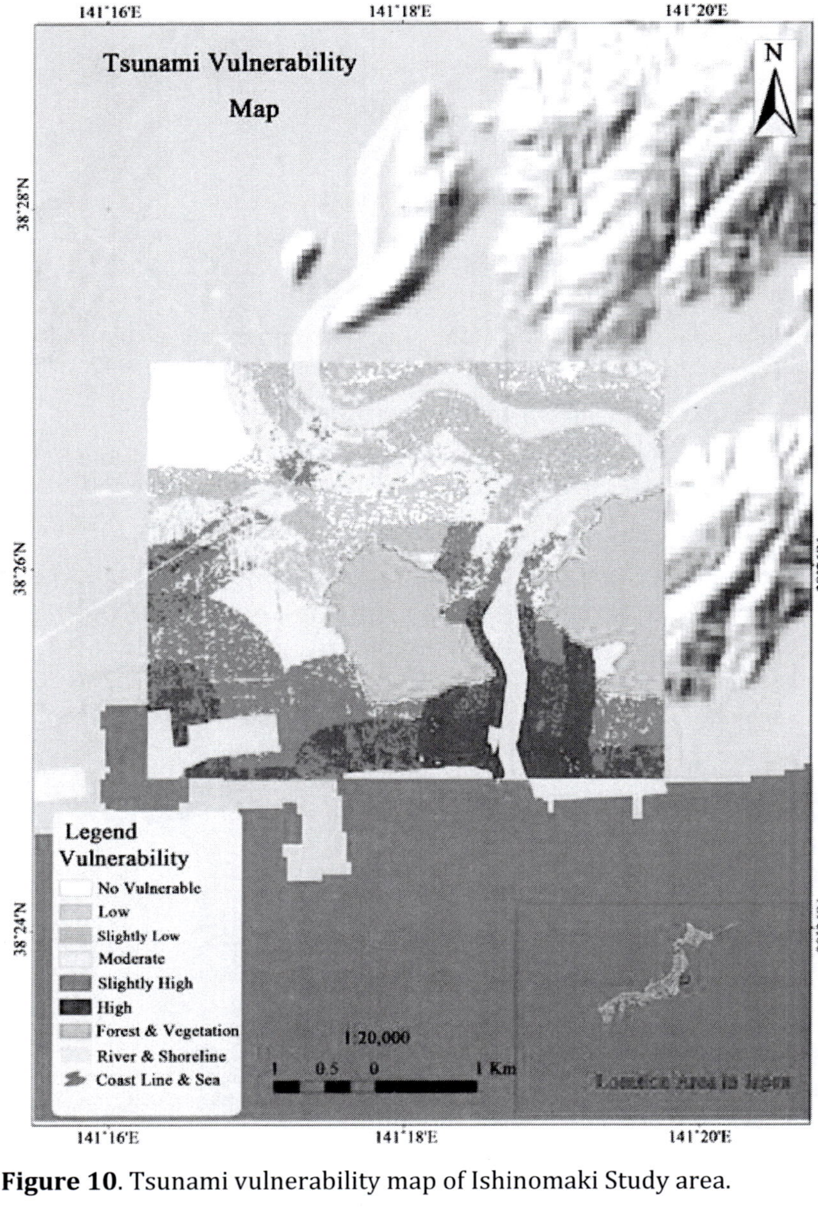

Figure 10. Tsunami vulnerability map of Ishinomaki Study area.

Table 5. Value of the Random Index (RI) [23] .

Random Consistency Index	Matrix Size									
	1	2	3	4	5	6	7	8	9	10
RI	0.00	0.00	0.52	0.89	1.11	1.25	1.35	1.40	1.45	1.49

We calculated each raster cell-by-cell basis of the factor to its weight. A Weighted Linear Combination (WLC) analysis is very straightforward in a raster GIS, and the factors that are being utilized are combined by applying a weight, to each followed by a summation of the results to proceed a suitability map. Equation (7) describes the suitability calculation [19] [20] and [23] :

$$S = \Sigma W_i \cdot X_i \qquad (7)$$

where:

S = Suitability;

W_i = Weight of factor i;

X_i = the criterion score of factor i (potential rating of the factor).

We estimated raster calculator in map algebra menu using the spatial analyst tools of ArcGIS 10.2.1 to produce vulnerability map by applying Equations (7).

Figure 10 shows the tsunami vulnerability map in the Ishinomaki area as output of this calculation. We estimated the statistics of the vulnerability map based on the vulnerability classification of five factors used in this study is shown in Table 6 while the vulnerability index of 91,437.35 grid cells ranged between 1 to 5, with a standard deviation of 0.792.The GIS produced tsunami vulnerability map is shown in Figure 10 which is in good agreement with the historical data recorded by "GSI" and "2011 Earthquake Tsunami Joint Survey Group". Based on the comparison most of the inundation areas are located in high and slightly high vulnerability areas.

Based on the result of our study, which is shown in Figure 11 the inundation area was 14.14 km^2, while GSI reported that the inundation area in Ishinomaki in the 2011 Japan tsunami was 13.44 km^2. On the study area the maximum inundation height (run-up) based on the 2011 Earthquake Tsunami Joint Survey Group was 8.6 m and found at Ayukawa in Ishinomaki City, Miyagi Prefecture.

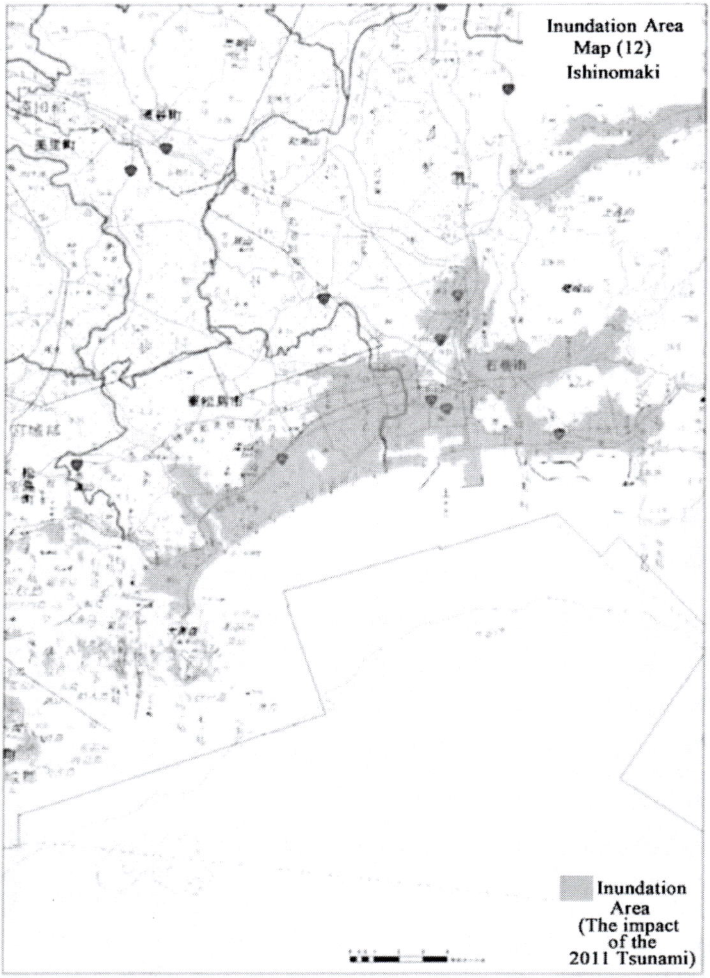

Figure 11. Tsunami Impact map in Ishinomaki (published by GSI) [24] .

Table 6. Vulnerability classification and tsunami inundated area.

Vulnerability index	Vulnerable Area Classification			
	Vulnerability class	Vulnerability Value in Tsunami Map	Area (km^2)	Area (%)
5	High	3.45 - 4.73	2.432	17.20
4	Slightly high	2.85 - 3.45	3.747	26.50
3	Medium	2.28 - 2.85	3.146	22.25
2	Slightly low	1.84 - 2.28	3.235	22.88
1	Low	1.57 - 1.84	1.579	11.17

5. DISCUSSION

Vulnerability is defined as the potential area that can be damaged by natural hazards. Physical vulnerability is often defined as the loss degree to a given factor or set of factors within the area affected by a hazard like tsunami. Physical factors, such as elevation, slope, shoreline distance, river distance and vegetation density could be used for vulnerability classification. Moreover, we know "Inundation" is the result of a tsunami traveling a long distance inland and is a horizontal measurement of the path of the tsunami. A tsunami effect is defined by several factors like height, run-up height and run-up distance [25] - [27] .

In this research, we used and analyzed satellite remote sensing (data), elevation data and field survey data followed by multi-criteria analysis through AHP raster overlay tools in GIS processing can be used as the basic operation for vulnerability mapping and inundation assessment due to tsunami hazard.

In this study a first attempt for assessing tsunami vulnerability was performed by using the factor of river distance besides elevation, vegetation, shoreline distance and slope and applying the AHP method combined with raster overlay tools through GIS processing in the Ishinomaki area. It is valuable step to estimate inundation area via this method and primary tsunami vulnerability mapping for impact assessment and output.

Remote sensing can be capable to achieve information about the input factors for tsunami vulnerability mapping and impact assessment. In the study area, several indicators of vulnerability can be achieved using a very high resolution satellite. It is very useful to derive data about the DEM in high spatial resolution so the output of the research will result in higher accuracy. Although obtaining the digital elevation model from GSI DEM needs some processing and ArcGIS is a powerful software for processing and combining spatial data of each factor and analyzing the result of AHP in order to generate a vulnerability map.

In this study, we used five classes of vulnerability for tsunami vulnerability map. Based on this, 2.432 km^2 of the study area was in high vulnerability, 3.747 km^2 was in slightly high vulnerability, 3.146 km^2 was in medium vulnerability, 3.235 km^2 was in slightly low vulnerability and 1.579 $km^2$of the area was in low vulnerability. The high-vulnerability areas were located in the coastal areas in low slopes. In high-vulnerability area and slightly high vulnerability inundation areas were predicted. In addition, the Kitakami River and water canal have the role to act as a flooding strip that transports inundation into the hinterland. The tsunami run-up comes up to the hinterland through the flat surface, urban areas and rivers. In this research we created the tsunami vulnerability map and inundation map which can be used for determining the priority for land-use planning related to tsunami hazard risk management.

In this research, we proposed the combined analysis of digital elevation data, very high resolution satellite images, tsunami historical data, AHP, and spatial multi-criteria processing via GIS to provide a tsunami vulnerability map and inundation map. Using the reflectance value of Geoeye-1 image before and after 2011 Japan, Tohoku tsunami was calculated as a preliminary study. In the next research, the height of the buildings and their materials in the inundation area can be considered in classifying the damaged buildings.

6. CONCLUSIONS

In recent years, vulnerable areas due to tsunami disaster have been studied by means of GIS application on satellite images and AHP method used in multi-criteria analysis. By means of GIS application, it is possible to manage hazards due to tsunami disasters. In the case of data limitation, very high resolution of the DEM and other factors, such as coastal type, relative direction of tsunami, coastal bathymetry and river morphological consideration is needed to achieve a better tsunami vulnerability mapping.

In this study, we proposed five factors in order to create a tsunami vulnerability map in accordance with the inundation area map of tsunami in Ishinomaki area of Miyagi Prefecture in Japan. As a result of this vulnerability map, although the importance of the shoreline areas in the creation of whole inundation pattern is observed, our results indicate an important role of rivers in inundation pattern far from the coast line. Furthermore, low level land areas far from the coast line are observed to be in inundated areas. The simulated tsunami pattern shows over eighty percent compatible with the inundation map.

This study used high resolution of DEM for the input factors of elevation and slope. We also calculated NDVI for the factor of vegetation index by using Geoeye-1 satellite image and the NDVI button on the Image Analysis window in the Arc GIS software.

In conclusion, by using multiple geospatial variables of topographic elevation, relation to tsunami direction, coastal proximity, and coastal shape incorporated by the AHP, an appropriate pair-wise comparison of AHP is proposed to construct a weighting scheme for the geospatial variables and assessing tsunami vulnerability.

Our research can be employed to evaluate the assessment of the vulnerable areas that could be affected by tsunami hazard in future natural disasters.

ACKNOWLEDGEMENTS

The technical and financial supports by Research Center for Urban Safety and Security (RCUSS) of Kobe University and Shahid Beheshti University are gratefully acknowledged. Authors are thankful to the GSI for providing DEM data and vector maps of the study area and 2011 Tohoku Earthquake Tsunami Joint Survey for the survey data in the study area.

CITE THIS PAPER

Mohammad Reza Poursaber, Yasuo Ariki (2016) Estimation of Tsunami Hazard Vulnerability Factors by Integrating Remote Sensing, GIS and AHP Based Assessment. *Open Access Library Journal*,**03**,1-11.

REFERENCES

1. Department of Regional Development and Environment Executive Secretariat for Economic and Social Affairs Organization of American States (1991) Primer on Natural Hazard Management in Integrated Regional Development Planning. Washington DC.
2. Roxana, L., Ciurean, D.S. and Glade, T. (2013) Conceptual Frameworks of Vulnerability Assessments for Natural Disasters Reduction. INTECH Published, 1-32.
3. FEMA (2015) Applications of GIS for Emergency Management, Lesson 1: Introduction and Course Overview, IS-922. Accessed on 16 August 2015. https://emilms.fema.gov/is922/GISsummary.htm
4. Stephen, J.C. (2007) Integrating Multi-Criteria Evaluation with Geographical Information Systems. Geographical Information Systems, 5, 321-339.
5. Islam, T. and Saman, S. (2006) Spatial Information-Based Approach for Integrated Coastal Resource Management in Asian Tsunami-Affected Countries. FAO Regional Office Asia and Pacific, The Regional Workshop on Information Managementand Coordination Mechanisms of Tsunami Emergency and Rehabilitation Operations in Agriculture, Fisheries and Forestry, 287-311.
6. James, C.A., Harriet, D.R., Hoag, D.L., McMaster, G.S., Vandenberg, B.C., Shaffer, M.J., Weltz, M.A. and Ahjua, L.R. (2002) Multi-Criteria Spatial Decision Support Systems: Overview, Applications, and Future Research Directions. 175-180.
7. Deichmann, U., Ehrlich, D., Small, C. and Zeug, G. (2011) Using High Resolution Satellite Data for the Identification of Urban Natural Disaster Risk. European Union and World Bank.
8. Sambah, A.B. and Miura, F. (2013) Remote Sensing, GIS, and AHP for Assessing Physical Vulnerability to Tsunami Hazard. World Academy of Science, Engineering

and Technology. International Journal of Environmental, Ecological, Geological and Mining Engineering, 7.

9. für Geowissenschaften, F. (2010) Assessing Building Vulnerability to Tsunami Hazard Using Integrative Remote Sensing and GIS Approaches. Dissertation Thesis.
10. Dall'Osso, F., Gonella, M., Gabbianelli, G., Withycombe, G. and Dominey-Howes, D. (2009) A Revised (PTVA) Model for Assessing the Vulnerability of Buildings to Tsunami Damage.
11. Dall'Osso, F., Bovio, L., Cavalletti, A., Immordino, F., Gonella, M. and Gabbianelli, G. (2010) A Novel Approach (the CRATER Method) for Assessing Tsunami Vulnerability at the Regional Scale Using ASTER Imagery. Italian Journal of Remote Sensing, 42, 55-74.
12. Sinaga, T.P.T., Nugroho, A., Lee, Y.-W. and Suh, Y. (2010) GIS Mapping of Tsunami Vulnerability: Case Study of the Jembrana Regency in Bali, Indonesia.
13. National Oceanic and Atmospheric Administration (2012) Ishinomaki Climate Normals 1961-1990.
14. Weih Jr., R.C. and Mattson, T.L. (2004) Modeling Slope in a Geographic Information System. Journal of the Arkansas Academy of Science, 58, 100-108.
15. Tanaka, H., Kayane, K., Adityawan, M.B. and Farid, M. (2013) The Effect of Bed Slope to the Tsunami Intrusion into Rivers. Coastal Dynamics, 1601-1610.
16. Adityawan, M.B., Roh, M., Tanaka, H. and Farid, M. (2012) The Effect of River Mouth Morphological Features on Tsunami Intrusion. 75-83.
17. Campbell, J.E. and Shin, M. (2012) Geographic Information System Basics. v. 1.0, Chap. 6, 133-163.
18. Podger, N.E., Colwell, W.B. and Taylor, M.H. (2011) GeoEye-1 Radiance at Aperture and Planetary Reflectance.
19. (2002) 3.3 Normalised Difference Vegetation Index (NDVI). NDVI: A Non-Technical Overview.
20. Singh, S.K., Chandel, V., Kumar, H. and Gupta, H. (2014) RS & GIS Based Urban Land Use Change and Site Suitability Analysis for Future Urban Expansion of Parwanoo Planning Area, Solan, Himachal Pradesh (India). International Journal of Development Research, 4, 1491-1503.
21. Saaty, T.L. (1977) A Scaling Method for Priorities in Hierarchical Structures. Journal of Mathematical Psychology, 15, 234-281.
22. Teknomo, K. (2006) Analytic Hierarchy Process (AHP).
23. Rao, R.V. (2013) Decision Making in the Manufacturing Environment Using Graph Theory and Fuzzy Multiple Attribute Decision Making Methods, Chapter 2, Improved Multiple Attribute Decision Making Methods. http://www. springer. com/978-1-4471-4374-1
24. Geospatial Authority of Japan (GSI) (2015) Map of Inundation Area Due to the 2011 Great East Japan Earthquake, Map Number 12. Accessed on 2 November 2015. http://www.gsi.go.jp/common/000059847.pdf
25. Chandrasekar, N., Immanuel, J.L., Sahayam, J.D., Rajamanickam, M. and Saravanan, S. (2007) Appraisal of Tsunami Inundation and Run-Up along the Coast of Kanyakumari District, India—GIS Analysis. Oceanologia, 49, 397-412.
26. Gencer, E.A. (2013) Natural Disasters, Urban Vulnerability, and Risk Management: A Theoretical Overview. In: Gencer, E.A., Ed., The Interplay between Urban Development, Vulnerability, and Risk Management, Mediterranean Studies, Vol. 7, Chap. 2, Springer-Verlag, Berlin, 7-43.
27. Papathoma-Köhle, M., Ulbrich, T., Keiler, M., Pedoth, L., Totschnig, R., Glade, T., Schneiderbauer, S. and Eidswig, U. (2014) Vulnerability to Heat Waves, Floods, and Landslides in Mountainous Terrain: Test Cases in South Tyrol. In: Birkmann, J., Kienberger, S. and Alexander, D., Eds., Assessment of Vulnerability to Natural Hazards: A European Perspective, Elsevier Publications, Amsterdam, 179-201.

CHAPTER 7

Probabilistic Earthquake–Tsunami Multi-Hazard Analysis: Application to the Tohoku Region, Japan

Raffaele De Risi* and Katsuichiro Goda

Department of Civil Engineering, University of Bristol, Bristol, UK

ABSTACT

This study develops a novel simulation-based procedure for the estimation of the likelihood that seismic intensity (in terms of spectral acceleration) and tsunami inundation (in terms of wave height), at a particular location, will exceed given hazard levels. The procedure accounts for a common physical rupture process for shaking and tsunami. Numerous realizations of stochastic slip distributions of earthquakes having different magnitudes are generated using scaling relationships of source parameters for subduction zones and then using a stochastic synthesis method of earthquake slip distribution. Probabilistic characterization of earthquake and tsunami intensity parameters is carried out by evaluating spatially correlated strong motion intensity through the adoption of ground motion prediction equations as a function of magnitude and shortest distance from the rupture plane and by solving non-linear shallow water equations for tsunami wave propagation and inundation. The minimum number of simulations required to obtain stable estimates of seismic and tsunami intensity measures is investigated through a statistical bootstrap analysis. The main output of the proposed procedure is the earthquake–tsunami hazard curves representing, for each mean annual rate of occurrence, the corresponding seismic and inundation tsunami intensity measures. This simulation-based procedure facilitates the earthquake–tsunami hazard deaggregation with respect to magnitude and distance. Results are particularly useful for multi-hazard mapping purposes, and the developed framework can be further extended to probabilistic earthquake–tsunami risk assessment.

Keywords: earthquake, tsunami, probabilistic hazard analysis, stochastic rupture models, scaling relationships of earthquake source parameters, mega-thrust subduction earthquake

INTRODUCTION

Earthquake and tsunami can be concurrent threats in many coastal regions around the world. In the last 2500 years, more than 2500 major tsunami events occurred globally (NGDC, 2016), and more than a half of those were triggered by seismic events. Other events were generated by volcanic eruptions (Latter, 1981), submarine landslides (Satake, 2001; Ward, 2001; Watts, 2004), or potentially by asteroid/meteorite impacts (Ward and Asphaug, 2000). Figure 1 shows the distribution of tsunami events triggered by seismic events at a global scale (NGDC, 2016). Tsunamis are particularly likely in active subduction zones surrounding the Pacific and Indian Oceans and are less expected in crustal seismogenic regions surrounding the Mediterranean Sea. Nonetheless, devastating tsunami disasters can occur in the Mediterranean areas, as exemplified by two historical disasters, i.e., the 1303 Crete Island tsunami (Guidoboni and Comastri, 1997) and the 1908 Messina event (Billi et al., 2008). It is therefore evident that simultaneous earthquake–tsunami hazard represents an urgent global issue and may cause catastrophic loss, affecting communities along coastal regions from economic and social viewpoints (Løvholt et al., 2014).

Seismic sources close to the shoreline can trigger tsunamis that cause devastating damage, especially due to the lack of sufficient reaction time (Monastersky, 2012). Thus, tsunamis triggered by near-field seismic sources can be regarded as the main contributors of the tsunami risk impact, and they should be studied in detail. Moreover, in comparison to local tsunamis, a simpler parameterization is usually sufficient for far-field tsunamis because seismic moment, source mechanism, and radiation pattern are more influential in comparison with slip distribution within a rupture plane (Geist and Parsons, 2006). For the above reasons, this work will focus on near-field scenarios by considering detailed features of the earthquake rupture.

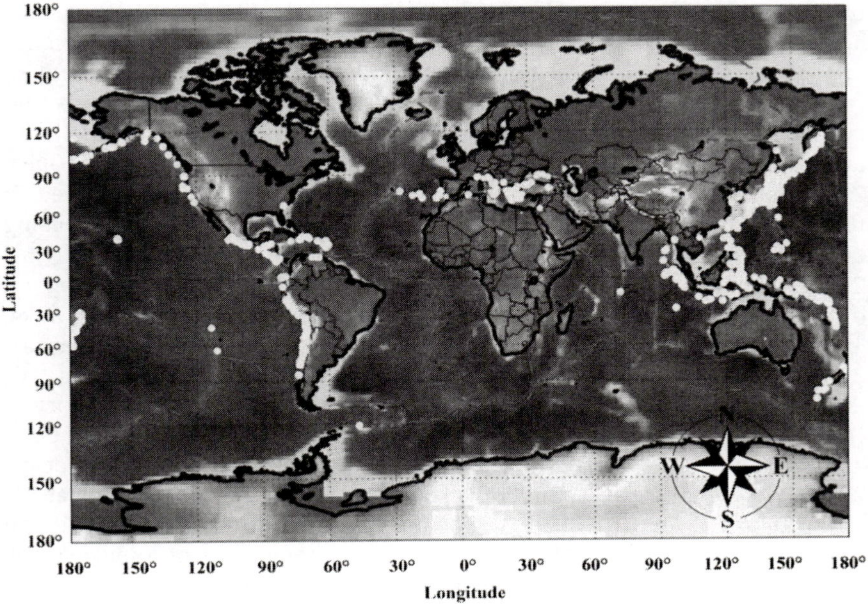

Figure 1. Locations of tsunamis triggered by seismic events. The red lines represent the global plate margins.

Probabilistic hazard analysis is the fundamental prerequisite for assessing disaster risk accurately and for deciding effective risk mitigation strategies. Both probabilistic earthquake and tsunami analyses involve various uncertain parameters that are related to geophysical processes and geological characteristics [e.g., slip rate, slip distribution, dip, and strike (Goda et al., 2014)], propagating media, local site conditions (e.g., soil type, roughness, and topography), and sea conditions [e.g., tidal level (Mofjeld et al., 2007)]. Conventional probabilistic seismic hazard analysis [PSHA (Cornell, 1968; McGuire, 2008)] can incorporate all major uncertain parameters in a comprehensive manner, with a potentially high computational effort. The computation becomes prohibitive when a logic tree with numerous branches (to capture full extent of epistemic uncertainty) is adopted for the assessment. In order to reduce this effort, a simulation-based probabilistic procedure can be implemented (Atkinson and Goda, 2013; Akkar and Cheng, 2015).

In the current probabilistic tsunami hazard analysis (PTHA), a comprehensive treatment of these uncertainties is rarely considered due to the lack of high-resolution/accuracy data and to the great computational effort involved in tsunami simulations. There are mainly three methodologies for tsunami hazard assessment in the literature (González et al., 2009): (a) probabilistic hazard analysis; (b) worst-case scenario approach, typically a deterministic method used for the development of practical emergency management products, such as evacuation maps and coastal infrastructure design (Cheung et al., 2011); and (c) sensitivity analysis, where the most influential model parameters are identified (Geist, 2002; Goda et al., 2014). The existing PTHA methods can be grouped in three broad categories. In the first category, PTHA is conducted by using tsunami catalogs (Burroughs and Tebbens, 2005; Tinti et al., 2005; Orfanogiannaki and Papadopoulos, 2007); in the second category, different scenario-based PTHA methods are suggested (Geist and Dmowska, 1999; Downes and Stirling, 2001; Farreras et al., 2007; Liu et al., 2007; Power et al., 2007; Yanagisawa et al., 2007; Burbidge et al., 2008; González et al., 2009; Løvholt et al., 2012). In the third category, a combination of the two previous categories is considered (Geist, 2005; Geist and Parsons, 2006; Annaka et al., 2007; Thio et al., 2007; Burbidge et al., 2008; Parsons and Geist, 2008; Grezio et al., 2010, 2012; Horspool et al., 2014; Fukutani et al., 2016). Specifically, for near-source subduction zones, Fukutani et al. (2016) extended the methodology of Annaka et al. (2007) for a Tohoku-type (M9) earthquake with fixed rupture geometry. They considered several cases for earthquake magnitude, slip pattern, and occurrence probability. However, the numbers of magnitudes and slip patterns are limited and thus not sufficient to capture a wide range of possible tsunami scenarios for this type of mega-thrust subduction earthquakes (Goda et al., 2014).

Building on the previous research, a new probabilistic earthquake and tsunami hazard assessment methodology for near-field seismic sources is presented. The novelty of the proposed methodology is the adoption of a single physical process for concurrent earthquake and tsunami threats; thus, dependency between ground-shaking and tsunami hazard parameters can be investigated probabilistically. On the one hand, the proposed methodology overcomes some of the previous limitations, such as inappropriate scaling relationships, simplistic slip distributions, subjective weights of the logic tree's branches, and simplified inundation models. On the other hand, some simplification, such as the adoption of discrete values of magnitude, and the fixed geometry and predefined meshing of the main subduction region, are maintained. This methodology can be extended to consider all possible sources in a region and can be applied to other subduction zones.

The first step is to define a suitable occurrence model; classical occurrence models in literature are the memory-less Poisson model, generally used for long-term hazard assessments, and the renewal model [e.g., Brownian passage time model (Matthews et al., 2002)] applied for short-term forecasting based on the seismic activity observed in the recent past. In this study, a classical Poisson model is adopted. Assuming a Poissonian inter-arrival time process, the probability of occurrence of an earthquake–tsunami event with specific characteristics in a given time window depends on the mean annual occurrence rate alone. A magnitude–frequency distribution of major seismic events that may potentially trigger tsunamis is then defined. For each value of earthquake magnitude, geometry of the rupture areas and other key source parameters (mean slip and spatial correlation parameters of slip distribution) are determined using new global scaling relationships for tsunamigenic earthquakes (Goda et al., 2016). In this step, both aleatory and epistemic uncertainties of the model parameters (i.e., position and geometry) are incorporated based on probabilistic information available in literature. Therefore, for each value of magnitude, multiple realizations of potential earthquake slip distribution are generated from a theoretical wavenumber spectrum model (Mai and Beroza, 2002). The incorporation of stochastic slip models in probabilistic earthquake–tsunami hazard analysis is another important novelty of this work with respect to the previous studies; conventionally, the slip distributions within a fault plane are considered as uniform or randomly distributed (without realistic spatial distribution of the slip).

Subsequently, simulations of the two hazard processes, i.e., ground shaking and tsunami, are carried out simultaneously. Specifically, for each slip distribution: (a) spatially correlated strong motion intensity measures are evaluated using ground motion prediction equations (GMPEs) for interface subduction events (Morikawa and Fujiwara, 2013; Abrahamson et al., 2016); and (b) the seafloor vertical displacement is calculated using analytical formulae (Okada, 1985; Tanioka and Satake, 1996), and tsunami simulation is performed by solving non-linear shallow water equations (Goto et al., 1997). By repeating the joint assessment of earthquake–tsunami hazards a sufficient number of times for each magnitude, a sample of spectral accelerations at multiple locations can be obtained by simulating a seismic intensity random field, while maximum tsunami wave heights/velocities can be obtained from tsunami hazard analysis. For each magnitude, the results obtained from the simulations are used to build the complementary cumulative distribution functions (CCDFs) for individual hazards, representing the conditional probability of reaching or exceeding a given intensity value. The CCDFs are provided with a confidence interval around the central estimates.

The site-specific earthquake–tsunami hazard curves can be derived by integrating the earthquake–tsunami simulation results and the magnitude–frequency distribution for the discrete values of magnitude, and by multiplying the result by the occurrence rate of earthquakes from the subduction zone. The result will be a triplet of CCDFs (central estimate and confidence interval curves), representing the mean annual rate of exceedance of specific values of seismic intensity or tsunami hazard parameters. The developed methodology is applied to the Tohoku region of Japan, where the subduction fault plane is well defined and information on regional seismicity is available. Finally, the hazards for a site in Sendai City, Miyagi Prefecture, are calculated.

METHODOLOGY

Formulation

The formulation presented herein is aimed at developing the earthquake–tsunami hazard curves for a specific location. Let **IM** represent the intensity measures of interest, such as spectral acceleration (S_a), inundation height (h), flow velocity (v), flux momentum, and tsunami force. Assuming a Poissonian arrival time process, the probability to observe an earthquake–tsunami sequence having intensity measure values **IM** equal to or greater than the specific values **im** in t years is

$$P(\mathbf{IM} \geq \mathbf{im}|t) = 1 - \exp[-\lambda(\mathbf{IM} \geq \mathbf{im}) \cdot t] \tag{1}$$

where $\lambda(\mathbf{IM} \geq \mathbf{im})$ is the mean annual rate at which the intensity measures **IM** will exceed specific values **im** at a given location. The rate $\lambda(\mathbf{IM} \geq \mathbf{im})$ can be expressed as a filtered Poisson process [e.g., Parsons and Geist (2008)]:

$$\lambda(\mathbf{IM} \geq \mathbf{im}) = \lambda(M \geq M_{\min}) \cdot \int P(\mathbf{IM} \geq \mathbf{im}|\boldsymbol{\theta}) \cdot S(\boldsymbol{\theta}|M) \cdot f(M) \cdot dM \tag{2}$$

$\lambda(M \geq M_{\min})$ is the mean annual rate of occurrence of the seismic events with magnitudes greater than the minimum magnitude considered in the magnitude–frequency distribution. $P(\mathbf{IM} \geq \mathbf{im}|\boldsymbol{\theta})$ is the probability that the joint intensity measures **IM** will exceed prescribed values **im** at a given coastal location for a given set of source parameters $\boldsymbol{\theta}$. $S(\boldsymbol{\theta}|M)$ represents the scaling relationships (or prediction models) of the uncertain earthquake source parameters conditioned on the magnitude. $f(M)$ is the magnitude–frequency distribution.

Five phases are defined: Phase 1 – fault model and earthquake occurrence, Phase 2 – source parameter characterization and stochastic slip synthesis, Phase 3 – earthquake simulation, Phase 4 – tsunami simulation, and Phase 5

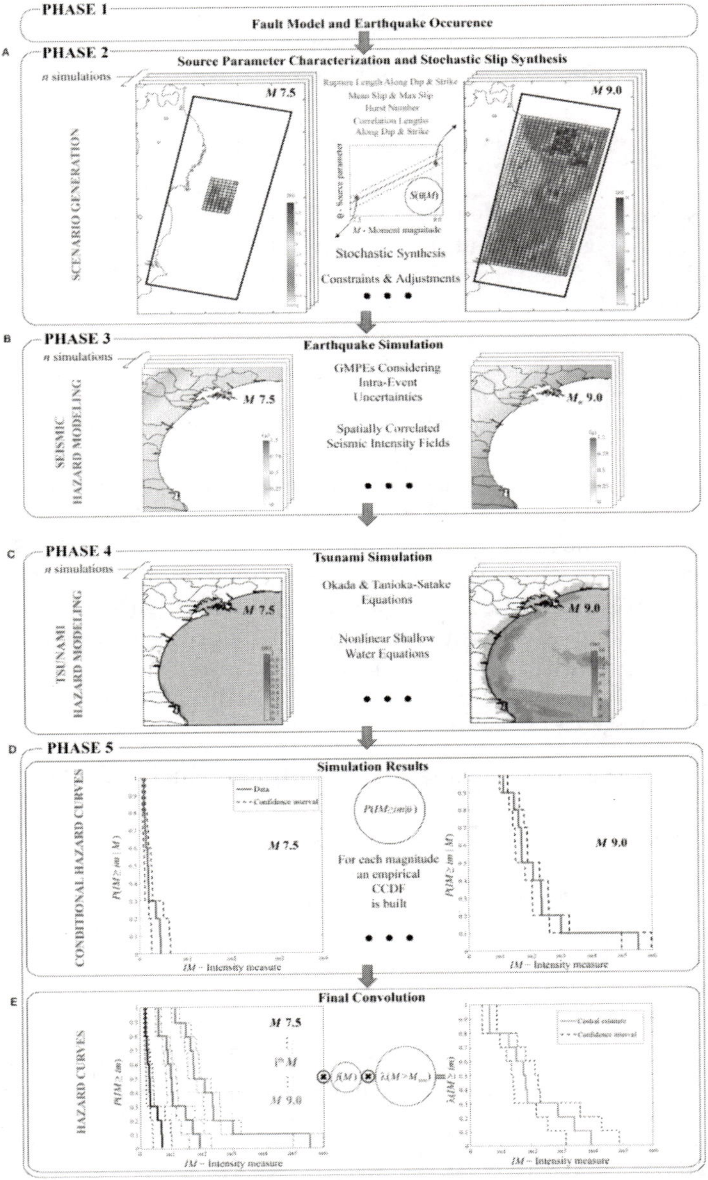

Figure 2. Computational framework for probabilistic earthquake–tsunami hazard analysis. **(A)** Scenario generation. **(B)** Seismic hazard modeling. **(C)** Tsunami hazard modeling. **(D)** Conditional hazard curves. **(E)** Hazard curves.

– development of earthquake–tsunami hazard curves. Detailed descriptions for each of these phases are presented in the following. Figure 2 shows the computational framework of the methodology.

Fault Model and Earthquake Occurrence

The first step is the identification of all seismic sources capable of producing damaging ground motions and tsunami inundation at a site. In this study, a curved surface is considered. Specifically, a 2011 Tohoku-type fault is analyzed with a source zone of 650 km along the strike and 250 km along the dip (Figure 3A); such geometry is capable of accommodating a *M*9 earthquake that is consistent with the maximum magnitude adopted for the magnitude–frequency distribution. The fault plane geometry is the extended version of the source model by Satake et al. (2013). Note that extremely large earthquakes that span across multiple seismotectonic segments are not considered (e.g., simultaneous rupture of the off-the-Tohoku subduction region and the off-the-Hokkaido subduction zone). To implement the stochastic synthesis method of earthquake slip distribution, the fault plane is discretized into many sub-faults; a 10-km mesh with variable dip based on Satake et al. (2013) is generated. Such a discretization allows simulating accurately the slip distribution corresponding to a seismic event with *M*7.5 (i.e., the smallest central magnitude value considered for the magnitude–frequency distribution, as shown later), involving at least 5-by-5 sub-faults.

To describe the earthquake sizes in the target region, i.e., the term *f* (*M*) in Eq. 2, a truncated Gutenberg–Richter relationship (Gutenberg and Richter, 1956) is adopted, and its CCDF is given by

$$G(M) = \frac{1 - 10^{-b(M - M_{\min})}}{1 - 10^{-b(M_{\max} - M_{\min})}} M_{\min} < M < M_{\max} \tag{3}$$

where $M_{\min}$ and $M_{\max}$ are the minimum and maximum moment magnitudes, respectively. For the simulation, it is convenient to convert the continuous distribution of magnitudes into a discrete set of values ($M_{\min}$, ..., M_i, ..., $M_{\max}$), assuming that they are the only possible magnitudes; such probabilities are computed as follows:

$$P(M_i) = G(M_i + 0.5 \cdot \Delta M) - G(M_i - 0.5 \cdot \Delta M) \tag{4}$$

where ΔM is the discretization interval. The discrete term presented in Eq. 4 is used in Eq. 2 instead of *f*(*M*).

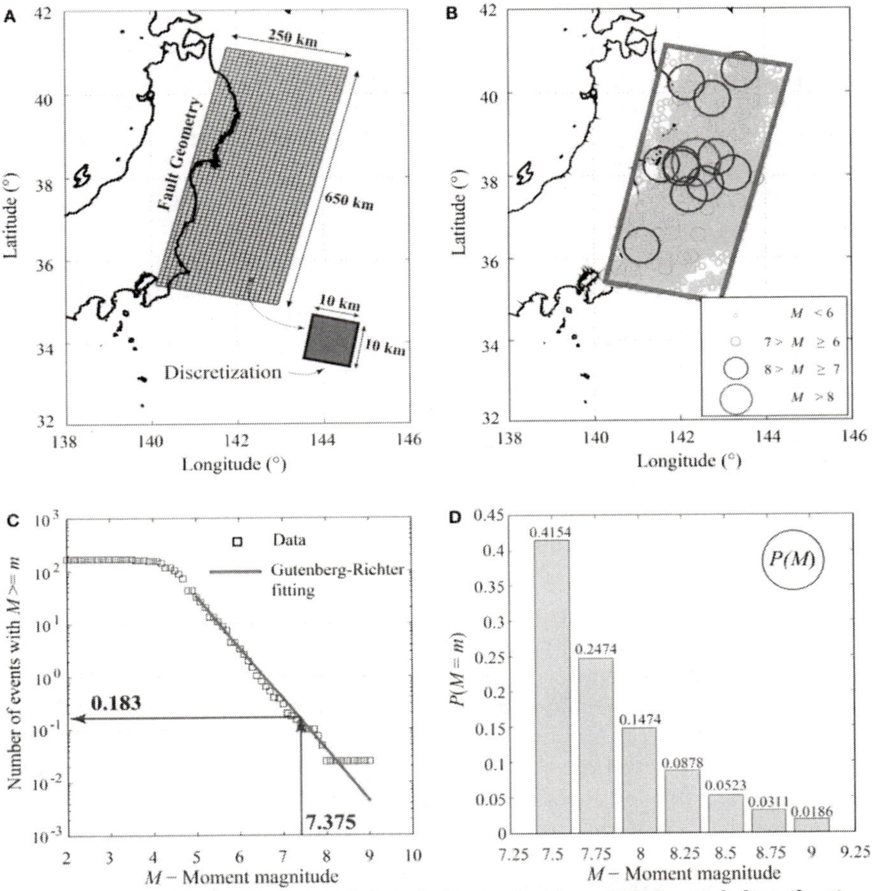

Figure 3. **(A)** Fault plane model and discretization. **(B)** Spatial distribution of earthquakes in the source region using the NEIC catalog. **(C)** Gutenberg–Richter relationship. **(D)** Discrete probability mass based on the fitted Gutenberg–Richter relationship.

For the analyses, $M_{\min}$ and $M_{\max}$ are set to 7.375 and 9.125, and a discretization interval of 0.25 is adopted. This means that seven central magnitude values, i.e., 7.5, 7.75, 8.0, 8.25, 8.5, 8.75, and 9.0, are considered to calculate the corresponding conditional probabilities as in Eq. 4. The minimum magnitude value is chosen, since small-to-moderate earthquakes rarely generate significant tsunamis, and their contributions to the tsunami hazard are negligible (Annaka et al., 2007). For the Tohoku case study, a *b*-value equal to 0.9 is adopted (Headquarters for Earthquake Research Promotion, 2013).

Once the magnitude interval is selected and the major source area containing all possible rupture scenarios is defined, the mean annual rate of occurrence of earthquakes with magnitudes greater than or equal to 7.375 falling in that area, i.e., the term $\lambda(M \geq M_{min})$ in Eq. 2, can be calculated. In order to perform such a calculation, the NEIC earthquake catalog (http://earthquake.usgs.gov/earthquakes/search/) is used. Figure 3B shows the events reported in the database that fall in the considered rupture area, recorded in the period between 1976 and 2012, having a depth varying between 0 km and 60 km, and considering a magnitude range between 5 and 9. According to the data analysis (Figure 3C), the estimated rate $\lambda(M \geq 7.375)$ is equal to 0.183. Figure 3D shows the occurrence probabilities for the discrete set of magnitude values (i.e., 7.5, 7.75, 8.0, 8.25, 8.5, 8.75, and 9.0). Note that the probability mass function shown in Figure 3D is normalized (conditional) with respect to the occurrence rate for the minimum magnitude event.

Source Parameter Characterization and Stochastic Slip Synthesis

To take into account uncertainties related to the rupture process, multiple random slip fields are simulated (Figure 2A). The simulation procedure is based on a spectral synthesis method (Goda et al., 2014; Fukutani et al., 2016), where the earthquake slip distribution is characterized by wavenumber spectra (Mai and Beroza, 2002; Lavallée et al., 2006). Scaling relationships that evaluate the source parameters as a function of moment magnitude are needed for stochastic tsunami simulation (e.g., rupture size and spectral characteristics of the rupture). In this study, new global scaling relationships for tsunamigenic earthquakes are employed. These relationships are obtained on the basis of 226 inverted source models in the SRCMOD database (Mai and Thingbaijam, 2014). The details of the adopted scaling laws can be found in Goda et al. (2016).

The following relationships are employed to obtain the rupture width (W) and length (L):

$$\log_{10} W = -0.4877 + 0.3125M + 0.1464 \times \varepsilon_W \tag{5}$$

$$\log_{10} L = -1.5021 + 0.4669M + 0.1717 \times \varepsilon_L \tag{6}$$

where the numbers multiplying by the ε terms are the SDs of the regression errors. The two geometrical dimensions are used to create the rupture area, which is randomly located inside the predefined subduction fault plane.

Subsequently, a slip distribution realization with desired properties is obtained using a stochastic synthesis method (Goda et al., 2014). First, a random field, having quasi-normal distribution with a desired spatial correlation structure, is generated using a Fourier integral method (Pardo-Iguzquiza and Chica-Olmo, 1993). The amplitude spectrum of the target slip distribution is specified by a theoretical power spectrum, while the phase spectrum is represented by a random phase matrix. For the amplitude spectrum, the von Kármán model is considered (Mai and Beroza, 2002):

$$P(k) \propto \frac{\mathrm{CL}_z \mathrm{CL}_x}{(1+k^2)^{\mathrm{HN}+1}} \tag{7}$$

where k is the wavenumber (i.e., reciprocal of the wavelength). The correlation lengths (CL_z along the dip and CL_x along the strike) are important source parameters that define the spatial heterogeneity of small wavenumber components in the spectrum and are determined from the following scaling relationships:

$$\log_{10}\mathrm{CL}_z = -1.0644 + 0.3093M + 0.1592 \times \varepsilon_{\mathrm{CL}_z} \tag{8}$$

$$\log_{10}\mathrm{CL}_x = -1.9844 + 0.4520M + 0.2204 \times \varepsilon_{\mathrm{CL}_x} \tag{9}$$

On the other hand, the Hurst number NH determines the spectral decay in the large wavenumber range and can be modeled as a bimodal random variable that takes a value of 0.99 with probability of 0.43 or a value sampled from the normal distribution with mean equal to 0.714 and SD equal to 0.172 with probability of 0.57 (Goda et al., 2016). The obtained complex Fourier coefficients are transformed into the spatial domain *via* 2-D inverse fast Fourier transform. The synthesized slip distribution is then scaled non-linearly to achieve suitable right-tail characteristics, in agreement with those observed in the finite-fault models, using the Box–Cox parameter λ (Box and Cox, 1964). It can be modeled as a normal random variable with mean equal to 0.312 and SD equal to 0.278.

Finally, the generated slip distribution is further adjusted in order to have a mean slip (D_a) and maximum slip (D_m), according to the values calculated from the scaling relationships for the given magnitude:

$$\log_{10}D_a = -5.7933 + 0.7420M + 0.2502 \times \varepsilon_{D_a} \tag{10}$$

$$\log_{10}D_m = -4.5761 + 0.6681M + 0.2249 \times \varepsilon_{D_m} \tag{11}$$

It is important to note that the error terms of the source parameters W, L, CL_z, CL_x, D_a, and D_m mentioned above are distributed according to a multivariate normal distribution (Goda et al., 2016). The linear correlation matrix of the regression errors ε is given by

$$\rho_\varepsilon = \begin{matrix} & \varepsilon_W & \varepsilon_L & \varepsilon_{CL_z} & \varepsilon_{CL_x} & \varepsilon_{D_a} & \varepsilon_{D_m} \\ \varepsilon_W \\ \varepsilon_L \\ \varepsilon_{CL_z} \\ \varepsilon_{CL_x} \\ \varepsilon_{D_a} \\ \varepsilon_{D_m} \end{matrix} \begin{bmatrix} 1.000 & 0.139 & 0.826 & 0.035 & -0.680 & -0.545 \\ & 1.000 & 0.249 & 0.734 & -0.595 & -0.516 \\ & & 1.000 & 0.288 & -0.620 & -0.564 \\ & & & 1.000 & -0.374 & -0.337 \\ & & & & 1.000 & 0.835 \\ & & & & & 1.000 \end{bmatrix} \quad (12)$$

Therefore, values of W, L, CL_z, CL_x, D_a, and D_m can be simulated jointly in the stochastic source simulation. The central estimates and the confidence interval (16th and 84th percentiles) of the scaling relationships are shown in Figure 4. The same figure also shows simulated data (green dots) and associated statistics (colored circles), which are obtained from the stochastic source modeling. Magnitude values for simulated data are not perfectly aligned at the seven discrete values; in fact, the simulation algorithm allows a tolerance band of ±0.05 around each magnitude value.

The preceding procedure of earthquake source characterization is innovative with respect to the literature. In particular, a common physical process for concurrent earthquake and tsunami threats is considered, i.e., the common fault rupture scenario that is modeled through the generic stochastic slip scenario on the subduction fault plane. The adoption of a common physical process facilitates the probabilistic investigation of the dependency between shaking and tsunami hazard parameters.

Earthquake Simulation

Ground motion prediction equations are extensively used as an effective way to predict seismic intensity measures for a given earthquake scenario (Wald et al., 2006). To account for seismic intensities at multiple locations that occur simultaneously for a given event, GMPEs together with spatial correlations in the regression residuals can be treated as statistical prediction models (Goda and Atkinson, 2010; Goda, 2011). This feature is particularly important in extending the seismic hazard assessment into a risk assessment of a portfolio of buildings/infrastructures. In this study, only the intra-event SD is propagated through the simulation procedure; such a choice is consistent with the simulation scenario of a single fault plane.

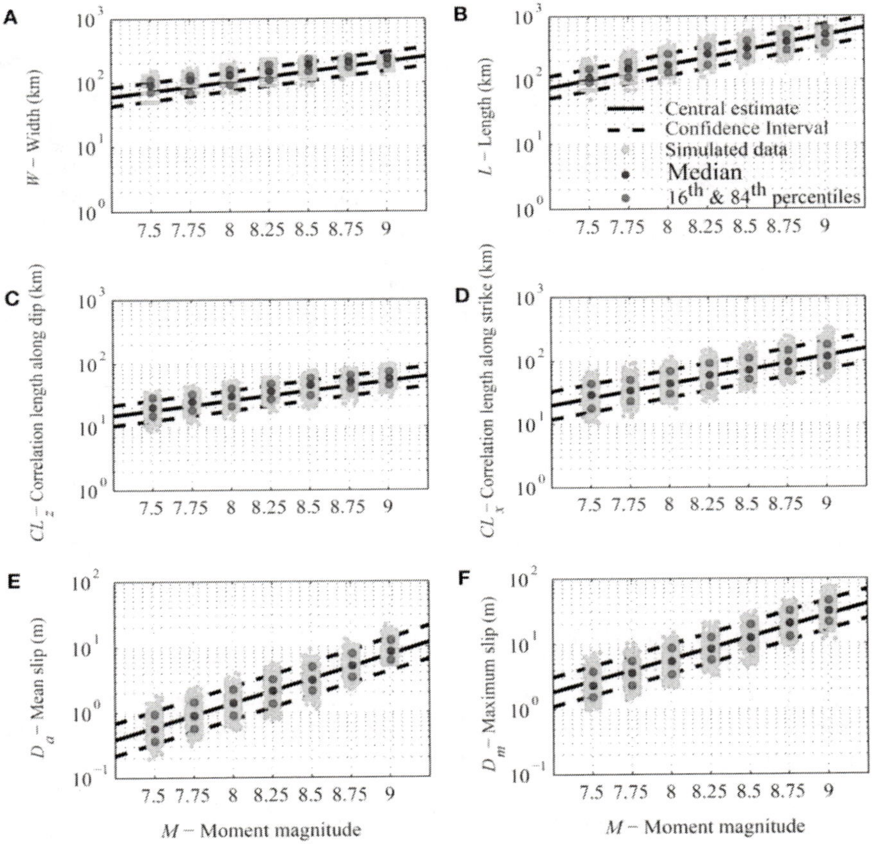

Figure 4. Scaling relationships for tsunamigenic earthquakes. **(A)** Rupture width versus moment magnitude. **(B)** Rupture length versus moment magnitude. **(C)** Correlation length along dip versus moment magnitude. **(D)** Correlation length along strike versus moment magnitude. **(E)** Mean slip versus moment magnitude. **(F)** Maximum slip versus moment magnitude. The simulated values (green dots) and the corresponding percentiles (colored circles) are also shown.

Two GMPEs that are applicable to subduction zones are used for the seismic simulations (Figure 2B). The first GMPE (Abrahamson et al., 2016) was developed with a global dataset of earthquakes in subduction zones and has been modified by adding the 2010 Maule Chile and 2011 Tohoku Japan earthquakes to the initial database. The basic functional form of the model for interface subduction events is

$$\ln(S_a) = \theta_1 + \theta_4 \cdot \Delta C_1 + [\theta_2 + \theta_3 \cdot (M - 7.8)] \cdot \ln\{R + C_4 \cdot \exp[\theta_9 \cdot (M - 6)]\} + \theta_6 \cdot R + f_{MAG}(M) + f_{FABA}(R) + f_{SITE}(PGA_{1000}, V_{S30}) + \sigma \cdot \varepsilon \quad (13)$$

where ln is the natural logarithm, R is the closest distance to the rupture area, V_{S30} is the shear wave velocity in the uppermost 30 m of soil column, PGA_{1000} is the median peak ground acceleration (PGA) value corresponding to $V_{S30} = 1000$ m/s, σ is the total SD, and ε is the Gaussian error term, represented by 0 mean and unit SD. The SD is period-dependent; it is obtained by the combination of intra-event (ϕ) and inter-event (τ) SDs. The magnitude function is

$$f_{MAG}(M) = \begin{cases} \theta_4 \cdot [M - (7.8 + \Delta C_1)] + \theta_{13} \cdot (10 - M)^2 & \text{for } M \leq 7.8 + \Delta C_1 \\ \theta_5 \cdot [M - (7.8 + \Delta C_1)] + \theta_{13} \cdot (10 - M)^2 & \text{for } M > 7.8 + \Delta C_1 \end{cases} \quad (14)$$

where ΔC_1 is the term representing the epistemic uncertainty in the break of the magnitude scaling and allows adjusting the GMPE for large interface events that were not originally considered in the earthquake database. $f_{FABA}(R)$ represents the forearc/backarc scaling term; it is equal to 0 for forearc or unknown site, and this is applicable to the case of this study. Finally, the model for site response scaling is given by

$$f_{SITE} = \theta_{12} \cdot \ln\left(\frac{\min(V_{S30}, 1000)}{V_{lin}}\right) - b \cdot \ln(PGA_{1000} + c) + b \cdot \ln\left[PGA_{1000} + c \cdot \left(\frac{\min(V_{S30}, 1000)}{V_{lin}}\right)^{n}\right] \quad V_{S30} < V_{lin}$$

$$f_{SITE} = \theta_{12} \cdot \ln\left(\frac{\min(V_{S30}, 1000)}{V_{lin}}\right) + b \cdot n \cdot \ln\left(\frac{\min(V_{S30}, 1000)}{V_{lin}}\right) \quad V_{S30} \geq V_{lin} \quad (15)$$

All the model coefficients for Eqs 13–15 can be found in Abrahamson et al. (2016).

The second GMPE by Morikawa and Fujiwara (2013) is suitable for $M9$ earthquakes in Japan. Two formulations are proposed: one is expressed with a quadratic magnitude term, while the other considers a linear magnitude term. In this study, the quadratic formulation is used since the correction factors (presented in the following) that are included in the quadratic formulation reduce the regression SD. The functional form for interface events is

$$\log(S_a) = a_1 \cdot (\min(M, 8.2) - M)^2 + b_1 \cdot R + c_1 - \log\left(R + d_1 \cdot 10^{e_1 \cdot \min(M, 8.2)}\right) + \sigma \cdot \varepsilon \quad (16)$$

where log is the base-10 logarithm, a_1, b_1, c_1, d_1, e_1, and σ are period-dependent regression coefficients and can be found in Morikawa and Fujiwara (2013). In Morikawa and Fujiwara (2013), no distinction is made between intra-event and inter-event SDs. The two SDs ϕ and τ can be determined by splitting the total variance into the intra-event and inter-event components based on the ratios of intra-event and inter-event variances to the total variance presented in Zhao et al. (2006). The estimate of the prediction equation is further modified using three additional correction terms: amplification due to the deep sedimentary layers (G_d), amplification due to shallow soft soils (G_s), and anomalous seismic intensity distribution due to the position of the site of interest with respect to the volcanic front (AI). In this study, only the second correction term is taken into account and is given by

$$\log(G_s) = p_s \cdot \log\left[\frac{\min(V_{S_{max}}, V_{S30})}{350}\right] \quad (17)$$

where VSmaxVSmax is a period-dependent regression parameter that can be found in Morikawa and Fujiwara (2013).

Figures 5A–D compare the two GMPEs for three seismic intensity parameters [i.e., PGA, $S_a(T = 0.3\text{ s})$, and $S_a(T = 3\text{ s})$], for M equal to 7.5 and 9.0 and for V_{S30} equal to 300 m/s. Figure 6 shows the acceleration response spectra obtained using the two GMPEs, considering two values of closest distance (i.e., 50 and 200 km), and the same values of magnitude and V_{S30} described before. Significant differences between the two GMPEs can be observed, especially for the large value of magnitude. Moreover, the Morikawa–Fuijwara GMPE tends to attenuate PGA and short-period spectral acceleration faster than the Abrahamson et al. GMPE with the distance, while the opposite trend occurs for the long-period spectral accelerations.

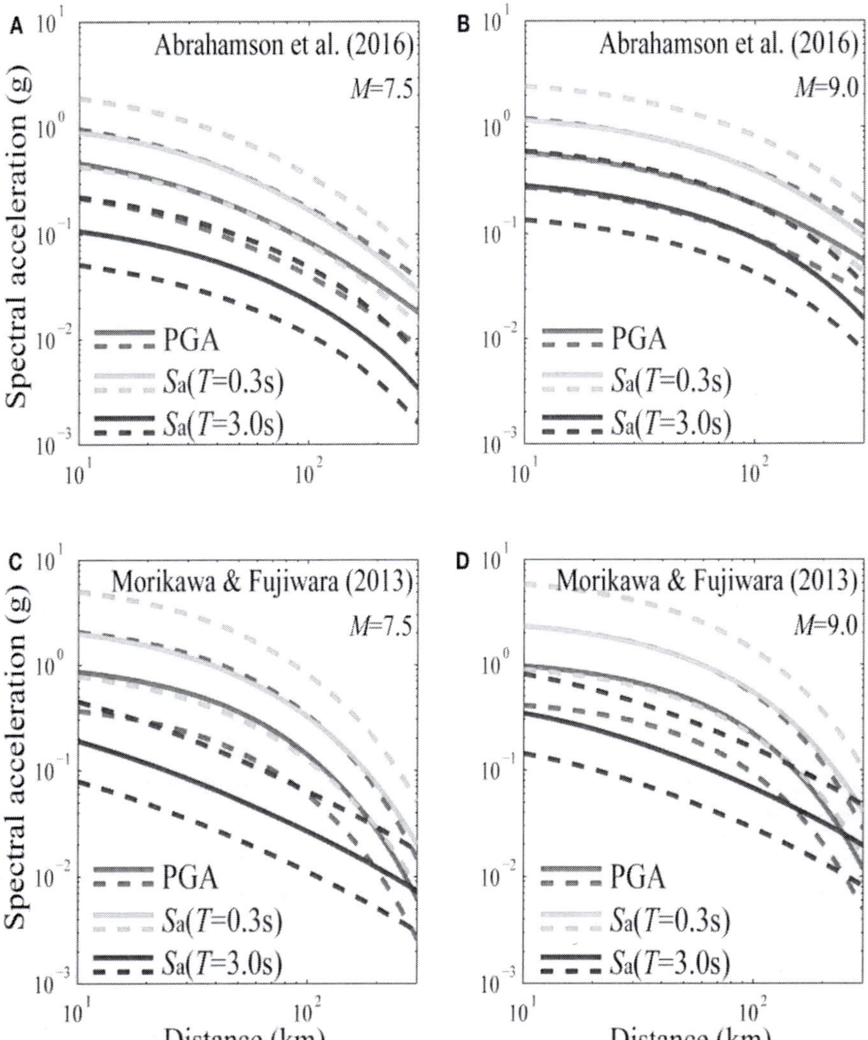

Figure 5. Comparison of GMPEs. (A) Abrahamson et al. GMPE for $M = 7.5$ and $V_{S30} = 300$ m/s. **(B)** Abrahamson et al. GMPE for $M = 9.0$ and $V_{S30} = 300$ m/s. **(C)** Morikawa–Fujiwara GMPE for $M = 7.5$ and $V_{S30} = 300$ m/s. **(D)** Morikawa–Fujiwara GMPE for $M = 9.0$ and $V_{S30} = 300$ m/s.

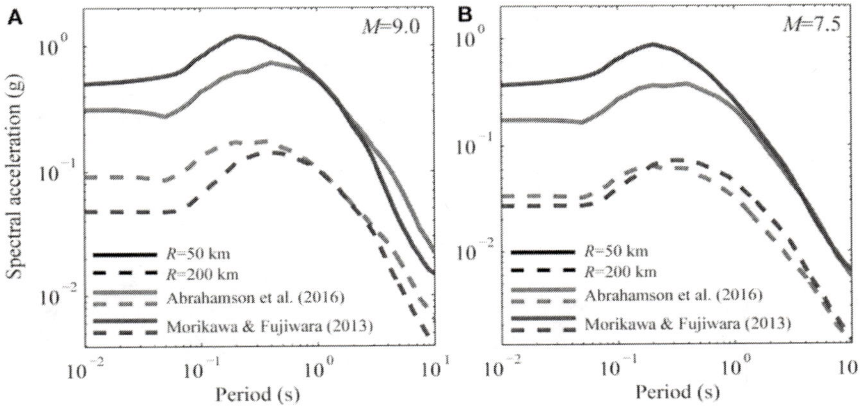

Figure 6. Response spectra obtained considering the Abrahamson et al. and Morikawa–Fujiwara GMPEs, V_{S30} = 300 m/s, and two shortest distances (*R* = 50 and 200 km). (A) *M* = 7.5. **(B)** *M* = 9.0.

For seismic simulations, three main inputs are required: event magnitude, distance from the rupture, and shear wave velocity for the considered site. Regarding the distance from the rupture, the GMPEs presented above are both based on the closest distance between the location of interest and the rupture area (Figure 7A). To optimize the computation of the shortest distance, the distances between the coastline location and each discretized element of the 2011 Tohoku-type fault are precomputed and stored (Figure 7B). As an example, Figure 7C shows the distances computed for 500 stochastic scenarios, which are considered in Section "Results." It is worth noting that the minimum distance is circa 50 km, corresponding to the depth of the fault plane under the considered location. As observed in Goda and Atkinson (2014), the source-to-site distance is affected by the location and size of the fault plane, which in turn is determined by the magnitude of the event. In Figure 7C, it can be observed that the greater magnitude value results in smaller variability of the closest distance (i.e., distribution function has a steeper slope). This is because the rupture plane can move more freely within the overall fault plane (Figure 2A) when the earthquake magnitude is small. For the shear wave velocity, the USGS global V_{S30} map server is used (Wald and Allen, 2007).

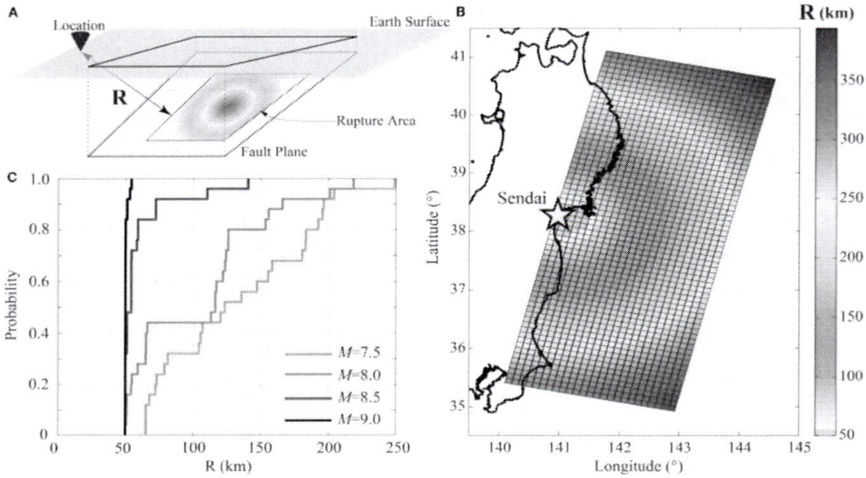

Figure 7. (A) Schematic representation of the shortest distance from the rupture. **(B)** Precalculated distance for each sub-fault from a site in Sendai City. **(C)** Distribution of the distances for 500 stochastic simulations associated with 4 values of magnitude.

Finally, to generate shake maps of intensity measures **IM**, the multivariate lognormal distribution can be adopted. The median values of **IM** at sites of interest are calculated from the GMPE, whereas their variances are based on the intra-event components. The prediction errors ε in the GMPE are spatially correlated; the correlation coefficient matrix has diagonal elements equal to 1 and off-diagonal elements equal to the correlation coefficient ρ. The correlation coefficient can be calculated using the following equation (Goda and Atkinson, 2010):

$$\rho_{i,j}(\Delta) = \max\left[\gamma \cdot \exp\left(-\alpha \cdot \Delta^{\beta}\right) - \gamma + 1, 0\right] \tag{18}$$

where Δ is the distance between the points *i* and *j*, while α, β, and γ are period-dependent model parameters that can be found in Goda and Atkinson (2010).

Tsunami Simulation

For each stochastic event, a tsunami simulation is carried out in order to compute the maximum inundation intensity measure (Figure 2C). To optimize the computational time, the subduction plane is discretized into sub-faults of 10 km × 10 km (Figure 3A), and for each sub-fault, the seafloor displacement corresponding to 1 m of slip is calculated using analytical equations by Okada (1985) and Tanioka and Satake (1996). Subsequently, for each simulated earthquake slip (i.e., event), the overall seafloor displacement field is estimated by scaling and summing the seafloor deformation fields of all individual sub-faults that make up the event.

Tsunami modeling is then carried out using a well-tested numerical code of Goto et al. (1997) that is capable of generating offshore tsunami propagation and inundation profiles by evaluating non-linear shallow water equations, with run-up using a leapfrog staggered-grid finite difference scheme. The run-up calculation is based on a moving boundary approach, where a dry/wet condition of a computational cell is determined based on total water depth relative to its elevation. The numerical tsunami calculation is performed for duration sufficient to model the most critical phases of tsunami waves (i.e., 2 h). The integration time step is determined by satisfying the CFL condition; it depends on the bathymetry/elevation data, and their grid sizes and is typically between 0.1 and 0.5 s. For the simulation, it is possible to obtain the maximum tsunami intensity measures of interest (i.e., tsunami height, tsunami velocity, etc.) for one or more specific locations along the coast. The results can also be used to evaluate aggregate tsunami hazard parameters, such as inundation areas above a certain depth.

A complete dataset of bathymetry/elevation, coastal/riverside structures (e.g., breakwater and levees), and surface roughness is obtained from the Miyagi prefectural government. The data are provided in the form of nested grids (1350 m–450 m–150 m– 50 m), covering the geographical regions of Tohoku. The ocean-floor topography data are based on the 1:50,000 bathymetric charts and JTOPO30 database developed by Japan Hydrographic Association and based on the nautical charts developed by Japan Coastal Guard. The tidal fluctuation is not taken into account in this study. The elevation data of the coastal/riverside structures are primarily provided by municipalities. In the tsunami simulation, the coastal/riverside structures are represented by a vertical wall at one or two sides of the computational cells. To evaluate the volume of water that overpasses these walls, Homma's overflowing formulae are employed. In the tsunami simulation, the bottom friction is evaluated using the Manning's formula. The Manning's coefficients are assigned to computational cells based on national land use data in Japan: 0.02 $m^{-1/3}s$ for agricultural land, 0.025 $m^{-1/3}s$ for ocean/water, 0.03 $m^{-1/3}s$ for forest vegetation, 0.04 $m^{-1/3}s$ for low-density residential areas, 0.06 $m^{-1/3}s$ for moderate-density residential areas, and 0.08 $m^{-1/3}s$ for high-density residential areas.

Development of Earthquake–Tsunami Hazard Curves

For each value of magnitude, the simulations are used to evaluate the term $P(\mathbf{IM} \geq \mathbf{im}|M)$ for the location of interest. Such probability is represented by the CCDF of the **IM** (Figure 2D). Specifically, the CCDF of the **IM** (i.e., spectral acceleration or tsunami inundation) is obtained as the Kaplan–Meier estimator (Kaplan and Meier, 1958), for which the variance can be calculated through the Greenwood's formula (Greenwood, 1926), and therefore, a confidence interval around the central estimate can be obtained. In this study, the 95% confidence interval is considered.

The curves obtained in the previous step for each magnitude are then multiplied by the probabilities corresponding to the related magnitude and eventually are summed up (Figure 2E). Also in this case, three curves are obtained, one corresponding to the central value and two for the confidence interval. The final hazard curves, representing the mean annual rate of occurrence of specific values of earthquake–tsunami intensity measures, are obtained by multiplying the previous three conditional curves (for each hazard) by the occurrence rate of events with magnitudes greater than the minimum magnitude considered in the magnitude–frequency distribution.

RESULTS

The developed methodology is applied to calculate the earthquake and tsunami hazard curves for a site along the coast line of Sendai City, Miyagi Prefecture (the yellow star in Figure 7B), for which V_{S30} = 240 m/s is obtained based on the USGS data. It is interesting to note that during the 2011 Tohoku earthquake, PGA of 0.8 g (Wald et al., 2006; USGS ShakeMap Archive, 2016) and tsunami wave height of 7 m [Ministry of Land, Infrastructure, and Transportation (MLIT), 2014] were observed in the vicinity of this site. The main results that are discussed in this section focus on (a) sensitivity of seismic and tsunami hazard estimates to the number of stochastic simulations, and (b) development of earthquake–tsunami multi-hazard curves for the target site and the deaggregation of the seismic and tsunami hazards. The former provides useful information regarding the stability of the simulation-based hazard assessments.

Sensitivity of Seismic and Tsunami Hazard Parameters to the Number of Simulations

Short or incomplete records lead to biased estimation of the hazard parameters, especially when conventional statistical methods are used

(Lamarre et al., 1992). To investigate the effect of the number of simulations on the final hazard estimation, a bootstrap procedure is carried out by randomly sampling *m* values from the original sample containing *n* elements (with $m \leq n$). This provides a pool of different samples of independent and identically distributed random variables, whose distribution function is the same as that of the original sample. For each generated sample, statistics of the parameter of interest (e.g., mean, median, and different percentiles) are then computed. The ensemble of such estimates can be used to quantify the uncertainty in the parameter value.

Figures 8 and 9 show five percentiles (i.e., 5th, 25th, 50th, 75th, and 95th) of the spectral acceleration and wave height, respectively; such intensity measures are calculated for the site in Sendai by considering different magnitude values (i.e., 7.5, 8.0, 8.5, and 9.0) as a function of the number of simulations. Moreover, Figure 8 shows bootstrap results for the seismic case considering two different GMPEs. The analysis is carried out for the maximum sample size of $n = 500$ simulations. The bootstrap procedure is then applied considering the number of simulations *m* varying between 1 and 500. For each trial number of simulations *m*, 1000 Monte Carlo samples are generated, and then the percentile curves are obtained as the mean value of such simulations. Based on Figure 8, it can be concluded that for the considered seismic case, 200 simulations is sufficient to observe stable percentiles for all magnitude values and for both GMPEs considered. Also in this case, for increasing values of magnitude, there is a decreasing trend of variability of the simulated values due to the reduction in variability of the closest distance (Figure 7C).

Tsunami simulation results show that the 50th percentile curves are stable after 100 simulations for all the considered magnitude values. To obtain stable estimates of the high percentiles, a larger number of simulations are needed (the red dotted line in Figure 9). In particular, 300 simulations are necessary for *M*7.5, 250 simulations for *M*8.0, and 200 simulations for *M*8.5 and *M*9.0. Such a decreasing trend with the magnitude is consistent with what was observed for the rupture distance (Figure 7C), i.e., when the magnitude is relatively small, the variability of the inundation intensity measures are increased because the rupture area can move more freely within the fault plane. In turn, when the magnitude is large, the fluctuation of the rupture area is more constrained.

Considering the results shown in Figures 8 and 9, 300 stochastic simulations are carried out for the final coupled multi-hazard simulation process in the following. The calculation can be completed in less than 1 week using a conventional workstation with parallel processing.

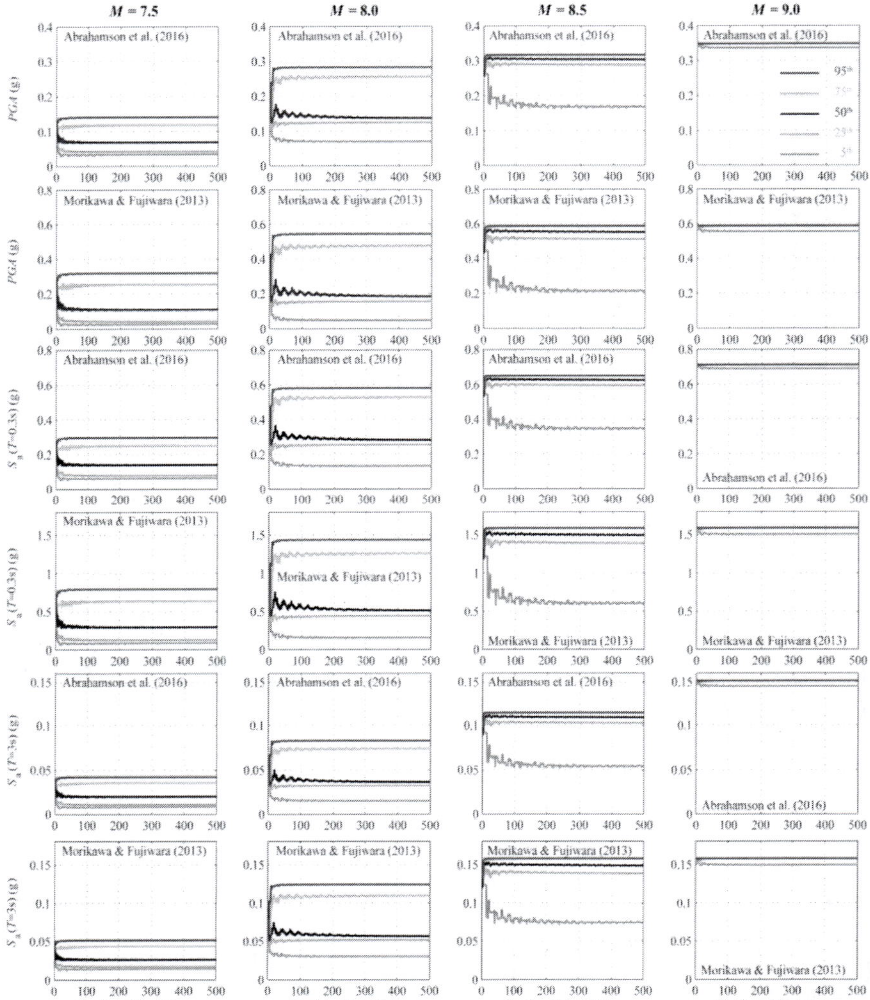

Figure 8. Convergence of estimated seismic intensity measures [i.e., PGA, $S_a(T = 0.3\text{ s})$ and $S_a(T = 3\text{ s})$] as a function of the number of simulations by considering the Abrahamson et al. and Morikawa–Fujiwara GMPEs and four moment magnitudes (M = 7.5, 8.0, 8.5, and 9.0). Different colors represent different percentiles.

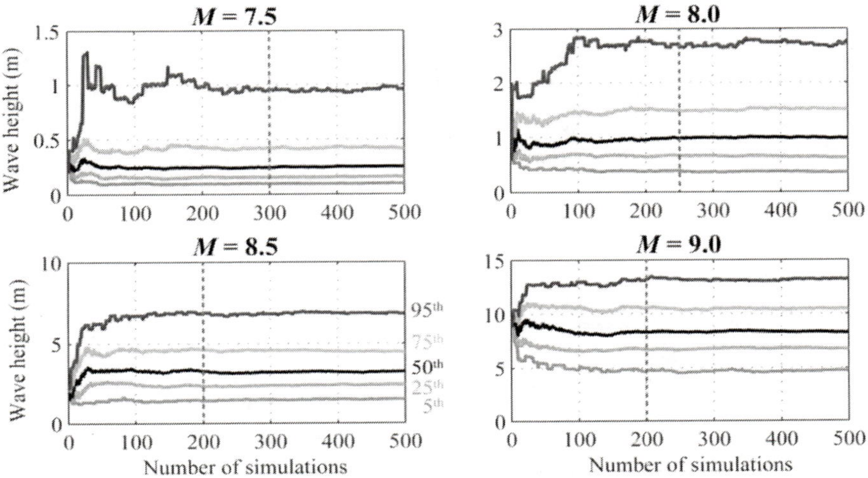

Figure 9. Convergence of estimated tsunami intensity measures (wave height) as a function of the number of simulations by considering four moment magnitudes (M = 7.5, 8.0, 8.5, and 9.0).

Earthquake–Tsunami Hazard Curves

For each value of 7 magnitudes (i.e., 7.5, 7.75, 8.0, 8.25, 8.5, 8.75, and 9.0), 300 sets of the source parameters $\boldsymbol{\theta}$ are generated using the scaling relationships by Goda et al. (2016). Figure 4 shows the simulated source parameters (the green dots). Simulated data are in agreement with the source parameter distributions (i.e., green dots are well clustered within the confidence interval of the scaling relationships). Then, 300 simulations are carried out for the earthquake hazard and tsunami hazard analysis, starting from the same stochastic source models.

The CCDFs (Figure 2D) in terms of PGA, $S_a(T = 0.3\text{ s})$, and $S_a(T = 3\text{ s})$ are shown in Figures 10A,C,E, respectively, for all the magnitude values analyzed. The analogous CCDFs for the tsunami wave height are presented in Figure 11A. Figures 10B,D,F and 11B show the CCDFs, weighted by the probability values obtained from the discretized Gutenberg–Richter relationship (Figure 3D).

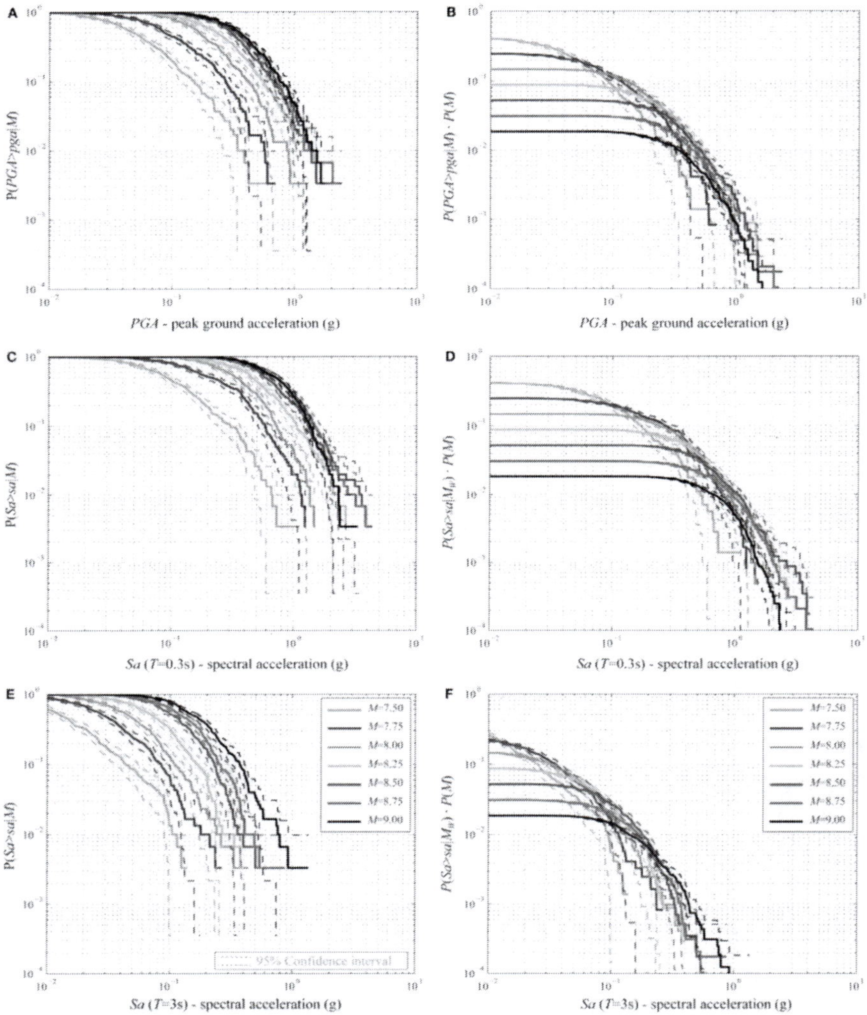

Figure 10. (A) Conditional hazard curve for PGA. **(B)** Weighted conditional hazard curves for PGA. **(C)**Conditional hazard curve for $S_a(T = 0.3$ s). **(D)** Weighted conditional hazard curves for $S_a(T = 0.3$ s). **(E)**Conditional hazard curve for $S_a(T = 3$ s). **(F)** Weighted conditional hazard curves for $S_a(T = 3$ s).

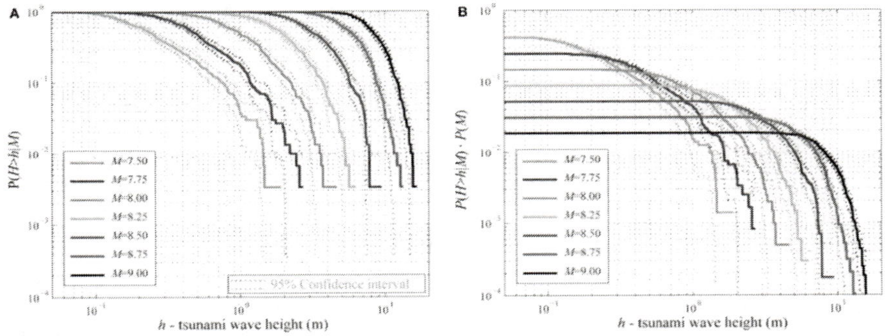

Figure 11. **(A)** Conditional hazard curve for tsunami wave height. **(B)** Weighted conditional hazard curves for tsunami wave height.

As shown in Figure 2E, for each IM [i.e., PGA, $S_a(T)$, h, etc.], the summation of the curves presented in Figures 10B,D,F and 11B, multiplied by $\lambda(M \geq 7.375) = 0.183$, leads to the final hazard curves. Figures 12A–C shows the final hazard curves, and the 95% confidence interval, for PGA, $S_a(T = 0.3$ s), and $S_a(T = 3$ s) obtained using the two GMPEs, i.e., Abrahamson et al. (2016), represented with blue lines, and Morikawa and Fujiwara (2013), represented with red lines. In the same figures, the mean seismic hazard curves are represented with black lines. Similarly, Figure 12D shows the final tsunami hazard curve and its 95% confidence interval that is very tight around the central estimate curve. It is noteworthy that the steep slope of the final tsunami hazard curve for wave heights greater than 10 m is because the tsunami height cannot be so high in the Sendai plain areas unlike ria-type coastal areas (e.g., Onagawa and Kesennuma), where the wave amplification due to topographical effects is significant.

Seismic Uniform Hazard Spectra and Earthquake–Tsunami Deaggregation

The proposed procedure also facilitates the construction of uniform hazard spectra (UHS) for seismic hazard. By repeating the seismic simulations for several spectral accelerations (Figure 13A) and considering specific values of the mean annual rate (e.g., 2, 5, and 10% in 50 years), it is possible to obtain the UHS, as shown in Figure 13B. The hazard curves and spectra are jagged because of the limited number of simulations (i.e., 300). Since the seismic simulations are less time consuming with respect to the tsunami simulations, it is possible to increase the number of simulations; in particular, for each scenario, several thousands of simulations can be conducted with a low additional computational effort. Figure 13C shows the seismic hazard curves obtained by performing 300 × 5000 simulations (i.e., 5000 simulations for each stochastic simulation). By increasing the number of simulations, the UHS become smooth (Figure 13D), and the confidence intervals around the central estimate hazard curves become narrow.

As a byproduct of the procedure, the earthquake–tsunami hazard deaggregation is obtained. Deaggregation shows the relative contributions of dominant seismic scenarios to the specified hazard levels and can be represented in terms of distance and magnitude. The deaggregation of the two hazards for the same mean annual rate of occurrence is demonstrated. Figure 14 shows the deaggregation results for Sendai by considering PGA (Figures 14A,C) and tsunami inundation height (Figures 14B,D) corresponding to two values of mean annual rate of occurrence (i.e., 63 and

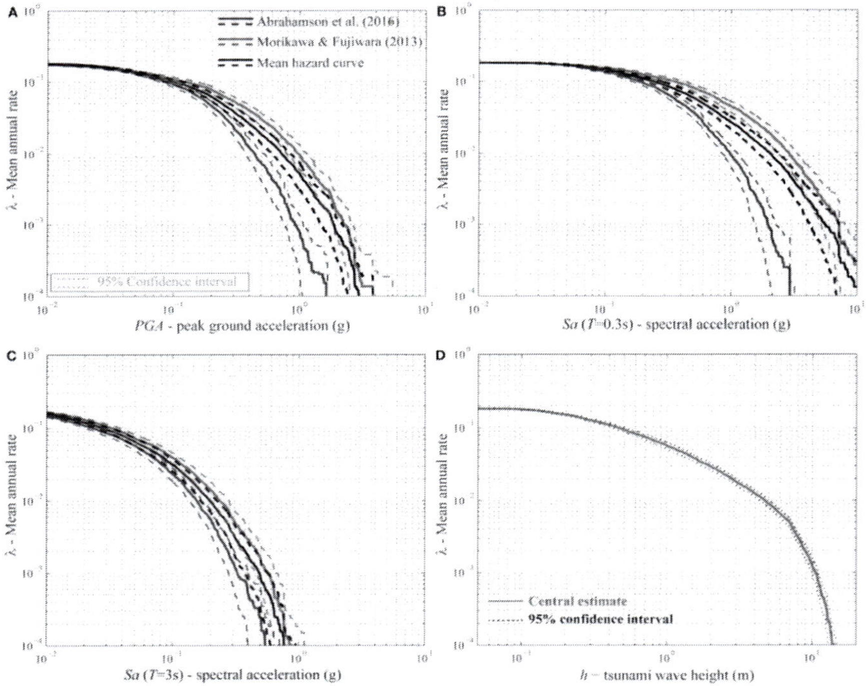

Figure 12. **(A)** Final seismic hazard curves for PGA. **(B)** Final seismic hazard curves for $S_a(T = 0.3$ s). **(C)** Final seismic hazard curves for $S_a(T = 3$ s). **(D)** Final tsunami hazard curves.

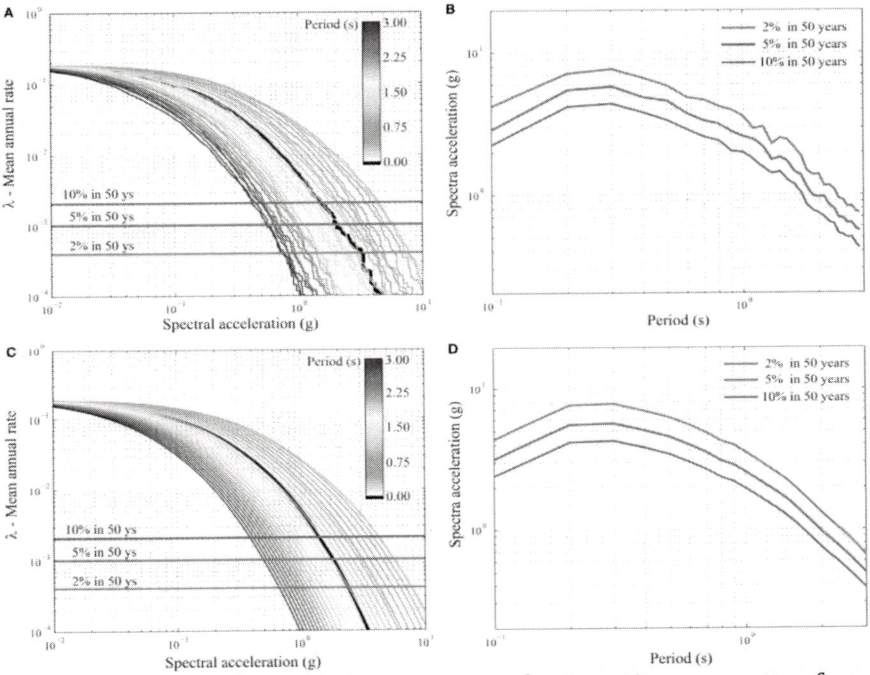

Figure 13. (A) Final seismic hazard curves for intensity parameters from PGA (the black line) to $S_a(T = 3$ s) based on 300 simulations. **(B)** Uniform hazard spectra based on 300 simulations. **(C)**Final seismic hazard curves for intensity parameters from PGA (the black line) to $S_a(T = 3$ s) based on 300 × 5000 simulations. **(D)** Uniform hazard spectra based on 300 × 5000 simulations.

10% in 50 years). The 10% in 50-year hazard level (corresponding to an event with 475-year return period) is commonly used to describe the life safety limit state, whereas the 63% in 50-year hazard level (corresponding to an event with 50-year return period) corresponds to the damage control limit state (CEN, 2004). It is worth noting that only large magnitude events contribute to higher values of **IM**s. Moreover, as observed before, the larger the magnitude is, the less the distance is influential. Finally, it is interesting to observe that the combinations of magnitude and distance that affect the seismic hazard and tsunami hazard differ significantly.

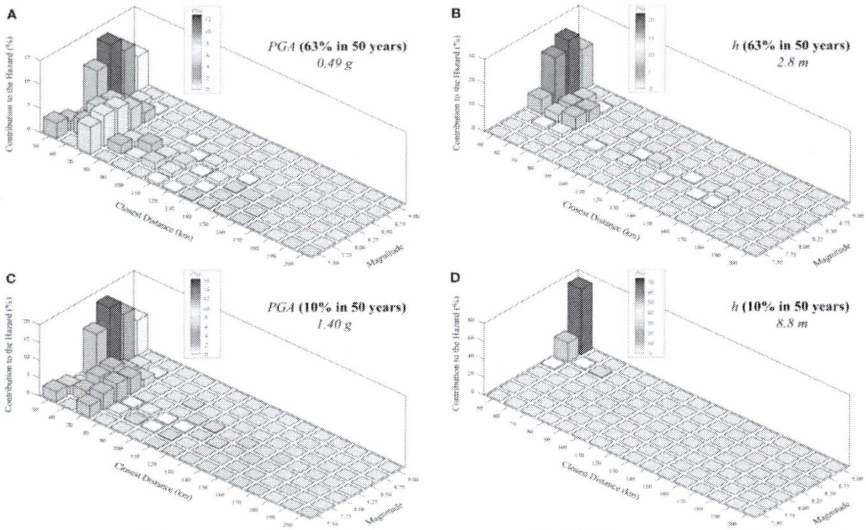

Figure 14. (A) Seismic hazard deaggregation for PGA corresponding to the hazard level of 63% in 50 years. **(B)** Tsunami hazard deaggregation for wave height corresponding to the hazard level of 63% in 50 years. **(C)** Seismic hazard deaggregation for PGA corresponding to the hazard level of 10% in 50 years. **(D)** Tsunami hazard deaggregation for wave height corresponding to the hazard level of 10% in 50 years.

CONCLUSION

A new simulation-based procedure to probabilistically calculate the earthquake–tsunami multi-hazard for specific locations was presented. The simulation framework allows implementing all potential sources of uncertainties, both epistemic and aleatory. The slip distribution on the fault plane was characterized in detail since it represents the major source of uncertainty. To generate a wide range of earthquake scenarios, new global scaling relationships of earthquake source parameters for tsunamigenic events were used. For each discrete magnitude value, multiple realizations of possible earthquake slip distributions were generated. The procedure was applied to the Tohoku region (Japan), and a single point located on the coastline in Sendai City was considered for assessing the concurrent earthquake–tsunami hazard. Three hundred simulations were performed for both seismic and tsunami intensity estimations at each discrete magnitude value. Data obtained from simulations were used to calculate the CCDFs of the considered intensity measures (i.e., spectral acceleration and tsunami inundation height) and their confidence intervals. Finally, such curves were combined with the magnitude–frequency distribution and were summed up in order to obtain the final triplets of earthquake–tsunami hazard curves: one representative of the central estimate and the others corresponding to the 95% confidence interval.

Based on the analysis results, the following conclusions can be drawn:

(a) For the considered case study, 300 simulations were sufficient to obtain a reliable and stable representation of both earthquake and tsunami hazard parameters at a single location, both in terms of central estimates and high percentiles.

(b) Given the same number of simulations and passing from small magnitude to large magnitude, a decrease in the dispersion of the simulation results was observed. This is due to the decreased variability of the earthquake location on the fault, and it also implies a reduction of the confidence interval.

(c) The procedure facilitates the calculation of UHS for seismic hazard and the deaggregation of both seismic and tsunami hazards.

The presented work can be considered a first step toward an earthquake–tsunami multi-hazard performance-based framework; in fact, a multi-risk assessment can be carried out by convoluting the obtained multi-hazard curves with seismic and tsunami fragility curves. The work can be extended using a Bayesian robust methodology (Cheung and Beck, 2010) to make more reliable estimations of earthquake–tsunami hazards. The proposed method can be further integrated into an operational tool for real-time earthquake–tsunami forecast (Tsushima et al., 2011) using data from offshore buoy and ocean-bottom pressure gauges. Furthermore, the methodology can be expanded to obtain the conditional tsunami or seismic hazard curve, given that a specific value of the counterpart hazard has been selected.

AUTHOR CONTRIBUTIONS

The two co-authors contributed equally to this work.

CONFLICT OF INTEREST STATEMENT

The authors declare that the research was conducted in the absence of any commercial or financial relationships that could be construed as a potential conflict of interest.

ACKNOWLEDGMENTS

This work is funded by the Engineering and Physical Sciences Research Council (EP/M001067/1).

REFERENCES

1. Abrahamson, N., Gregor, N., and Addo, K. (2016). BC hydro ground motion prediction equations for subduction earthquakes. *Earthquake Spectra* 32, 23–44.
2. Akkar, S., and Cheng, Y. (2015). Application of a Monte-Carlo simulation approach for the probabilistic assessment of seismic hazard for geographically distributed portfolio. *Earthquake Eng. Struct. Dyn.* 45, 525–541.
3. Annaka, T., Satake, K., Sakakiyama, T., Yanagisawa, K., and Shuto, N. (2007). Logic-tree approach for probabilistic tsunami hazard analysis and its applications to the Japanese coasts. *Pure Appl. Geophys.* 164, 577–592.
4. Atkinson, G. M., and Goda, K. (2013). "Probabilistic seismic hazard analysis of civil infrastructure: chapter 1, 3–28," in *Handbook of Seismic Risk Analysis and Management of Civil Infrastructure Systems*, eds S. Tesfamariam and K. Goda (Cambridge: Woodhead Publishing), 912.
5. Billi, A., Funiciello, R., Minelli, L., Faccenna, C., Neri, G., Orecchio, B., et al. (2008). On the cause of the 1908 Messina tsunami, southern Italy. *Geophys. Res. Lett.* 35, L06301.
6. Box, G. E. P., and Cox, D. R. (1964). An analysis of transformations. *J. R. Stat. Soc. Ser. B* 26, 211–252.
7. Burbidge, D., Cummins, P. R., Mleczko, R., and Thio, H. K. (2008). A probabilistic tsunami hazard assessment for Western Australia. *Pure Appl. Geophys.* 165, 2059–2088.
8. Burroughs, S. M., and Tebbens, S. F. (2005). Power-law scaling and probabilistic forecasting of tsunami runup heights. *Pure Appl. Geophys.* 162, 331–342.
9. CEN. (2004). *Eurocode 8: Design of Structures for Earthquake Resistance – Part 1: General Rules, Seismic Actions and Rules for Buildings*. Brussels: Comite International de Normalization.
10. Cheung, K. F., Wei, Y., Yamazaki, Y., and Yim, S. C. (2011). Modeling of 500-year tsunamis for probabilistic design of coastal infrastructure in the Pacific Northwest. *Coast. Eng. J.* 58, 970–985.
11. Cheung, S. H., and Beck, J. L. (2010). Calculation of posterior probabilities for Bayesian model class assessment and averaging from posterior samples based on dynamic system data. *Comput. Aided Civil Infrastruct. Eng.* 25, 304–321.
12. Cornell, C. A. (1968). Engineering seismic risk analysis. *Bull. Seismol. Soc. Am.* 58, 1583–1606.
13. Downes, G. L., and Stirling, M. W. (2001). "Groundwork for development of a probabilistic tsunami hazard model for New Zealand," in *Proceedings of International Tsunami Symposium, Session 1* (Seattle, USA: National Oceanic and Atmospheric Administration), 293–301.
14. Farreras, S., Ortiz, M., and Gonzalez, J. I. (2007). Steps towards the implementation of a tsunami detection, warning, mitigation and preparedness program for southwestern coastal areas of Mexico. *Pure Appl. Geophys.* 164, 605–616.
15. Fukutani, Y., Suppasri, A., and Imamura, F. (2016). Stochastic analysis and uncertainty assessment of tsunami wave height using a random source parameter model that targets a Tohoku-type earthquake fault. *Stoch. Environ. Res. Risk Assess.* 29, 1763–1779.

16. Geist, E. L. (2002). Complex earthquake rupture and local tsunamis. *J. Geophys. Res. Solid Earth* 107, ESE 2-1–ESE 2-15.
17. Geist, E. L. (2005). *Local Tsunami Hazards in the Pacific Northwest from Cascadia Subduction Zone Earthquakes*. Reston: U.S. Geological Survey, 17. Professional Paper 1661-B.
18. Geist, E. L., and Dmowska, R. (1999). Local tsunamis and distributed slip at the source. *Pure Appl. Geophys.* 154, 485–512.
19. Geist, E. L., and Parsons, T. (2006). Probabilistic analysis of tsunami hazards. *Nat. Hazards* 37, 277–314.
20. Goda, K. (2011). Interevent variability of spatial correlation of peak ground motions and response spectra. *Bull. Seismol. Soc. Am.* 101, 2522–2531.
21. Goda, K., and Atkinson, G. M. (2010). Intraevent spatial correlation of ground-motion parameters using SK-net data. *Bull. Seismol. Soc. Am.* 100, 3055–3067.
22. Goda, K., and Atkinson, G. M. (2014). Variation of source-to-site distance for megathrust subduction earthquakes: effects on ground motion prediction equations. *Earthquake Spectra* 30, 845–866.
23. Goda, K., Mai, P. M., Yasuda, T., and Mori, N. (2014). Sensitivity of tsunami wave profiles and inundation simulations to earthquake slip and fault geometry for the 2011 Tohoku earthquake. *Earth Planets Space* 66:105.
24. Goda, K., Yasuda, T., Mori, N., and Maruyama, T. (2016). New scaling relationships of earthquake source parameters for stochastic tsunami simulation. *Coast. Eng. J.* 58:1650010.
25. González, F. I., Geist, E. L., Jaffe, B., Kânoglu, U., Mofjeld, H., Synolakis, C. E., et al. (2009). Probabilistic tsunami hazard assessment at Seaside, Oregon, for near- and far-field seismic sources. *J. Geophys. Res. Oceans* 114, C11023.
26. Goto, C., Ogawa, Y., Shuto, N., and Imamura, F. (1997). *Numerical Method of Tsunami Simulation with the Leap-Frog Scheme (IUGG/IOC Time Project), IOC Manual*. Paris, France: UNESCO.
27. Greenwood, M. (1926). *The Natural Duration of Cancer. Reports on Public Health and Medical Subjects*, Vol. 33. London: Her Majesty's Stationery Office, 1–26.
28. Grezio, A., Gasparini, P., Marzocchi, W., Patera, A., and Tinti, S. (2012). Tsunami risk assessments in Messina, Sicily – Italy. *Nat. Hazards Earth Syst. Sci.* 12, 151–163.
29. Grezio, A., Marzocchi, W., Sandri, L., and Gasparini, P. (2010). A Bayesian procedure for probabilistic tsunami hazard assessment. *Nat. Hazards* 53, 159–174.
30. Guidoboni, E., and Comastri, A. (1997). The large earthquake of 8 August 1303 in Crete: seismic scenario and tsunami in the Mediterranean area. *J. Seismol.* 1, 55–72.
31. Gutenberg, B., and Richter, C. F. (1956). Magnitude and energy of earthquakes. *Ann. Geophys.* 9, 1–15.
32. Headquarters for Earthquake Research Promotion. (2013). *Investigations of Future Seismic Hazard Assessment*. Tokyo: Japanese Govern, 217.
33. Horspool, N., Pranantyo, I., Griffin, J., Latief, H., Natawidjaja, D. H., Kongko, W., et al. (2014). A probabilistic tsunami hazard assessment for Indonesia. *Nat. Hazards Earth Syst. Sci.* 14, 3105–3122.
34. Kaplan, E. L., and Meier, P. (1958). Nonparametric estimation from incomplete observations. *J. Am. Stat. Assoc.* 53, 457–481.
35. Lamarre, M., Townshend, B., and Shah, H. C. (1992). Application of the bootstrap method to quantify uncertainty in seismic hazard estimates. *Bull. Seismol. Soc. Am.* 82, 104–119.
36. Latter, J. H. (1981). Tsunamis of volcanic origin: summary of causes, with particular reference to Krakatoa, 1883. *Bull. Volcanol.* 44, 467–490.
37. Lavallée, D., Liu, P., and Archuleta, R. J. (2006). Stochastic model of heterogeneity in earthquake slip spatial distributions. *Geophys. J. Int.* 165, 622–640.
38. Liu, Y., Santos, A., Wang, S. M., Shi, Y., Liu, H., and Yuen, D. A. (2007). Tsunami hazards along Chinese coast from potential earthquakes in South China Sea. *Phys. Earth Planet. Inter.* 163, 233–244.

39. Løvholt, F., Glimsdal, S., Harbitz, C. B., Horspool, N., Smebye, H., de Bono, A., et al. (2014). Global tsunami hazard and exposure due to large co-seismic slip. *Int. J. Disaster Risk Reduct.* 10, 406–418.
40. Løvholt, F., Glimsdal, S., Harbitz, C. B., Zamora, N., Nadim, F., Peduzzi, P., et al. (2012). Tsunami hazard and exposure on the global scale. *Earth-Sci. Rev.* 110, 58–73.
41. Mai, P. M., and Beroza, G. C. (2002). A spatial random field model to characterize complexity in earthquake slip. *J. Geophys. Res. Solid Earth* 107, ESE 10-1–ESE 10-21.
42. Mai, P. M., and Thingbaijam, K. K. S. (2014). SRCMOD: an online database of finite-fault rupture models. *Seismol. Res. Lett.* 85, 1348–1357.
43. Matthews, M. V., Ellsworth, W. L., and Reasenberg, P. A. (2002). A Brownian model for recurrent earthquakes. *Bull. Seismol. Soc. Am.*92, 2233–2250.
44. McGuire, R. K. (2008). Probabilistic seismic hazard analysis: early history. *Earthquake Eng. Struct. Dyn.* 37, 329–338.
45. Ministry of Land, Infrastructure, and Transportation (MLIT). (2014). *Survey of Tsunami Damage Condition*. Available at: http://www.mlit.go.jp/toshi/toshi-hukkou-arkaibu.html
46. Mofjeld, H. O., González, F. I., Titov, V. V., Venturato, A. J., and Newman, J. C. (2007). Effects of tides on maximum tsunami wave heights: probability distributions. *J. Atmos. Oceanic Technol.* 24, 117–123.
47. Monastersky, R. (2012). The next wave. *Nature* 483, 144–146.
48. Morikawa, N., and Fujiwara, H. (2013). A new ground motion prediction equation for Japan applicable up to M9 mega-earthquake. *J. Disaster Res.* 8, 878–888.
49. NGDC. (2016). *National Geophysical Data Center (NGDC WDS) Global Historical Tsunami Database*. Boulder, CO, USA. Available at: https://www.ngdc.noaa.gov/hazard/tsu_db.shtml
50. Okada, Y. (1985). Surface deformation due to shear and tensile faults in a half-space. *Bull. Seismol. Soc. Am.* 75, 1135–1154.
51. Orfanogiannaki, K., and Papadopoulos, G. A. (2007). Conditional probability approach of the assessment of tsunami potential: application in three tsunamigenic regions of the Pacific Ocean. *Pure Appl. Geophys.* 164, 593–603.
52. Pardo-Iguzquiza, E., and Chica-Olmo, M. (1993). The Fourier integral method: an efficient spectral method for simulation of random fields. *Math. Geol.* 25, 177–217.
53. Parsons, T., and Geist, E. L. (2008). Tsunami probability in the Caribbean region. *Pure Appl. Geophys.* 165, 2089–2116.
54. Power, W., Downes, G., and Stirling, M. (2007). Estimation of tsunami hazard in New Zealand due to South American earthquakes. *Pure Appl. Geophys.* 164, 547–564.
55. Satake, K. (2001). "Tsunami modeling from submarine landslides," in *Proceedings of the International Tsunami Symposium, Session 6*(Seattle, USA: National Oceanic and Atmospheric Administration), 665–674.
56. Satake, K., Fujii, Y., Harada, T., and Namegaya, Y. (2013). Time and space distribution of coseismic slip of the 2011 Tohoku earthquake as inferred from tsunami waveform data. *Bull. Seismol. Soc. Am.* 103, 1473–1492.
57. Tanioka, Y., and Satake, K. (1996). Tsunami generation by horizontal displacement of ocean bottom. *Geophys. Res. Lett.* 23, 861–864.
58. Thio, H. K., Somerville, P., and Ichinose, G. (2007). Probabilistic analysis of strong ground motion and tsunami hazards in Southeast Asia. *J. Earthquake Tsunami* 1, 119–137.
59. Tinti, S., Armigliato, A., Tonini, R., Maramai, A., and Graziani, L. (2005). Assessing the hazard related to tsunamis of tectonic origin: a hybrid statistical-deterministic method applied to southern Italy coasts. *ISET J. Earthquake Technol.* 42, 189–201.
60. Tsushima, H., Hirata, K., Hayashi, Y., Tanioka, Y., Kimura, K., Sakai, S. I., et al. (2011). Near-field tsunami forecasting using offshore tsunami data from the 2011 off the Pacific coast of Tohoku earthquake. *Earth Planets Space* 63, 821–826.
61. USGS ShakeMap Archive. (2016). *The 2011 Tohoku Earthquake*. Available at: http://earthquake.usgs.gov/earthquakes/shakemap/global/shake/c0001xgp/

62. Wald, D. J., Worden, B. C., Quitoriano, V., and Pankow, K. L. (2006). *ShakeMap Manual: Technical Manual, User's Guide, and Software Guide.* Reston: United States Geological Survey, 156.
63. Wald, D. J., and Allen, T. I. (2007). Topographic slope as a proxy for seismic site conditions and amplification. *Bull. Seismol. Soc. Am.*97, 1379–1395.
64. Ward, S. N. (2001). Landslide tsunami. *J. Geophys. Res. Solid Earth* 106, 11201–11215.
65. Ward, S. N., and Asphaug, E. (2000). Asteroid impact tsunami: a probabilistic hazard assessment. *Icarus* 145, 64–78.
66. Watts, P. (2004). Probabilistic predictions of landslide tsunamis off Southern California. *Mar. Geol.* 203, 281–301.
67. Yanagisawa, K., Imamura, F., Sakakiyama, T., Annaka, T., Takeda, T., and Shuto, N. (2007). Tsunami assessment for risk management at nuclear power facilities in Japan. *Pure Appl. Geophys.* 164, 565–576.
68. Zhao, J. X., Zhang, J., Asano, A., Ohno, Y., Oouchi, T., Takahashi, T., et al. (2006). Attenuation relations of strong ground motion in Japan using site classification based on predominant period. *Bull. Seismol. Soc. Am.* 96, 898–913.

CHAPTER 8

Advances in Earthquake and Tsunami Sciences and Disaster Risk Reduction Since the 2004 Indian Ocean Tsunami

Kenji Satake

Earthquake Research Institute, The University of Tokyo, Yayoi, Bunkyo-ku, Tokyo 113-0032, Japan

ABSTRACT

The December 2004 Indian Ocean tsunami was the worst tsunami disaster in the world's history with more than 200,000 casualties. This disaster was attributed to giant size (magnitude $M \sim 9$, source length >1000 km) of the earthquake, lacks of expectation of such an earthquake, tsunami warning system, knowledge and preparedness for tsunamis in the Indian Ocean countries. In the last ten years, seismology and tsunami sciences as well as tsunami disaster risk reduction have significantly developed. Progress in seismology includes implementation of earthquake early warning, real-time estimation of earthquake source parameters and tsunami potential, paleoseismological studies on past earthquakes and tsunamis, studies of probable maximum size, recurrence variability, and long-term forecast of large earthquakes in subduction zones. Progress in tsunami science includes accurate modeling of tsunami source such as contribution of horizontal components or "tsunami earthquakes", development of new types of offshore and deep ocean tsunami observation systems such as GPS buoys or bottom pressure gauges, deployments of DART gauges in the Pacific and other oceans, improvements in tsunami propagation modeling, and real-time inversion or data assimilation for the tsunami warning. These developments have been utilized for tsunami disaster reduction in the forms of tsunami early warning systems, tsunami hazard maps, and probabilistic tsunami hazard assessments. Some of the above scientific developments helped to reveal the source characteristics of the 2011 Tohoku earthquake, which caused devastating tsunami damage in Japan and Fukushima Dai-ichi Nuclear Power Station accident. Toward tsunami disaster risk reduction, interdisciplinary and trans-disciplinary approaches are needed for scientists with other stakeholders.

Keywords: Earthquake Tsunami Disaster risk reduction Tsunami warning system The 2004 Indian Ocean tsunami The 2011 Tohoku tsunami

INTRODUCTION

On 26 December 2004, five months after the inauguration of the Asia Oceania Geoscience Society (AOGS), the countries around the Indian Ocean suffered from the devastating tsunami. This tsunami, generated by the Sumatra-Andaman earthquake (magnitude *M* 9.1), was the worst tsunami disaster in the world's written history, and the casualties were not only from the Indian Ocean countries but also extended to European countries because many tourists were spending their Christmas vacations in Asian countries.

This disaster was attributed to several factors. The earthquake was huge and such a giant earthquake was not expected in the Indian Ocean; thus tsunami warning system did not exist in the Indian Ocean; and consequently the coastal residents, tourists and governments did not have knowledge on tsunamis and were not prepared for such a disaster.

In the last decade, significant improvements have been made in earthquake and tsunami sciences as well as in their applications for disaster risk reduction. Scientific developments include real-time estimation of earthquake and tsunami source parameters, implementation of early warning of earthquakes and tsunamis, historical and geological studies of past earthquakes and tsunamis, examination of probable maximum earthquake size, long-term forecast of large earthquakes, new types of tsunami observations in the open ocean and on the coast, and accurate tsunami modeling and inversion. The disaster risk reduction includes delivery of tsunami early warning messages to officials and coastal residents, making tsunami hazard maps or probabilistic hazard assessments, construction of infrastructure such as speakers to disseminate the warning messages, seawalls, evacuation signs, and designated evacuation areas, as well as public education. During the time period of such developments, the 2011 Tohoku earthquake and tsunami occurred and caused devastating tsunami damage in Japan and the Fukushima Daiichi Nuclear Power Station accident. Some of the above scientific developments helped to reveal the source characteristics of this giant earthquake and tsunami, yet they could not prevent the disaster.

In this review paper, I first describe the 2004 Indian Ocean tsunami in section 2, then review developments in seismology in section 3, followed by those in tsunami science in section 4. The 2011 Tohoku earthquake and tsunami are described in section 5. I then discuss efforts and issues that show how scientific developments can be utilized for disaster risk reduction in section 6.

THE 2004 INDIAN OCEAN TSUNAMI

The source of the Indian Ocean tsunami was the Sumatra-Andaman earthquake on 26 December 2004 (Figure 1). The earthquake size, expressed by a moment magnitude scale (M_w), was 9.1 (according to United States Geological Survey: USGS), the largest in the world in the past 40 years. Moment magnitude is derived from seismic moment, which shows a physical size of the earthquake. The seismic moment of this single event was comparable to cumulative moment from global earthquakes in the preceding decade [1]. Only few earthquakes of this size ($M_w \sim 9$) occurred in the 20th century, and they were all around the Pacific Ocean. The 2004 Sumatra-Andaman earthquake was the first instrumentally-recorded event of this size in the Indian Ocean.

The 2004 earthquake was an interplate earthquake between the Indo-Australian plate and the Andaman (or Burma) microplate, a part of the Eurasian plate (Figure 1). The Indo-Australian plate subducts along the Sunda Trench at a rate of approximately 5 cm per year, and the direction of subduction changes from normal to the trench to oblique toward north. This subduction causes upper plate to be dragged and deformed up to a certain limit, then suddenly rebound to cause an interplate earthquake. The focal mechanism solution, estimated by Centroid Moment Tensor inversion [2] indicates a thrust faulting, or low-angle reverse fault, mechanism. The epicenter of the 2004 earthquake was located off Sumatra Island, but the source area extended northward through Nicobar to Andaman Islands.

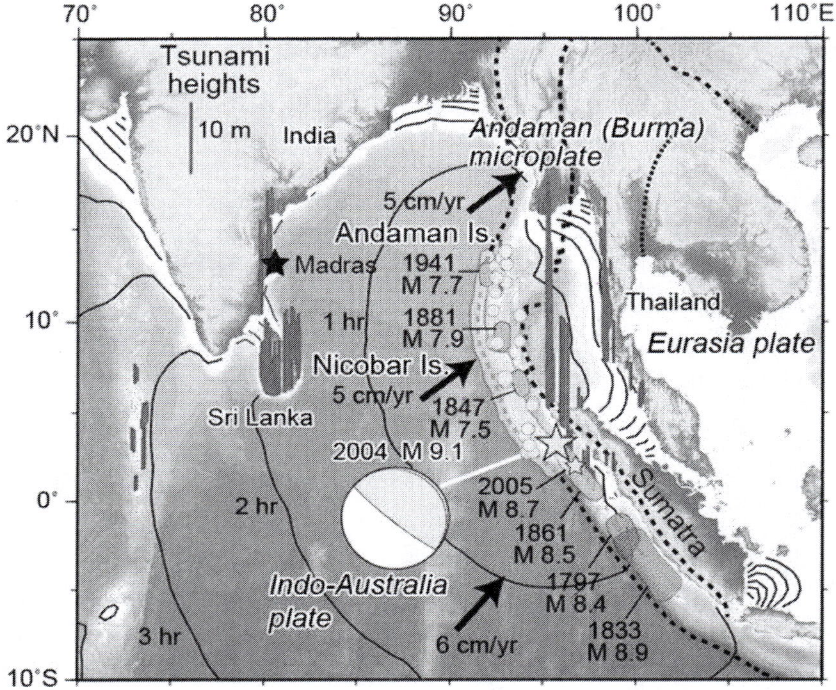

Figure 1. THE 2004 SUMATRA-ANDAMAN EARTHQUAKE. The tsunami heights measured by field surveys are shown by red bars (NOAA NGDC Tsunami Database). The yellow circles and beach ball show the one-day aftershocks and the focal mechanism. Computed tsunami fronts are shown for every hour. Black arrows indicate the direction and speed of Indo-Australian plate. Past earthquakes with their occurrence year and magnitudes are also shown with filled polygons and ovals.

The source lengths of the 2004 earthquake estimated from various data are somewhat different. Seismological analyses indicate that the 2004 rupture started at the epicenter off the west coast of Sumatra, then propagated toward north through Nicobar and Andaman Islands in about 500 seconds with a total length of 1200 to 1300 km [1],[3]. The fault slip was largest, 20 to 30 m, off the coast of northern Sumatra, followed by ~5 m slip off Nicobar Island. The fault slip around the Andaman Islands was estimated to be small and was speculated to be slow. The analysis of tsunami waveforms recorded on tide gauge stations [4] showed a shorter, up to 900 km, source of the tsunami. Satellite image analyses and ground-truth field investigations [5]–[7] indicated that the coseismic coastal sea level change extended from Sumatra through North Andaman Island with a total length of 1600 km. Some of the northern slip is attributed to afterslip on the fault plane which occurred up to 40 days [8].

This earthquake generated a tsunami which devastated the shores of the Indian Ocean. Within 30 minutes of the earthquake, the tsunami first attacked Banda Aceh and other coastal villages of Sumatra Island in Indonesia causing 160,000 casualties. The tsunami then reached the coasts of Thailand (casualty 8000), Sri Lanka (35,000) and India (16,000) within approximately two hours. About a half of tsunami victims in Thailand were foreign tourists. The tsunami further propagated and reached the east coast of Africa where it caused 300 casualties in Somalia. The total casualties of the Indian Ocean tsunami were more than 200,000.

The distribution of the 2004 tsunami heights, measured by scientists and engineers from many countries, looks proportional to the damage distribution (Figure 1). The tsunami heights were mostly larger than 20 m with a maximum height above 30 m near Sumatra Island, particularly in the Aceh province. The tsunami heights along the Andaman Sea coast varied greatly; 5 to 15 m near Thailand but less than 3 m near Myanmar. The tsunami heights were up to 5 m in Andaman Islands. In Sri Lanka, the tsunami heights were 5 to 15 m.

The tsunami was instrumentally recorded by coastal tide gauges in the Indian Ocean as well as in the Atlantic and Pacific Oceans [9]. The tsunami propagation in deep water was captured by deep-sea pressure gauges [10], satellite altimeters [11], hydrophones [12] and horizontal components of broad-band seismographs [13].

At the time of the 2004 tsunami, the tsunami warning system existed only in the Pacific Ocean. The Pacific Tsunami Warning Center (PTWC), located in Hawaii, issued the first information bulletin only 15 minutes after the earthquake. The earthquake was located off the west coast of Northern Sumatra, and the magnitude was initially estimated to be 8.0. The second bulletin was issued at 69 minutes after the earthquake, but still before the tsunami arrivals at the coasts of Thailand, Sri Lanka and India. The earthquake size was updated to 8.5 and the possibility of a local tsunami was included in the bulletin. However, these messages did not reach the governments or coastal communities around the Indian Ocean [14].

The 2004 tsunami caused slight damage to Madras Atomic Power Station at Kalpakkam, near Chennai, on the east coast of India. This was the first tsunami damage to a nuclear power plant in the world. At about 3 hours after the earthquake, the tsunami arrived at the nuclear power station with 4.5 m height, and caused flooding of the seawater pump house and construction site of a new reactor. The switchboard of the pump house was submerged, but the reactors were safely shut down. After this accident, International Atomic Energy Agency revised their safety guide [15].

REVIEW OF DEVELOPMENTS IN SEISMOLOGY

Can we forecast earthquakes and tsunamis in advance? Earthquake source is a fault motion, which is movement or rupture across a plane within the earth. Sudden fault motion generates seismic waves which cause ground shaking and seafloor displacement which becomes the source of tsunami. If we can forecast future earthquakes, or tell in advance where, when and how big they will be, it would benefit to reduce damage from earthquakes and tsunamis. Earthquake and tsunami forecast is made at various time scales; in seconds or minutes between fault rupture and arrival of seismic waves (called Earthquake Early Warning, EEW), in minutes to hours between the earthquake occurrence and the first tsunami arrival (Tsunami Warning), in hours, days or months before the earthquake (Short-term earthquake prediction), and in years to decades before earthquake (Long-term earthquake forecast).

The EEW system forecasts ground shaking after the earthquake occurrence but before the arrival of seismic waves, based on quick analysis of seismic data recorded near the earthquake source [16],[17]. The EEW was developed before 2004 but has been implemented and in operation in the last decade in several countries such as Japan [18]. Typical lead time between the announcement and start of large ground shaking is from several to several tens of seconds, yet providing useful information through TV, radio or cell phones.

The current tsunami warning system also relies on quick analysis of seismic data. The recent deployment of advanced seismological analysis methods for rapid determination of earthquake source parameters, such as the *W*phase analysis [19], makes it possible to quickly assess an earthquake's size with acceptable accuracy and to estimate the potential tsunami size, in order to issue tsunami warnings in less than half an hour for global earthquakes. For example, during the 2012 Sumatra earthquake, PTWC issued bulletins with not only earthquake parameters but also tsunami amplitudes predicted by simulation-based empirical formula [20]. Thus tsunami warning is practically possible at least for far-field tsunamis. Accurate near-filed tsunami warning is still challenging as discussed in next sections.

Earthquake prediction depends on monitoring reliable precursory phenomena which are yet to be discovered. In the rest of this section, we limit our discussion on long-term forecast, which are commonly expressed as future probabilities of occurrence.

Probabilities of future earthquakes can be estimated from past earthquake data. Earthquake probabilities in a certain time window, for example in the next 30 years, can be calculated by fitting inter-earthquake times with a probabilistic density function. If earthquakes occur randomly in time, or a fault does not have any memory of past earthquakes, the Poisson process is assumed to compute the time-independent probabilities; i.e., the probability of the next earthquake is constant through time, depending solely on the average recurrence interval. Alternatively, earthquake probabilities may increase with time, if similar size earthquakes recur more or less regularly (called characteristic earthquakes). The elastic rebound theory explains that an earthquake occurs when the accumulated stress at the plate boundary reaches certain limit. In such a case, statistical distributions such as log-normal distribution or Brownian passage model [21],[22], with the average recurrence interval and the date of most recent events, are used to calculate the time-dependent probabilities.

Was the 2004 Sumatra-Andaman earthquake the first mega-event in the region? Seismological data indicate that earthquakes with *M* 7.5 and 7.9 occurred in the Nicobar Islands and an *M* 7.7 earthquake occurred in the Andaman Islands in 1941 [23]. These past earthquakes had been considered as the maximum earthquakes in the Andaman and Nicobar Islands. Instrumental seismological data are available since the last century. Historical records of damage from past earthquakes or tsunamis are kept for more than 1000 years in some countries like China or Japan [24],[25]. In other places, such historical earthquake data exist only for less than a few centuries, which may not be long enough to record the history of large earthquakes. Geological records such as traces of coastal sea level change or deposits brought by tsunami, called tsunami deposit, are used to study older earthquakes. Such a study area is called paleoseismology.

Paleoseismological studies of tsunami deposits have been conducted since 2004 in Sumatra Island [26], Thailand [27]–[29], the Andaman and Nicobar Islands [30],[31] and India [32]. These studies have shown geological evidence of past tsunamis in the regions. The last earthquake was estimated to have occurred around AD 1300–1450 in Thailand, AD 1290–1400 in Sumatra, AD 1250–1450 near the Andaman and Nicobar Islands, post AD 1600 in South Andaman Island, and around AD 1020–1160 along the Indian coast. These various dates may indicate that the last great earthquake was not exactly the same as the 2004 Sumatra-Andaman earthquake.

Besides the studies of past earthquakes in particular regions, seismologists have attempted to make global assessments of probable maximum earthquake size. Because of infrequent nature of such giant earthquakes, global collection of data is needed to increase our sample and knowledge on such large earthquakes. A comparative studies of subduction zones [33] showed that there are two end members of subduction zones, i.e., Chilean type and Mariana type, among which only the former types can produce great earthquakes. Subsequent studies proposed that the age of subducting plate and plate convergence rate may control the maximum size of earthquakes [34]; larger earthquakes occur in subduction zones where younger plate subducts at a higher convergence rate. However, re-examination of the relation among the plate age, convergence rate and the maximum earthquake size, made after the 2004 earthquake, showed that such a relationship is not as strong as it was believed before [35].

One way to calculate earthquake probability is to assume that the maximum earthquake size is *M* 9.5, which is the size of the 1960 Chile earthquake, the largest earthquake in the 20th century (Figure 2a). For example, McCaffrey [36] proposed that any subduction zone in the world could produce an $M \sim 9$ earthquake. But was the 1960 Chile earthquake really the maximum earthquake? It should be noted that the size of the 1960 Chile earthquake was estimated in the 1970's [37]. Recently, Matsuzawa [38] proposed that we should prepare for an $M \sim 10$ earthquake, although the maximum size of an earthquake on the earth could be $M \sim 11$.

Variability in size and recurrence interval is likely a characteristic nature of great earthquakes in subduction zones [39]. Historical and geological data in other subduction zones indicate that recurrence patterns of past great earthquakes are highly variable (Figure 2b). For example, in southern Chile, historical records indicate that past earthquakes occurred in 1575, 1737, 1837 and 1960, with an average recurrence interval of 130 years. However, the geological evidence or tsunami deposits were found only from the 1960 and 1575 earthquakes as well as older earthquakes, yielding the recurrence interval of ~300 years based on paleoseismological studies [40]. The 2011 Tohoku earthquake added another example of such variability.

REVIEW OF DEVELOPMENTS IN TSUNAMI SCIENCE

Tsunamis are generated by submarine earthquakes, volcanic eruptions or landslides. For the earthquake source, vertical displacement due to subsurface faulting, which can be computed from earthquake source parameters [41] are usually considered as the tsunami source. For the case of the 2004 Sumatra-Andaman earthquake, seafloor was uplifted on the western edge and subsided on the eastern edge of the source area. This asymmetric seafloor deformation yielded initial receding wave on the east (e.g., Thailand) whereas the water level initially rose on the west (e.g., Indian or Sri Lankan coasts). When the tsunami source is on a steep seafloor slope and the horizontal displacement is large, the vertical

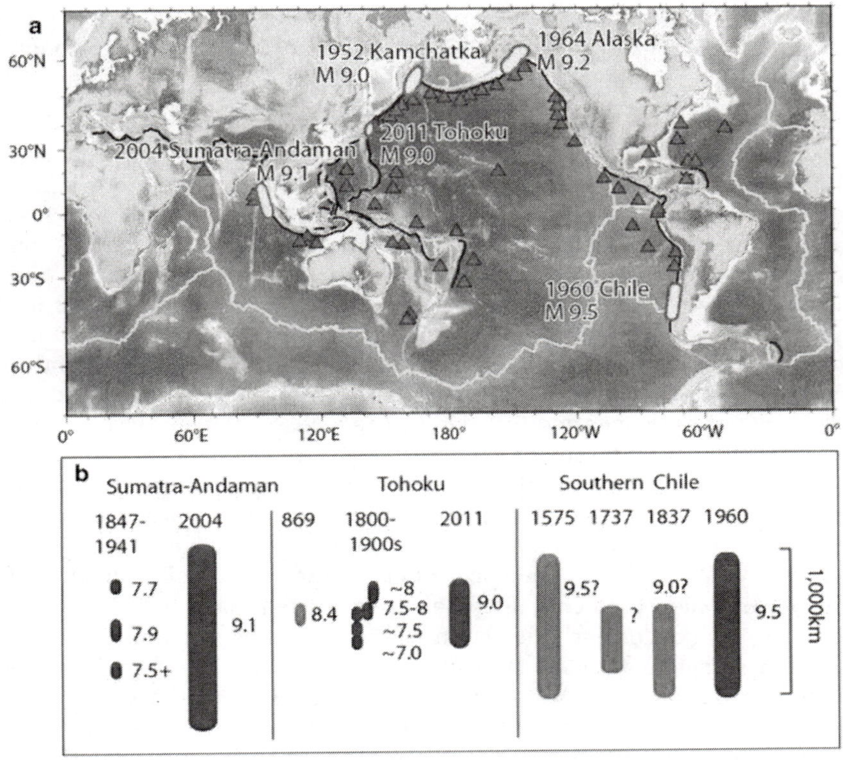

Figure 2. GIANT EARTHQUAKES IN THE WORLD. **(A)** The locations of M ~ 9 earthquakes since the 20th century are shown by yellow ovals. Red triangles indicate DART stations. Black, yellow and green colors indicate three kinds of plate boundaries, i.e., subduction zones, mid-oceanic ridges and transform faults. **(B)** Variability of earthquake size in subduction zones. The colored bars represent simplified rupture area: blue, inferred solely from instrumental and written records; green, includes paleoseismological evidence. Numerals denote moment magnitudes. Updated from Satake and Atwater [39].

movement of water due to the horizontal displacement of the slope must be also considered [42]. For the 2011 Tohoku tsunami, the large horizontal displacement of seafloor slope was responsible for 20–40% of the observed tsunami amplitudes [43]. While tsunami is generally larger for larger earthquake, notable exceptions are "tsunami earthquakes" which generate much larger tsunamis than expected from seismic waves [44],[45]. Typical examples are the 1896 Sanriku earthquake which produced much smaller ground shaking than the 2011 Tohoku earthquake, but the tsunami heights on Sanriku coasts from these earthquakes were similar [46]. More recent examples of "tsunami earthquake", such as the 1992 Nicaragua earthquake and the 2010 Mentawai earthquakes, indicate that very large slip near the trench axis is responsible for the large tsunami and smaller seismic waves [47],[48].

Tsunamis are instrumentally recorded by sea level recorders such as coastal tide gauges, near-shore wave and GPS buoys, and deep-ocean bottom pressure gauges (Figure 3). Coastal tide gauges have various types such as mechanical type with a float, and pressure, acoustic or radar sensors. After the 2004 Indian Ocean tsunami, more coastal tide gauges have been installed in the Indian Ocean region. Currently, sea level data at several hundreds of stations are available in real-time (e.g., http://www.ioc-sealevelmonitoring.org/). Near-shore gauges include wave gauges using ultrasonic waves and GPS buoys. They measure offshore sea levels at water depths of 50 to 200 m, and can detect tsunamis before their coastal arrivals giving some lead time for issuing tsunami warnings. Tsunami waveforms are much simpler in deep oceans, where they are free from the effects of coastal reflection or refraction due to bathymetry. A kinematic GPS analysis of a ship in open ocean detected the 2010 Chile tsunami [49]. Deep-ocean measurements of tsunamis have been made by using bottom pressure gauges for early detection and warnings of tsunamis. The Deep-ocean Assessment and Reporting of Tsunamis (DART), developed by NOAA (National Oceanic and Atmospheric Administration) of the USA, records water levels using bottom pressure gauges, and sends signals to a surface buoy via acoustic telemetry in the ocean, then via satellites to a land station in real time [50]. After the 2004 Indian Ocean tsunami, the total number of DART stations in the Pacific as well as Indian Ocean increased from 6 in 2004 to about 60 in 2013 [51]. An alternative way to retrieve data from deep ocean bottom pressure data is to use submarine cables. Around Japan, more than 10 bottom pressure gauges were installed at the time of the 2011 Tohoku earthquake, and more cabled networks, DONET along the Nankai trough [52] and S-net along the Japan Trench [53] are being deployed.

Tsunami can be hydrodynamically considered as shallow-water (or long) waves, whose phase velocity is given as a square root of product of water depth and the gravitational acceleration. Because the ocean depth, or bathymetry, is globally surveyed and mapped, the tsunami propagation can be simulated using the actual bathymetry data. A popular method of tsunami numerical simulation is a finite-difference computation of equation of motion for shallow-water waves (momentum conservation) and the equation of continuity (mass conservation) [54]. For deep ocean, a typical grid size is a few to several kilometers. Near the coasts with shallow depths, non-linear effects and bottom friction need to be included. In addition, effects of local topography and bathymetry, such as reflection or refraction, also play important role, hence the smaller grid, typically with several tens of meter interval, is adopted. For computation of tsunami inundation on land, topographic data are also used with moving boundary conditions [55].

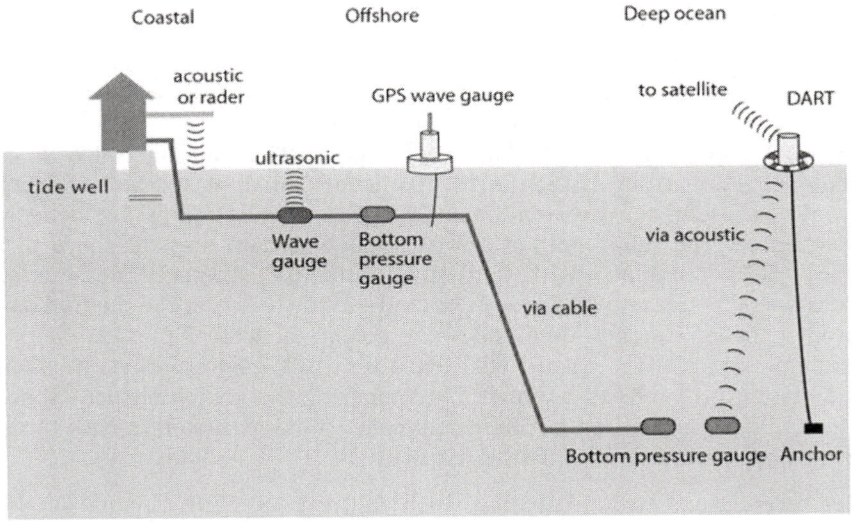

Figure 3. Various types of instruments designed to measure tsunami.

The tsunami waveform data are used to estimate the water height, or fault slip, distribution at the source. In this method, called tsunami waveform inversion [56], the tsunami source or fault plane is divided into subfaults, and tsunami waveforms, or the Green's functions, are computed for each of the subfault with unit amount of slip. Assuming that the observed tsunami waveforms are linear superposition of the Green's function, the distribution of displacement or fault slip can be estimated using a least-square method. The tsunami waveform inversion is used to study the tsunami sources. For the 2011 Tohoku earthquake, because high-quality and high-sampling offshore tsunami waveforms were available, the temporal change as well as spatial distribution of the slips on subfaults was estimated [43].

The tsunami waveform inversion has been also used for tsunami warning systems, both far-field and near-field. The current tsunami warning system, based on seismic monitoring, first determines location, depth, and magnitude of earthquake, then predict tsunami arrival time and coastal heights using database of pre-computed tsunami waveforms for various earthquake sources [57],[58]. For far-field tsunami warnings, the tsunami waveforms at the DART locations are computed for numerous tsunami sources around the Pacific Ocean. When the DART stations record tsunami waveforms, they are compared with pre-computed tsunami waveforms to first estimate the tsunami sources. Then, the estimated sources are used to predict tsunami arrival times and the amplitudes at more distant locations. The predicted tsunami waveforms from the real-time data assimilation, without assuming earthquake source parameters, show good agreement with the observed waveforms [59].

For the near-field tsunami warning, tsunami waveforms recorded on cabled bottom pressure gauges can be used to estimate the sea surface displacement rather than the fault slip [60],[61]. This method, called tFISH (tsunami Forecasting based on Inversion for initial sea-Surface Height), would be useful for tsunami warning if the cabled stations are densely distributed. The same method can be applied to data on near-shore GPS buoys [62]. Combined with Real-time Automatic detection method for Permanent Displacement (RAPiD) of land-based GPS data, the method can predict tsunami arrival time and wave heights at least 5 minutes before tsunami arrival for near-field tsunamis [63]. Forecasting tsunami inundation on land can be made by comparing the computed near-shore tsunami waveforms from actual earthquake source parameters with those pre-computed and stored in the database [64].

For recent trans-Pacific tsunamis, such as the 2010 Chile tsunami or the 2011 Tohoku tsunami, discrepancies (a few percent) in the travel time between the observed waveforms recorded at DART stations and the computed waveforms based on linear shallow water have been reported. A small reduction of the tsunami phase velocity at very long period (>1000 seconds), caused by the coupling of seawater and self-gravitating elastic Earth, is considered to be responsible for these delays [65].

THE 2011 TOHOKU EARTHQUAKE AND TSUNAMI

A giant earthquake occurred off the northern coast of Honshu, Japan, on 11 March 2011. With the largest magnitude (M_w 9.0) in the history of Japan, it caused a devastating tsunami disaster and serious damage to the nearby Fukushima Dai-ichi Nuclear Power Station. The earthquake and subsequent tsunami caused approximately 18,500 dead and missing persons. The 2011 Tohoku earthquake occurred at the Japan Trench where the Pacific plate subducts beneath northern Honshu at a rate of approximately 8 cm per year. This earthquake was also an interplate earthquake with a thrust-type fault motion.

Very dense geophysical measurements both on land and offshore Japan made this event the best recorded subduction-zone earthquake in the world (Figure 4a). The nation-wide land-based GPS network with more than 1000 stations recorded large coseismic movements with a maximum amount of 5.3 m eastward and 1.2 m downward [66]. The repeated marine geophysical measurements, such as GPS/acoustic positioning, bottom pressure gauge, or multi-beam swath bathymetry survey, started before the 2011 Tohoku earthquake, detected huge seafloor displacement, approximately 50 m in horizontal direction [67]–[69].

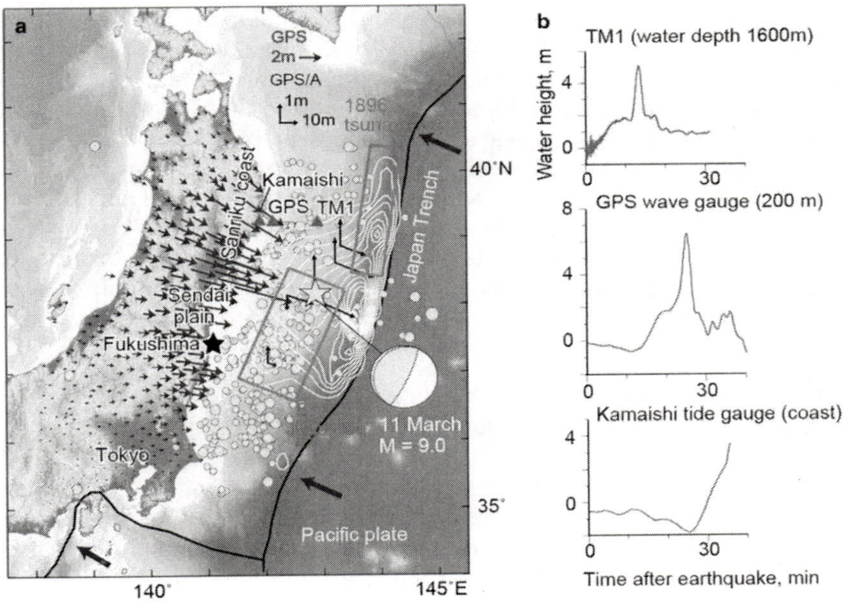

Figure 4. THE 2011 TOHOKU EARTHQUAKE. (A) The source region of the 2011 Tohoku earthquake. Thick black curves are plate boundaries and the arrows show the relative plate motions. The mainshock (yellow star) and earthquakes with $M \geq 5.0$ occurred within a week (yellow circles). Blue arrows are coseismic movements of land-based GPS stations [66] and black arrows are marine GPS/acoustic measurements [69]. The white contours show the slip distribution with 4 m interval estimated from the tsunami waveforms [43]. The red triangles show the locations of the bottom pressure (TM1), GPS wave and coastal tide gauges. The proposed fault models (orange rectangles) of the 1896 Sanriku tsunami earthquake [70] and the 869 Jogan earthquake [71] are shown. Black star shows the location of Fukushima Dai-ichi Nuclear Power Station. **(B)** The tsunami waveforms recorded by the bottom pressure gauge (TM1), GPS wave gauge and coastal tide gauge [72].

The 2011 tsunami was first detected on ocean bottom pressure and GPS wave gauges. A cabled bottom pressure gauge about 76 km off Sanriku coast at a 1600 m water depth recorded ~2 m water rise in about 6 minutes starting immediately after the earthquake, followed by an impulsive wave with additional 3 m rise within 2 minutes [72]. Similar two-stage tsunami waveforms were also recorded on a GPS wave gauge near the coast 12 minutes later, just before tsunami arrival on the coast (Figure 4b).

The Japan Meteorological Agency (JMA) issued the first tsunami warning 3 minutes after the earthquake. The estimated tsunami heights were at maximum 6 m, significantly smaller than the actual tsunami heights with a maximum of 40 m [73]. The smaller expected coastal tsunami heights were due to the initial underestimation of the earthquake magnitude (M =7.9). Nevertheless, very strong ground shaking and the tsunami warning urged many coastal residents to evacuate to high ground and thus saved their lives. After detecting the large offshore tsunami on GPS wave gauges, JMA upgraded the tsunami warning messages to a higher level of estimated tsunami heights at 28 minutes after the earthquake. Although it was announced before the actual tsunami arrival to the nearest coast, the updated warning messages did not reach all the coastal communities due to power failures and the fact that coastal residents had already started evacuation.

The occurrence of an $M \sim 9$ earthquake near the Japan Trench was another surprise to the global seismological communities. Off Miyagi prefecture, near the epicenter of the 2011 Tohoku earthquake, large ($M \sim 7.5$) earthquakes have repeatedly occurred since 1793 with an average interval of 37 years. On the basis of this recurrence, the Earthquake Research Committee of the Japanese government estimated the probability of a great ($M \sim 8$) earthquake occurring between 2010 and 2040 as 99% [74]. The long-term forecast failed to predict the size (M) of the Tohoku earthquake [75].

The Sanriku coast of Tohoku had been devastated by previous tsunamis. The 1896 Sanriku earthquake, a typical "tsunami earthquake," caused large tsunami, with the maximum height of 38 m, despite its weak ground shaking. The 2011 tsunami heights along the Sanriku coast were as high as nearly 40 m, roughly similar to the 1896 tsunami heights [46]. The 1896 tsunami caused about 22,000 casualties, slightly more than the 2011 tsunami. Study of tsunami waveforms indicate that the 1896 earthquake was generated from a fault motion near the trench axis [70], similar to other "tsunami earthquakes."

The predecessor of the 2011 Tohoku earthquake is considered to be the 869 Jogan earthquake [71],[76]. A national history book depicts strong ground shakings, collapse of houses, kilometers of tsunami flooding with

1000 drowned people in Sendai plain in AD 869 in Jogan era. In addition, paleoseismological studies found tsunami deposits in coastal lowlands, more than 4 km from the current coast in the Sendai plain. Older tsunami deposits were also found. From the distribution of the tsunami deposits and computed inundation area, the size (M = 8.3 to 8.4), location and fault models of the Jogan earthquake were proposed with a recurrence interval between 500 and 1000 years [71],[77].

Tsunami models indicate that the source of the 2011 earthquake appear to be a combination of the 1896 Sanriku "tsunami earthquake" and a Jogan-type deeper interplate earthquake [43],[72]. A huge slip near the trench axis, similar to the 1896 Sanriku tsunami earthquake, caused the first impulsive tsunami waves recorded by the bottom pressure and GPS wave gauges and large tsunami runup heights along Sanriku coast [46]. The fault motion and large slip along the deeper plate interface, similar to a proposed model of the Jogan earthquake [71], produced a long-wavelength seafloor displacement which caused the first gradual rise at offshore gauges (Figure 4b) and the large tsunami inundation in the Sendai plain [78],[79].

The 2011 tsunami also impacted four nuclear power stations located near the source area. At these stations, the strong ground shaking automatically shut down the reactors, and the diesel power generators started to cool down the reactors. At the Fukushima Dai-ichi Nuclear Power Station, strong ground shaking damaged the external power supply system, and the 15 m tsunami damaged the diesel generator. The Fukushima Dai-ichi Station thus failed to cool down the reactors, which led to melt down of three reactors, hydrogen explosions and release of radioactive materials into atmosphere [80], ocean [81], and land/soil [82]. At the other three nuclear power stations, the reactors were cooled down by using external or backup power and succeeded to avoid fatal accident.

The estimated maximum tsunami, or design tsunami height, at the Fukushima Dai-ichi Nuclear Power Station was 6.1 m, based on the M 7.5 earthquake which occurred in 1938. The long-term forecast of Earthquake Research Committee indicated that a "tsunami earthquake" may occur anywhere along the Japan Trench. The tsunami height at the Fukushima Dai-ichi Station from the Jogan earthquake model was estimated as 8.9 m, and that from a "tsunami earthquake" off Fukushima was calculated as 15.7 m by Tokyo Electronic Power Company, but no preventive actions were taken. For critical facilities such as a nuclear power plant, the seismological progress should be closely monitored and reflected in safety preparedness.

TOWARD TSUNAMI DISASTER RISK REDUCTION

Despite advances in natural science on hazards, why do disaster losses continue to increase? This is a motivation to initiate an international and interdisciplinary program called IRDR (Integrated Research for Disaster Reduction) under ICSU (International Council for Science), together with ISSC (International Social Science Council) and UN International Strategy for Disaster Reduction (UN-ISDR). Disaster risk consists of hazard and vulnerability. Natural hazard such as earthquake or tsunamis cannot be controlled, but may be forecasted. However, disaster risk can be reduced by minimizing vulnerability or exposure to hazards. The disaster risk reduction is therefore closely related to how science can be utilized for safety of the society.

In January 2005, immediately after the 2004 Indian Ocean tsunami, Hyogo Framework of Action (HFA) for 2005–2015 was adopted at the World Disaster Reduction Conference and later endorsed by UN General Assembly. It is for building the resilience of nations and communities to disasters, and consists of five action items. (1) Make disaster risk reduction a national and local priority; (2) Identify, assess and monitor disaster risks and enhance early warning; (3) Use knowledge, innovation and education to build understanding and awareness; (4) Reduce risk factors; and (5) Be prepared and ready to act. We will review how the scientific developments can contribute to these actions.

The tsunami early warning systems have been implemented in the Indian Ocean and other ocean basins. In the Pacific Ocean, an international tsunami warning system was established after the 1960 Chile tsunami, and International Coordination Group was formed under UNESCO Intergovernmental Oceanographic Commission. Following the 2004 Indian Ocean tsunami, similar groups were formed for the Indian Ocean, for the North-eastern Atlantic and Mediterranean Sea, and for the Caribbean Sea. In coordination with the UNESCO group, three regional Tsunami Warning Centers were recently established in Australia [83], India [84] and Indonesia [85]. These centers are staffed 24 hours a day and 7 days a week to monitor seismic activity and the possibility of a tsunami occurrence. The warning systems rely on the most advanced seismic and sea-level monitoring, a database of past tsunami events, and pre-made numerical simulations. These systems can issue tsunami warning messages typically about 5 minutes after an earthquake.

Once the coastal residents receive tsunami warning message, they need to know what it means, and where to evacuate. Tsunami is a Japanese word meaning "harbor wave," but few people around the Indian Ocean knew the word before the 2004 Indian Ocean tsunami. It has been used internationally since the 2004 Indian Ocean tsunami. An effective tool to guide coastal residents for evacuation is a hazard map showing the tsunami risk zones. Possible flooding zones and safe evacuation places such as tsunami shelters can be shown in the hazard maps. One of the lessons learned from the 2004 Indian Ocean tsunami is that not only coastal residents but also foreign tourists need to be informed about tsunami hazards. In the Hawaiian Islands, tsunami hazard maps are prepared and published in the local phone books that are available at every hotel room. Those in high-rise hotel buildings are advised to move vertically to the third or higher floors, rather than horizontally moving out of the possible flooding area.

Tsunami hazard maps are constructed for past tsunamis or for the most likely tsunami source. In the Sendai plain before the 2011 Tohoku tsunami, tsunami hazard maps and other countermeasures were prepared for an $M \sim 8$ earthquake, which was estimated to occur with 99% probability in the next 30 years (see The 2011 Tohoku earthquake and tsunami section). The predicted inundation area was, for the most part, within 1 km from the coast, and much smaller than the actual tsunami inundation area of the 2011 $M = 9.0$ earthquake which was up to 5 km. The distribution of the 869 Jogan tsunami deposits, however, was similar to the inundation area of the 2011 Tohoku tsunami. The hazard maps need to consider such infrequent gigantic earthquake and tsunamis.

One of the developments in tsunami hazard assessment in the last decade is Probabilistic Tsunami Hazard Assessment (PTHA) [86],[87]. Results of the PTHA are typically displayed as hazard curves that show the annual frequency of exceedance of tsunami heights. The hazard from a large number of possible sources including non-earthquake source can be aggregated together to develop a tsunami hazard curve. In addition, multiple sources of uncertainty related to the source parameters and tsunami numerical computations can be considered in the PTHA. Uncertainty can be classified into two types: aleatory and epistemic. Aleatory uncertainty, or random variability, relates to the natural or stochastic uncertainty inherent in a physical system, and cannot be reduced but can be estimated from repeated observations or experiments. Epistemic uncertainty is due to incomplete knowledge and data, and can be reduced by the collection of new data. Epistemic uncertainty can be treated as logic trees [88]. A single hazard curve is obtained by integration over the aleatory uncertainties, and a large number of hazard curves are obtained for different branches of a logic-tree representing epistemic uncertainty. The PTHA for Fukushima Dai-ichi Nuclear Power Station estimated that the annual exceedance of 10 m high tsunami was an order of 1×10^{-5}, or return period of around 100,000 years [89].

To reduce social vulnerability, various infrastructures, or hardwares, have been implemented since the 2004 Indian Ocean and 2011 Tohoku tsunamis. Speakers to broadcast tsunami warning messages have been installed in coastal areas of Indonesia or Thailand. Sign boards showing the altitude and route to evacuation places have been installed at numerous coastal communities in the world. New and higher seawalls have been constructed for coastal cities, particularly to protect critical facilities such as nuclear power plants. These kinds of infrastructure have their lifetime and may not be maintained until the next tsunami disaster.

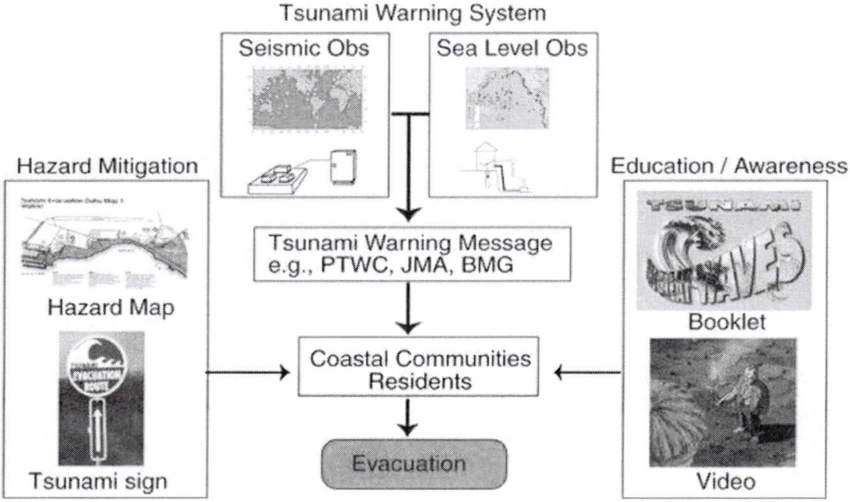

Figure 5. ELEMENTS OF A TSUNAMI-RESILIENT SOCIETY. Tsunami warning system (center), hazard assessment (left) and education systems (right) to reduce tsunami disaster risks.

Soft measures to reduce vulnerability include education and awareness efforts. Numerous books and videos have been published and used for education. In Japan, a famous story, called "Inamura-no-hi" (fire of rice sheaves), has been used for tsunami education. After a strong earthquake was felt at a coastal village in 1854, the village chief put fire on his just-harvested rice crops to guide villagers to high ground and to save their lives. Another concept, "Tsunami tendenko," which calls for a quick tsunami evacuation without waiting for others, not even one's parents or children, became famous and popular after the 2011 tsunami [90],[91]. Periodic practice and drills are also important to keep the tsunami warning and mitigation system functional. In Indonesia, tsunami evacuation drills have been carried out in many communities including Banda Aceh, Padang and Bali in the last decade.

Interdisciplinary studies of natural, social and human sciences, as well as trans-disciplinarity of science, that is cooperation between scientist and society, are important for disaster risk reduction. For the latter, results of scientific developments need to be shared with and utilized by various stakeholders such as national government, local government or communities (Figure 5).

CONCLUSIONS

(1) The 2004 Sumatra-Andaman earthquake, the largest event in the last 40 years, caused the worst tsunami disaster in countries around the Indian Ocean. The main factors for the disaster were unexpected occurrence of a gigantic earthquake in the region and lacks of tsunami warning system, education and awareness for tsunamis in the Indian Ocean.

(2) Seismological developments since 2004 include early detection and estimation of tsunami occurrence, paleoseismological studies on evidence of similar tsunamis in the past, and global studies of recurrence nature of large earthquakes in subduction zones. The assessments of probable maximum size and long-term forecast of great subduction zone earthquakes still appear to be complicated, because of variability of recurrent earthquakes.

(3) Developments in tsunami science include accurate modeling of tsunami source such as contribution of horizontal components or "tsunami earthquakes", instrumental developments for offshore and deep-ocean tsunami observation, deployments of DART gauges in the Pacific and other oceans, improvements in tsunami propagation modeling, and real-time inversion of various kinds of data for the tsunami warning.

(4) At the time of the 2011 Tohoku earthquake, the tsunami warning issued in 3 minutes of the earthquake saved many lives yet resulted in 18,500 casualties. The long-term earthquake forecast made before 2011 estimated the 30 year probability of 99 % in the source region with smaller ($M \sim 8$) size. The source of the 2011 earthquake was modeled as a combination of the 1896 "tsunami earthquake" and the 869 Jogan earthquake.

(5) Towards tsunami disaster reduction, the development of seismology and tsunami science can be implemented as tsunami early warning systems, tsunami hazard maps, and probabilistic tsunami hazard assessments. In

addition, interdisciplinary and trans-disciplinary approaches are needed for scientists with other stakeholders.

ACKNOWLEDGEMENTS

The author acknowledges Drs. Shingo Watada and Mohammad Heidarzadeh for their reviews of the manuscript before submission. Valuable comments and suggestions by two anonymous reviewers also improved the paper. This work was partially supported by KAKENHI (24241080).

COMPETING INTERESTS

The author declares that he has no competing interests.

AUTHORS' CONTRIBUTIONS

This is a single-authored paper, and the no other person has contributed to this manuscript.

REFERENCES

1. Lay T, Kanamori H, Ammon CJ, Nettles M, Ward SN, Aster RC, Beck SL, Bilek SL, Brudzinski MR, Butler R, DeShon HR, Ekstrom G, Satake K, Sipkin S: The great Sumatra-Andaman earthquake of 26 december 2004. *Science* 2005, 308(5725): 1127–1133.
2. Ekström G: 4.16 - global seismicity: results from systematic waveform analyses, 1976–2005. In *Treatise on geophysics*. Edited by: Schubert G. Elsevier, Amsterdam; 2007:473–481.
3. Ammon CJ, Ji C, Thio HK, Robinson D, Ni SD, Hjorleifsdottir V, Kanamori H, Lay T, Das S, Helmberger D, Ichinose G, Polet J, Wald D: Rupture process of the 2004 Sumatra-Andaman earthquake. *Science* 2005, 308: 1133–1139.
4. Fujii Y, Satake K: Tsunami source model of the 2004 Sumatra-Andaman earthquake inferred from tide gauge and satellite data. *Bull Seism Soc Am* 2007, 97: S192-S207.
5. Kayanne H, Ikeda Y, Echigo T, Shishikura M, Kamataki T, Satake K, Malik JN, Basir SR, Chakrabortty GK, Roy AKG: Coseismic and postseismic creep in the Andaman islands associated with the 2004 Sumatra-Andaman earthquake. *Geophys Res Lett* 2007, 34: L01310.
6. Meltzner AJ, Sieh K, Abrams M, Agnew DC, Hudnut KW, Avouac J-P, Natawidjaja DH: Uplift and subsidence associated with the great Aceh-Andaman earthquake of 2004. *J Geophys Res Solid Earth* 2006, 111: B02407.

7. Tobita M, Suito H, Imakiire T, Kato M, Fujiwara S, Murakami M: Outline of vertical displacement of the 2004 and 2005 Sumatra earthquakes revealed by satellite radar imagery. *Earth Planets Space* 2006, 58: e1-e4.
8. Chlieh M, Avouac J-P, Hjorleifsdottir V, Song T-RA, Ji C, Sieh K, Sladen A, Hebert H, Prawirodirdjo L, Bock Y, Galetzka J: Coseismic slip and afterslip of the great Mw 9.15 Sumatra–Andaman earthquake of 2004. *Bull Seism Soc Am* 2007, 97(1A):S152-S173.
9. Titov V, Rabinovich AB, Mofjeld HO, Thomson RE, Gonzalez FI: The global reach of the 26 december 2004 Sumatra tsunami. *Science* 2005, 309(5743):2045–2048.
10. Rabinovich AB, Woodworth PL, Titov VV: Deep-sea observations and modeling of the 2004 Sumatra tsunami in drake passage. *Geophys Res Lett* 2011, 38(16):L16604.
11. Smith WHF, Scharroo R, Titov VV, Arcas D, Arbic BK: Satellite altimeters measure tsunami, early model estimates confirmed. *Oceanography* 2005, 18(2):10–12. 10.5670/oceanog.2005.62
12. Hanson JF, Reasoner C, Bowman JR: High frequency tsunami signals of the great Indonesian earthquakes of 26 december and 28 march 2005. *Bull Seism Soc Am* 2007, 97: S232-S248.
13. Yuan XH, Kind R, Pedersen HA: Seismic monitoring of the Indian ocean tsunami. *Geophys Res Lett* 2005, 32: L15308.
14. NOAA (2005) NOAA and the Indian ocean tsunami. . Accessed October 28 2015, [http://www.noaanews.noaa.gov/stories2004/s2358.htm]
15. Meteorological and hydrological hazards in site evaluation for nuclear installations. IAEA, Vienna; 2011.
16. Grecksch G, Kümpel H-J: Statistical analysis of strong-motion accelerograms and its application to earthquake early-warning systems. *Geophys J Intl* 1997, 129(1):113–123.
17. Wu Y-M, Teng T-l: A virtual subnetwork approach to earthquake early warning. *Bull Seism Soc Am* 2002, 92(5):2008–2018.
18. Kamigaichi O, Saito M, Doi K, Matsumori T, Sy T, Takeda K, Shimoyama T, Nakamura K, Kiyomoto M, Watanabe Y: Earthquake early warning in Japan: warning the general public and future prospects. *Seism Res Lett* 2009, 80(5):717–726.
19. Kanamori H, Rivera L: Source inversion of W phase: speeding up seismic tsunami warning. *Geophys J Intl* 2008, 175(1):222–238.
20. Wang D, Becker NC, Walsh D, Fryer GJ, Weinstein SA, McCreery CS, Sardiña V, Hsu V, Hirshorn BF, Hayes GP, Duputel Z, Rivera L, Kanamori H, Koyanagi KK, Shiro B: Real-time forecasting of the april 11, 2012 Sumatra tsunami. *Geophys Res Lett*2012, 39(19):L19601.
21. Matthews MV, Ellsworth WL, Reasenberg PA: A Brownian model for recurrent earthquakes. *Bull Seism Soc Am* 2002, 92(6):2233–2250.
22. Nishenko SP, Buland R: A generic recurrence interval distribution for earthquake forecasting. *Bull Seism Soc Am* 1987, 77: 1382–1399.
23. Bilham R, Engdahl R, Feldl N, Satyabala SP: Partial and complete rupture of the indo-Andaman plate boundary 1847–2004. *Seism Res Lett* 2005, 76: 299–311.
24. Guidoboni E, Ebel JE: *Earthquyakes and tsunamis in the past.* Cambridge University Press, Cambridge; 2009.
25. Ishibashi K: Status of historical seismology in Japan. *Ann Geophys* 2004, 47(2/3):339–368.
26. Monecke K, Finger W, Klarer D, Kongko W, McAdoo BG, Moore AL, Sudrajat SU: A 1,000-year sediment record of tsunami recurrence in northern Sumatra. *Nature* 2008, 455(7217):1232–1234.
27. Fujino S, Naruse H, Matsumoto D, Jarupongsakul T, Sphawajruksakul A, Sakakura N: Stratigraphic evidence for pre-2004 tsunamis in southwestern Thailand. *Mar Geol* 2009, 262: 25–28.
28. Jankaew K, Atwater BF, Sawai Y, Choowong M, Charoentitirat T, Martin ME, Prendergast A: Medieval forewarning of the 2004 Indian ocean tsunami in Thailand. *Nature* 2008, 455(7217):1228–1231.

29. Prendergast AL, Cupper ML, Jankaew K, Sawai Y: Indian ocean tsunami recurrence from optical dating of tsunami sand sheets in Thailand. *Mar Geol* 2012, 295–298: 20–27.
30. Malik JN, Shishikura M, Echigo T, Ikeda Y, Satake K, Kayanne H, Sawai Y, Murty CVR, Dikshit O: Geologic evidence for two pre-2004 earthquakes during recent centuries near port Blair, south Andaman island, India. *Geology* 2011, 39(6):559–562.
31. Rajendran CP, Rajendran K, Andrade V, Srinivasalu S: Ages and relative sizes of pre-2004 tsunamis in the Bay of Bengal inferred from geologic evidence in the Andaman and Nicobar islands. *J Geophys Res Solid Earth* 2013, 117: 1345–1362.
32. Rajendran CP, Rejendran K, Machado T, Satyamurthy T, Aravazhi P, Jaiswal M: Evidence of ancient sea surges at the mamallapuram coast of India and implications for previous Indian ocean tsunami events. *Curr Sci* 2006, 91(9):1242–1247.
33. Uyeda S, Kanamori H: Back-arc opening and the mode of subduction. *J Geophys Res Solid Earth* 1979, 84(B3):1049–1061.
34. Ruff L, Kanamori H: Seismicity and the subduction process. *Phys Earth Planet Inter* 1980, 23: 240–252.
35. Stein S, Okal E: Ultralong period seismic study of the december 2004 Indian ocean earthquake and implications for regional tectonics and the subduction process. *Bull Seism Soc Am* 2007, 97: S279-S295.
36. McCaffrey R: Global frequency of magnitude 9 earthquakes. *Geology* 2008, 36(3):263–266.
37. Kanamori H: The energy release in great earthquakes. *J Geophys Res Solid Earth* 1977, 82: 2981–2987.
38. Matsuzawa T: The largest EarthquakesWe should prepare for. *J Disaster Res* 2014, 9(3):248–251.
39. Satake K, Atwater BF: Long-term perspectives on giant earthquakes and tsunamis at subduction zones. *Annu Rev Earth Planet Sci* 2007, 35: 349–374.
40. Cisternas M, Atwater BF, Torrejon F, Sawai Y, Machuca G, Lagos M, Eipert A, Youlton C, Salgado I, Kamataki T, Shishikura M, Rajendran CP, Malik JK, Rizal Y, Husni M: Predecessors of the giant 1960 Chile earthquake. *Nature* 2005, 437(7057):404–407
41. Okada Y: Surface deformation due to shear and tensile faults in a half-space. *Bull Seism Soc Am* 1985, 75: 1135–1154.
42. Tanioka Y, Satake K: Tsunami generation by horizontal displacement of ocean bottom. *Geophys Res Lett* 1996, 23(8):861–864.
43. Satake K, Fujii Y, Harada T, Namegaya Y: Time and space distribution of coseismic slip of the 2011 Tohoku earthquake as inferred from tsunami waveform data. *Bull Seism Soc Am* 2013, 103(2B):1473–1492.
44. Kanamori H: Mechanism of tsunami earthquakes. *Phys Earth Planet Inter* 1972, 6: 246–259.
45. Satake K, Tanioka Y: Sources of tsunami and tsunamigenic earthquakes in subduction zones. *Pure Appl Geophys* 1999, 154(3–4):467–483.
46. Tsuji Y, Satake K, Ishibe T, Harada T, Nishiyama A, Kusumoto S (2014) Tsunami heights along the pacific coast of northern Honshu recorded from the 2011 Tohoku and previous great earthquakes. Pure Appl Geophys 1–33,
47. Satake K: Mechanism of the 1992 Nicaragua tsunami earthquake. *Geophys Res Lett* 1994, 21(23):2519–2522.
48. Satake K, Nishimura Y, Putra PS, Gusman AR, Sunendar H, Fujii Y, Tanioka Y, Latief H, Yulianto E: Tsunami source of the 2010 mentawai, Indonesia earthquake inferred from tsunami field survey and waveform modeling. *Pure Appl Geophys* 2013, 170: 1567–1582.
49. Foster JH, Brooks BA, Wang D, Carter GS, Merrifield MA: Improving tsunami warning using commercial ships. *Geophys Res Lett* 2012, 39(9):L09603.
50. González FI, Bernard EN, Meinig C, Eble MC, Mofjeld HO, Stalin S: The NTHMP tsunameter network. *Nat Hazards* 2005, 35(1):25–39.

51. Mungov G, Eblé M, Bouchard R: DART® tsunameter retrospective and real-time data: a reflection on 10 years of processing in support of tsunami research and operations. *Pure Appl Geophys* 2013, 170(9–10):1369–1384.
52. Nakano M, Nakamura T, Kamiya S, Ohori M, Kaneda Y: Intensive seismic activity around the nankai trough revealed by DONET ocean-floor seismic observations. *Earth Planets Space* 2013, 65: 5–15.
53. Uehira K, Kanazawa T, Noguchi S, Aoi S, Kunugi T, Matsumoto T, Okada Y, Sekiguchi S, Shiomi K, Yamada T (2012) Ocean bottom seismic and tsunami network along the Japan trench. AGU Fall Meeting Abstracts:OS41C-1736 presented at 2012 Fall Meeting, AGU, San Francisco, California, 3–7 December
54. Satake K: Linear and nonlinear computations of the 1992 Nicaragua earthquake tsunami. *Pure Appl Geophys* 1995, 144(3–4):455–470.
55. Imamura F: Tsunami modeling: calculating inundation and hazard maps. In *Tsunamis, vol 15*. Edited by: Bernard EN, Robinson AR. Harvard University Press, Cambridge; 2009:321–332.
56. Satake K: Inversion of tsunami waveforms for the estimation of heterogeneous fault motion of large submarine earthquakes - the 1968 tokachi-Oki and 1983 Japan Sea earthquakes. *J Geophys Res Solid Earth* 1989, 94(B5):5627–5636.
57. Kamigaichi O: Tsunami forecasting and warning. In *Encyclopedia of complexity and systems science*. Edited by: Meyers RA. Springer, New York; 2009:9592–9613.
58. Tatehata H: The new tsunami warning system of the Japan meteorological agency. In *Perspectives on tsunami hazards reduction*. Edited by: Hebenstreit G. Kluwer Academic Publishers, Dordrecht; 1997:175–188.
59. Tang L, Titov VV, Chamberlin CD: Development, testing, and applications of site-specific tsunami inundation models for real-time forecasting. *J Geophys Res Oceans* 2009, 114(C12):C12025.
60. Tsushima H, Hino R, Fujimoto H, Tanioka Y, Imamura F: Near-field tsunami forecasting from cabled ocean bottom pressure data. *J Geophys Res Solid Earth* 2009, 114(B6):B06309.
61. Tsushima H, Hino R, Tanioka Y, Imamura F, Fujimoto H: Tsunami waveform inversion incorporating permanent seafloor deformation and its application to tsunami forecasting. *J Geophys Res Solid Earth* 2012, 117(B3):B03311.
62. Yasuda T, Mase H: Real-time tsunami prediction by inversion method using offshore observed GPS buoy data: nankaido. *J Waterway Port Coastal Ocean Engin* 2013, 139(3):221–231.
63. Tsushima H, Hino R, Ohta Y, Iinuma T, Miura S: TFISH/RAPiD: rapid improvement of near-field tsunami forecasting based on offshore tsunami data by incorporating onshore GNSS data. *Geophys Res Lett* 2014, 41(10):L059863.
64. Gusman AR, Tanioka Y, MacInnes BT, Tsushima H (2014) A methodology for near-field tsunami inundation forecasting: application to the 2011 Tohoku tsunami. J Geophys Res Solid Earth 2014JB010958,
65. Watada S, Kusumoto S, Satake K: Traveltime delay and initial phase reversal of distant tsunamis coupled with the self-gravitating elastic earth. *J Geophys Res Solid Earth* 2014, 119(5):4287–4310.
66. Ozawa S, Nishimura T, Suito H, Kobayashi T, Tobita M, Imakiire T: Coseismic and postseismic slip of the 2011 magnitude-9 tohoku-oki earthquake. *Nature* 2011, 475(7356):373–376.
67. Fujiwara T, Kodaira S, No T, Kaiho Y, Takahashi N, Kaneda Y: The 2011 tohoku-oki earthquake: displacement reaching the trench axis. *Science* 2011, 334(6060):1240.
68. Kido M, Osada Y, Fujimoto H, Hino R, Ito Y: Trench-normal variation in observed seafloor displacements associated with the 2011 tohoku-oki earthquake. *Geophys Res Lett* 2011, 38: L24303.
69. Sato M, Ishikawa T, Ujihara N, Yoshida S, Fujita M, Mochizuki M, Asada A: Displacement above the hypocenter of the 2011 tohoku-oki earthquake. *Science* 2011, 332(6036):1395.

70. Tanioka Y, Satake K: Fault parameters of the 1896 sanriku tsunami earthquake estimated from tsunami numerical modeling. *Geophys Res Lett* 1996, 23(13):1549–1552.
71. Sawai Y, Namegaya Y, Okamura Y, Satake K, Shishikura M: Challenges of anticipating the 2011 Tohoku earthquake and tsunami using coastal geology. *Geophys Res Lett* 2012., 39:
72. Fujii Y, Satake K, Sakai S, Shinohara M, Kanazawa T: Tsunami source of the 2011 off the pacific coast of Tohoku earthquake. *Earth Planets Space* 2011, 63(7):815–820.
73. Ozaki T: Outline of the 2011 off the pacific coast of Tohoku earthquake (Mw 9.0) - tsunami warnings/advisories and observations-. *Earth Planets Space* 2011, 63: 827–830.
74. Long-term evaluation of earthquakes from sanriku-oki to boso-oki. Headquarters for Earthquake Research Promotion, Tokyo; 2009.
75. Satake K, Fujii Y: Review: source models of the 2011 Tohoku earthquake and long-term forecast of large earthquakes. *J Disaster Res* 2014, 9(3):272–280.
76. Minoura K, Nakaya S: Traces of tsunami preserved in inter-tidal lacustrine and marsh deposits - some examples from northeast Japan. *J Geology* 1991, 99(2):265–287.
77. Minoura K, Imamura F, Sugawara D, Kono Y, Iwashita T: The 869 jogan tsunami deposit and recurrence interval of large-scale tsunami on the pacific coast of norheast Japan. *J Natural Disaster Sci* 2001, 23: 83–88.
78. Goto K, Fujima K, Sugawara D, Fujino S, Imai K, Tsudaka R, Abe T, Haraguchi T: Field measurements and numerical modeling for the run-up heights and inundation distances of the 2011 tohoku-oki tsunami at Sendai plain. *Earth Planets Space*2012, 64: 1247–1257.
79. Nakajima H, Koarai M: Assessment of tsunami flood situation from the great east Japan earthquake. *Bull Geospatial Info Authority Japan* 2011, 59: 55–66.
80. Stohl A, Seibert P, Wotawa G, Arnold D, Burkhart JF, Eckhardt S, Tapia C, Vargas A, Yasunari TJ: Xenon-133 and caesium-137 releases into the atmosphere from the Fukushima Dai-Ichi nuclear power plant: determination of the source term, atmospheric dispersion, and deposition. *Atmos Chem Phys* 2012, 12: 2313–2343.
81. Miyazawa Y, Masumoto Y, Varlamov SM, Miyama T, Takigawa M, Honda M, Saino T: Inverse estimation of source parameters of oceanic radioactivity dispersion models associated with the Fukushima accident. *Biogeosciences* 2013, 10(4):2349–2363.
82. Kato H, Onda Y, Gomi T: Interception of the Fukushima reactor accident-derived137Cs, 134Cs and 131I by coniferous forest canopies. *Geophys Res Lett* 2012, 39(20):L20403.
83. Allen SCR, Greenslade DJM: Developing tsunami warnings from numerical model output. *Nat Hazards* 2008, 46(1):35–52.
84. Kumar TS, Nayak S, Kumar P, Yadav RBS, Kumar A, Sunanda MV, Devi EU, Shenoi SSC: Performance of the tsunami forecast system for the Indian ocean. *Curr Sci* 2012, 102(1):110–114.
85. Munch U, Rudloff A, Lauterjung J: Postface "the GITEWS project - results, summary and outlook". *Nat Hazards Earth Syst Sci*2011, 11(3):765–769.
86. Geist E, Parsons T: Probabilistic analysis of tsunami hazards. *Nat Hazards* 2006, 37: 277–314.
87. González FI, Geist EL, Jaffe B, Kânoğlu U, Mofjeld H, Synolakis CE, Titov VV, Arcas D, Bellomo D, Carlton D, Horning T, Johnson J, Newman J, Parsons T, Peters R, Peterson C, Priest G, Venturato A, Weber J, Wong F, Yalciner A: Probabilistic tsunami hazard assessment at seaside, Oregon, for near- and far-field seismic sources. *J Geophys Res Oceans* 2009, 114(C11):C11023.
88. Annaka T, Satake K, Sakakiyama T, Yanagisawa K, Shuto N: Logic-tree approach for probabilistic tsunami hazard analysis and its applications to the Japanese coasts. *Pure Appl Geophys* 2007, 164(2–3):577–592.
89. Sakai T, Takeda T, Soraoka H, Yanagisawa K, Annaka T: Development of a probabilistic tsunami hazard analysis in Japan. In *Proceedings of ICONE 14,*

international conference on nuclear engineering. ASME (American Society of Mechanical Engineers), Miami, Florida, USA; 2006:1–7.

90. Kodama S (2013) Tsunami-tendenko and morality in disasters. J Medical Ethics

91. Yamori K: Revisiting the concept of tsunami tendenko: tsunami evacuation behavior in the great east Japan earthquake. *J Disaster Res* 2013, 8: 115–116.

CHAPTER 9

Tsunami Waves Extensively Resurfaced the Shorelines of an Early Martian Ocean

J.Alexis P. Rodriguez[1,2], AlbertoG. Fairén[3,4], Kenneth L.Tanaka[5], Mario Zarroca[6], Rogelio Linares[6], Thomas Platz[1], Goro Komatsu, Hideaki Miyamoto[9], Jeffrey S. Kargel , JianguoYan, VirginiaGulick[2], Kana Higuchi[3], Victor R. Baker & NatalieGlines[2]

[1]Planetary Science Institute, 1700 East Fort Lowell Road, Suite 106, Tucson, AZ 85719-2395, USA.

[2]NASA Ames Research Center, Mail Stop 239-20, Moffett Field, CA, 94035, USA.

[3] Department of Planetology and Habitability, Centro de Astrobiología (CSIC-INTA), Madrid 28850, Spain.

[4] Department of Astronomy, Cornell University, Ithaca, NY 14850, USA.

[5]Astrogeology Science Center, U.S. Geological Survey, Flagstaff, AZ 86001, USA.

[6]External Geodynamics and Hydrogeology Group, Department of Geology, Autonomous University of Barcelona, 08193 Bellaterra, Barcelona, Spain.

ABSTRACT

It has been proposed that ~3.4 billion years ago an ocean fed by enormous catastrophic floods covered most of the Martian northern lowlands. However, a persistent problem with this hypothesis is the lack of definitive paleoshoreline features. Here, based on geomorphic and thermal image mapping in the circum-Chryse and northwestern Arabia Terra regions of the northern plains, in combination with numerical analyses, we show evidence for two enormous tsunami events possibly triggered by bolide impacts, resulting in craters ~30 km in diameter and occurring perhaps a few million years apart. The tsunamis produced widespread littoral landforms, including run-up water-ice-rich and bouldery lobes, which extended tens to hundreds of kilometers over gently sloping plains and boundary cratered highlands, as well as backwash channels where wave retreat occurred on highland-boundary surfaces. The ice-rich lobes formed in association with the younger tsunami, showing that their emplacement took place following a transition into a colder global climatic regime that occurred after the older tsunami event. We conclude that, on early Mars, tsunamis played a major role in generating and resurfacing coastal terrains.

INTRODUCTION

The existence of an early Mars northern ocean[1,2,3,4,5,6,7] remains a fundamental mystery[8,9]. During the Hesperian Period (~3.71 to 3.37 Ga; ages herein based on Neukum chronology as given in Michael)[10], Mars' ancient hydrosphere was apparently cold-trapped within vast systems of subsurface aquifers underneath a thick, ice-rich permafrost zone[7]. Groundwater outbursts at the end of the Hesperian may have generated catastrophic floods that produced an ocean in the northern lowlands, as evidenced by a deposit that covers most of this region and generally exhibits a roughly topographically equipotential margin[1,2,3,4,5,6,7,11,12]. Radar-sounding data are consistent with the deposit being comprised of mostly water-ice[13]. This deposit is identified as the Late Hesperian lowland unit (*lHl*) on the latest geologic map of Mars[14]. However, until now, the lack of wave-cut paleoshoreline features[9] and the presence of lobate margins[8,12] appeared to be inconsistent with the Late Hesperian paleo-ocean hypothesis. Our new geologic mapping in Chryse Planitia and northwestern Arabia Terra regions reveals previously undistinguished, older and younger members of the unit (lHl_1 and lHl_2, respectively, Fig. 1A). Both members are bounded by south-facing lobes that are typically tens of kilometers in length and width; however, in Chryse Planitia these dimensions reach a few hundred kilometers in scale (Fig. 1B, Fig. S1). The lobes reach upland boundary surfaces distributed between approximately −4087 m and −3191 m of elevation (Fig. S2). These deposits embay dozens of streamlined promontories scattered over a surface area of ~570,000 km^2 (Fig. 1B,C).

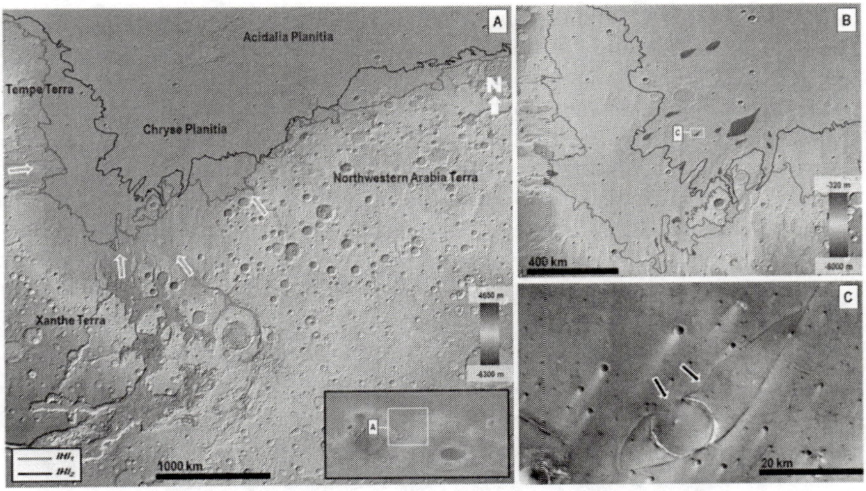

Figure 1.(**A**) View of the circum-Chryse highland-lowland boundary region, which is made up of the Chryse and Acidalia Planitiae lowlands and the Tempe, Xanthe, and Arabia Terrae highlands (inset shows region within the planetwide context). The boundary is breached by the planet's largest outflow channels (blue arrows). The red and black lines trace the margins of Late Hesperian lowland members lHl_1 and lHl_2, respectively. **(B)** View of Chryse and Acidalia Planitiae showing the distribution of streamlined promontories (green) buried by these deposits. The base images for panels **(A,B)** are color-coded shaded-relief MOLA digital elevation models (460 m/pixel). Credit: MOLA Science Team, MSS, JPL, NASA. **(C)** View of a streamlined promontory that is embayed, and partly buried (e.g., black arrows), by member lHl_2 materials. Part of HRSC image H1436_0000_ND3 (12.5 m/pixel) centered at 30.3°N, 35.9°W, (http://hrscview.fu-berlin.de/cgi-bin/ion-p?page=product.ion&mage=1436_0000), which is licensed under Attribution-ShareAlike 3.0 IGO license. The license terms can be found on the following link: https://creativecommons.org/licenses/by-sa/3.0/igo/. Credit: ESA/DLR/FU Berlin. We produced the mosaics and maps in this figure using Esri's ArcGIS®10.3 software (http://www.esri.com/software/arcgis).

In THEMIS night-time infrared images, the upper reaches of the older deposit that were emplaced along Arabia and Tempe Terrae (member lHl_1 (Fig. 1A)) appear thermally bright (i.e., rocky exposures)15 and abruptly transition upland-ward into thermally dark (i.e., fine-grained sediments)15 surfaces (e.g., Figs 2A,B and 3B). Close-up views show that the bright surfaces consist of boulder deposits, with individual boulders typically meters in diameter (Figs 2C and 3E, Fig. S3D). Exhumation of the boulder deposit from beneath ejecta blanket materials along impact crater rims (black arrows in Fig. 2B,C), as well as distinct onlapping contacts (e.g., Fig. S3C), show that the deposit overlies the thermally dark surfaces consisting of finer-grained materials (e.g., Figs 2A,B and 3B). Throughout spatially disconnected locations in the eastern part of northwestern Arabia Terra, the marginal parts of member lHl_1 cover low-slope ramps that are extensively dissected by NNW-trending (Fig. S4A) sets of aligned channels (e.g., Fig. 3A–D). These channels were first identified in Viking data (but only locally along Arabia Terra in association with an older "lowland unit A")1.

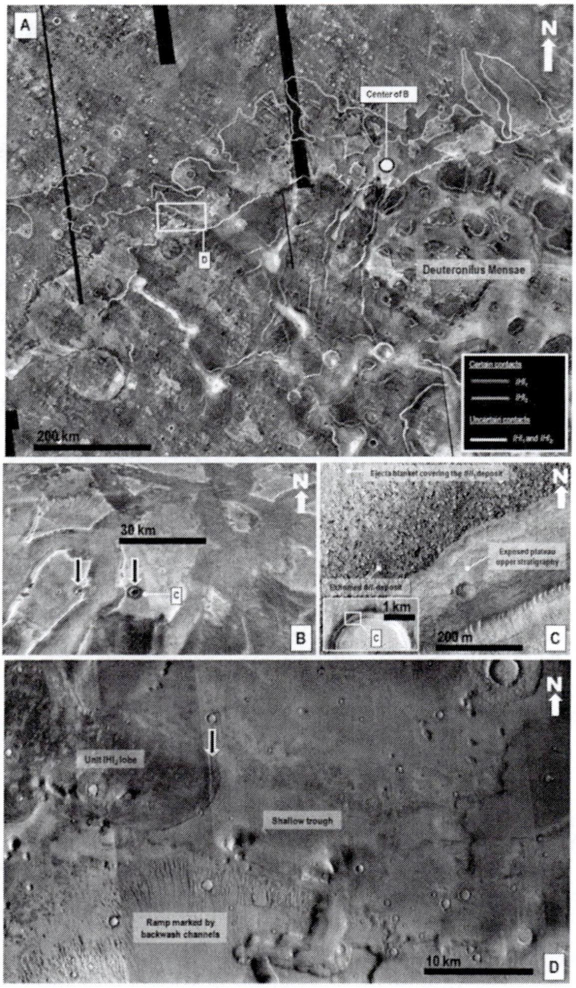

Figure 2. **(A)**View of highland surfaces along the highland-lowland boundary in northwestern Arabia Terra, which are embayed by the upper reaches of members *lHl*$_1$ and *lHl*$_2$. The red (*lHl*$_1$) and orange (*lHl*$_2$) lines show contact sections mapped as certain. The yellow lines are zones where the contacts are obscured due to localized resurfacing or mantling. Out of the total 4124 km length of these mapped contacts, 76% is certain and 23% is uncertain. Part of THEMIS night-time infrared image mosaic (100 m/pixel) centered at 43°26′N, 12°29′E. Credit: Christensen *et al.*32. **(B)** Perspective close-up view on a system of mesas, which are partly covered by the upper reaches of member *lHl*$_1$. These deposits consist of thermally bright materials that pass upland-wards into thermally dark surfaces. The black arrows show locations where the thermally dark materials are exhumed from beneath the thermally bright materials by impact craters. Center location indicated by dot in panel A. Part of THEMIS night-time infrared mosaic (100 m/pixel) centered at 45°45′N, 16°41′E. Credit: Christensen *et al.*32. **(C)** Visible light close-up view on one of the craters identified in panel **(B)**. The impact crater's inner wall shows the boulder deposit (member *lHl*$_1$) overlying an exhumed non-bouldery upper stratigraphy. The boulder deposit is also locally covered by a non-bouldery ejecta blanket. The inset shows the view's location along the crater's northern margin. Part of HiRISE image ESP_017355_2260 centered at 45.82°N 16.47°E (50 cm/pixel). Credit: NASA/JPL/University of Arizona. **(D)** Close-up view of northeast Deuteronilus Mensae showing a low-angle ramp marked by backwash channels (part of member *lHl*$_1$), which is cut by a shallow trough. A lobe of member *lHl*$_2$ can be observed to run up the trough's surface. The black arrow identifies a marginal compressional fold. Location is shown in panel **(A)**. Part of a CTX mosaic centered at 44°49′N, 11°10′E. Credit: NASA/JPL. The license terms can be found on the following link: pds-imaging.jpl.nasa.gov/portal/mro_mission.html. We produced the mosaics and maps in this figure using Esri's ArcGIS® 10.3 software (http://www.esri.com/software/arcgis).

Upslope flows leading to the emplacement of the *lHl* unit are implied by the highland-facing orientation of the deposits' lobes as well as their relief gains, which commonly are a few hundred meters (e.g., Fig. 1A,B, Figs. S5, S6). These characteristics rule out emplacement by gravity-driven downslope moving flows such as debris, flood, glacier and lava flows. Uphill unidirectional winds can generate elongate aeolian deposits known as wind streaks. However, these deposits are largely composed of saltating sand-sized lithic particles that are deposited in scattered patches on the lee sides of topographic obstacles (typically impact craters), exhibit surface bedforms, generally cover hills and mesas situated along their paths, and

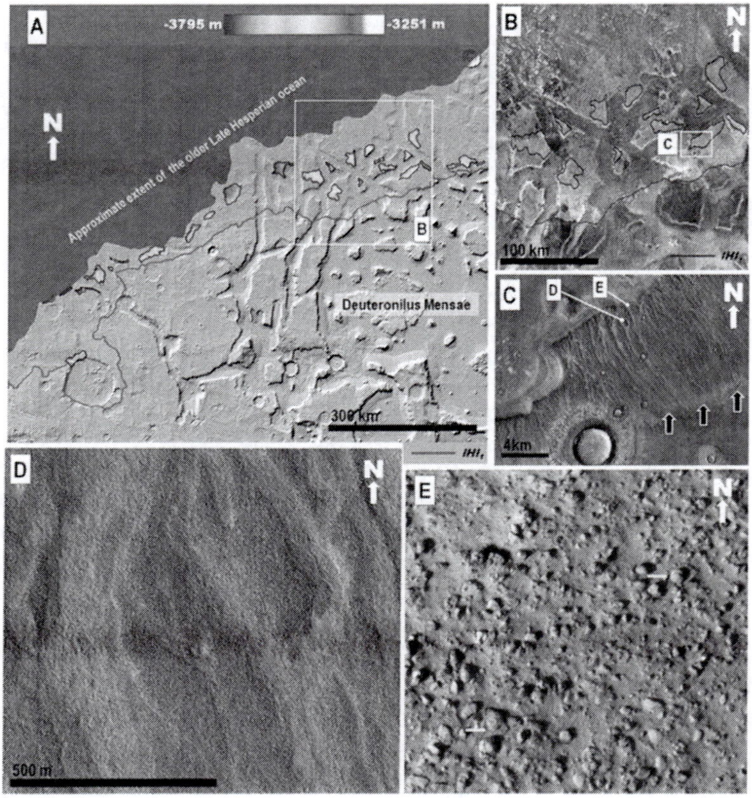

Figure 3.(A) View of Deuteronilus Mensae in northwestern Arabia Terra showing the distribution and elevation ranges of ramp surfaces marked by backwash channels as well as the reconstructed location of the paleoshoreline from which the older tsunami (member *lHl*$_1$) propagated (see Fig. 4 and Fig. S4 for details on paleoshoreline reconstructions). Shaded-relief MOLA digital elevation model (460 m/pixel). Credit: MOLA Science Team, MSS, JPL, NASA. **(B)** Close-up view on panel (**A**) showing the distribution of some of the channel-scoured, north-sloping highland mesas (blue areas). Channels occur on bright (i.e., higher thermal inertia) surfaces, which abruptly transition (red line) in lower part of image to uplands covered by dark (lower thermal inertia) materials. THEMIS night-time infrared image mosaic, context and location in panel **A**. Credit: Christensen *et al.*32. Example of channeled surface (**C**) displaying streamlined bars (**D**) made up of rounded to angular boulders as much as ~10 m in diameter (yellow bars are 10 m in length) (**E**). (**C–E)**, Parts of HiRISE image ESP_028537_2270, 25 cm/pixel. Credit: NASA/JPL/University of Arizona.) We produced the mosaics and maps in this figure using Esri's ArcGIS® 10.3 software (http://www.esri.com/software/arcgis).

mostly have length-to-width ratios >1 (ref. 16). In contrast, the lobes of member lHl_1 include boulders several meters in diameter (Figs 2C and 3E, Fig. S3D), and those of member lHl_2 appear to be mostly composed of water-ice[6,12,13,14]. In addition, the lobes in both members diverge around numerous mesas (e.g., Fig. 1C) as well as broad rises (e.g., Fig. S3), and have length-to-width ratios mostly <1 (Fig. S1) (which is consistent with uphill flow along with substantial lateral spreading). Therefore, we propose that the two unit *lHl* members represent deposits emplaced by highly energetic, sediment-rich tsunami waves that originated from a Late Hesperian paleo-ocean.

In Deuteronilus Mensae, extensive troughs cut the boundary scarps covered by member lHl_1. The troughs are locally intruded by member lHl_2 run-up lobes (e.g., Fig. 2D), indicating that they formed during the time interval separating the two tsunami events. Active resurfacing leading to the formation of these troughs likely lasted a few million years and could have been the result of Late Hesperian glacial erosion[17]. Crater-count statistics show that, while the deposits formed during the Late Hesperian Epoch, their absolute ages could differ as much as several tens of millions of years (see supplementary crater statistics).

The boulder deposits of member lHl_1 drape over, and therefore postdate, the incision of adjoining highland channels (e.g., orange arrow in Fig. S3A), ruling out upland fluvial systems as possible discharge sources. Highly energetic, boulder-rich tsunami fronts on Earth show diversion around topographic obstacles as they propagate onshore[18]. Similarly, member lHl_1 boulder deposits exhibit well-defined landward lobate margins around broad promontories (Fig. S3A–C). Member lHl_1 boulders range from rounded to angular and are as much as ~10 m in diameter (Figs 2C and 3E, Fig. S3D), which are also characteristics of some terrestrial tsunami deposits[18]. Thus, we interpret the member lHl_1 lobes as made up of lowland and boundary clastic materials that were captured and transported by a tsunami wave, then beached farther inland as the wave lost its momentum.

Subsequently, we suggest that rapid gravity-forced backwash of the tsunami wave into the paleo-ocean dissected the channel systems on the marginal parts of member lHl_1 in the eastern part of northwestern Arabia Terra (Fig. 3A–D). These channels have remarkable similarities to terrestrial tsunami backwash channels; including the presence of aligned channel heads[19] (black arrows in Fig. 3C), perpendicular orientations to the reconstructed paleoshoreline[19] (Fig. S4A), streamlined bars composed of reworked boulders[20,21] (Fig. 3D,E), and widths ranging between ~50 and ~200 m (refs 19,22) (Fig. 3C,D). Parker *et al.*[23] observed a few of these parallel channel systems in Arabia Terra using lower-resolution image data, and they also interpreted them as tsunami backwash channels.

The lower terminations of the proposed backwash channels are generally truncated by younger scarps (Figs 2D and 3C, Fig. S4A). However, the identification of a possibly subaqueously emplaced sedimentary lobe adjoining the lower reaches of a set of these channels located at ~−3795 m in elevation (Fig. S4B) provides an approximate upper boundary to the paleoshoreline from which the older tsunami propagated (Fig. 4A). The lowest margins of the mapped lHl_2lobes are at ~−4100 m in elevation (Fig. S2), which we have used as an upper bound to the paleoshoreline elevation from which the younger tsunami propagated (Fig. 4B). The elevation difference between the two paleoshorelines implies a decrease in ocean level of ~300 m, which could have taken place via evaporation/sublimation within several million years6.

Based on these paleo-oceanographic reconstructions, we estimate that the areas inundated by the older and younger tsunamis within the study region were $\sim 8 \times 10^5$ km^2 and $\sim 1 \times 10^6$ km^2, respectively (Fig. S5). Measured typical run-up distances are tens to a few hundred kilometers for both the older and younger tsunamis, and their respective maxima reach ~529 km and ~650 km (Fig. S6). Overall, the morphometric characterizations of both tsunamis are strikingly similar. The slightly larger inundation area that was apparently covered during the younger event is consistent with the tsunami extending from a lower shoreline, and therefore, flowing over relatively smooth, older ocean and tsunami deposits. These run-up distances and inundation areas are enormous by terrestrial standards, which explain why the backwash channels exhibit lengths of ~20 km, while some terrestrial examples of backwash channel lengths produced by much smaller tsunamis range between ~200 and ~300 m in length22.

Our mapping (Fig. 1, Fig. S6) shows comparatively shorter run-up distances along the rougher and steeper cratered topography of the Arabia Terra boundary terrains, indicative of relatively lower wave heights and velocities, as predicted by tsunami numerical simulations24. These simulations also indicate that as the waves overflowed the Arabia Terra cratered boundary, their velocities would have abruptly dropped below the ~1 m/s threshold required to move multi-meter-scale boulders, explaining the occurrence of the boulder deposits in the region (Figs 2C and 3E, Fig. S3D). On the other hand, the more gentle slopes in Chryse Planitia would have resulted in a more gradual decrease in wave velocity, leading to the emplacement of more sorted sedimentary lobes, with their distal-most areas primarily consisting of finer-grained sediments. In addition, prior to their inundation by tsunami waves, the highland boundary surfaces were likely covered by extensive boulder fields, which would have been captured and redistributed by the waves, which is also another important factor accounting for the regional prevalence of boulder-rich lobes. In Chryse Planitia the tsunamis would have mostly propagated over gently-sloping

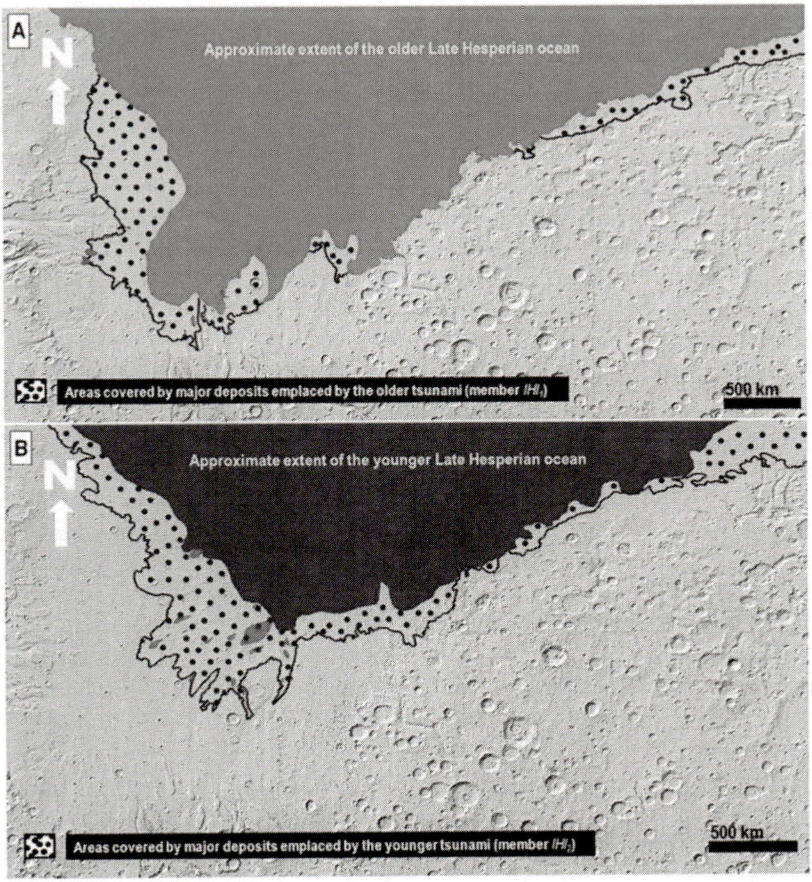

Figure 4. **(A)** Late Hesperian ocean with a paleoshoreline elevation close to −3795 m (light blue), and estimated extent of the older tsunami (dotted surface). **(B)** Late Hesperian ocean with a paleoshoreline elevation close to −4100 m (dark blue), and estimated extent of the younger tsunami (dotted surface). The ocean margins were reconstructed by tracing the estimated paleoshoreline elevations, while carefully interpolating across topography generated by clearly younger resurfacing processes. The tsunami margins were produced using the *lHl*unit's lobes extending from the projected paleoshoreline elevations. Areas marked in red represent the locations of streamlined promontories in Chryse Planitia, which were embayed, and/or buried, by materials emplaced during each of the tsunami events. Topography in **(A,B)** from shaded-relief MOLA digital elevation model (460 m/pixel). Credit: MOLA Science Team, MSS, JPL, NASA. We produced the mosaics and maps in this figure using Esri's ArcGIS® 10.3 software (http://www.esri.com/software/arcgis).

plains that were largely made up of less bouldery outflow channel sedimentary deposits[14].

The simulations also show that bolide impacts causing craters ~30 km in diameter would have generated tsunami waves with typical onshore heights of ~50 m and local variations from ~10 m to as much as ~120 m (ref. 24). Using run-up distances measured in 71 topographic profiles (Fig. S6), we have calculated the tsunami wave heights and find that they reasonably match the simulations' predicted ranges[24] (see supplementary calculations). In addition, whereas the simulations do not describe the hydrodynamic behavior of the backwash stage, the formation of several marine impact craters on Earth has also resulted in documented tsunami backwash channels[25].

Using the surface area of the paleo-ocean's region included in the numerical simulation by Iijima *et al.*[24] (i.e., ~4.5×10^6 km^2) and the crater production function of Ivanov[26], we find that ~23 marine impact craters ≥30 km in diameter would have formed within this part of Mars during the Late Hesperian Epoch (3.61–3.37 Ga)[10,27]. Of these, 7 fit in the diameter range of 30–35 km, which was used in the tsunami simulations[24]. The prediction is that, within the particular region of the ocean analyzed here, on average about 2 impact craters ~30 km in diameter formed every 30 million years during this time period. Therefore, within statistical constraints for the deposits' surface ages and for crater production rates, impacts can account for generation of both *lHl* members as tsunami deposits (see supplementary crater statistics).

Briny aqueous chemistry models show that the ocean could have remained in liquid form over millions of years, and consequently mostly free of an ice cover even during cryogenic climatic conditions[28]. Another geologic scenario invokes the formation of an ice-covered ocean soon after the ocean's emplacement[6]. However, no numerical simulations have been performed to detail the behavior of impact-related tsunamis[24] on these types of Martian marine environments.

High rates of marine, and subsequent periglacial[6,12,14] resurfacing, likely reduced the topography of the tsunami-generating crater structures. Such resurfacing can also explain the lack of well-preserved impact craters predating the Amazonian Period in the northern lowlands[12]. The frequency rate of ~30 km in diameter impact craters for the entire ocean's surface area (~24×10^6 km^2, as determined by Head *et al.*[3]) is one every 2.7 million years during the Late Hesperian. Although we have only identified evidence for two tsunami events in our study area, other regions in the northern plains likely experienced similar tsunami-related coastal resurfacing, perhaps associated with other impacts, huge landslides, or large marsquakes. Older but less extensive tsunami deposits may have been completely resurfaced by more recent events with similar run-up distances. Thus, the mapped tsunami margins comprise only the largest magnitude tsunami events located at the highest elevations.

Many of the *lHl_1* lobes are mostly made up of lithic deposits and exhibit backwash modifications. In contrast, the landward-facing lobate termini of unit *lHl_2* lack evidence indicative of a backwash phase subsequent to their emplacement. Like on Earth, the absence of backwash features associated with these flows could have been the result of the waves transitioning into sub-aerial sediment-laden slurry flows extending over low gradient surfaces[29,30] (supplementary calculations), which can also explain the presence of possible contractional folds along the margins of some of the member's lobes (e.g., black arrow in Fig. 2D). However, the *lHl_2*lobes appear to be mostly composed of water-ice[6,12,13,14], suggesting that the transition into slurry likely involved the formation and incorporation of a significant proportion of ice particles. In May 2013, the Saskatchewan Water Security Agency filmed an ice surge in the Codette Reservoir near Nipawin, Saskatchewan, Canada. The surge comprises a spectacular terrestrial analog of rarely observed catastrophic ice-rich flows leading to the emplacement of enormous lobate fronts, which are remarkably similar to those of member *lHl_2*(video link included in ref. 31).We propose that these morphologic differences might be linked to colder environmental conditions following the first tsunami event.

Our mapping of two unit *lHl* members as tsunami lobes is consistent with the occurrence of two paleoshoreline levels of a receding Martian northern plains ocean during the Late Hesperian (Fig. 4, Fig. S5). However, resurfacing by the tsunami waves has obscured the paleoshorelines, thus making rigorous testing of their equipotentiality impossible.

MAPPING METHODOLOGY

Mapping in this investigation was performed using Esri's ArcGIS® 10.3 software (http://www.esri.com/software/arcgis). Embayment and overlapping relationships leading to the recognition of the outer margins of members *lHl_1* and *lHl_2* involved an integrated analysis of (1) thermal infrared image data (i.e., Mars Odyssey Thermal Emission Imaging System (THEMIS) night-time and day-time infrared image mosaics (100 m per pixel)), (2) visible image data (i.e., Mars Reconnaissance Orbiter Context Camera (CTX, (5.15–5.91 m/pixel)) images, and (3) Mars Global Surveyor Mars Orbital Laser Altimeter (MOLA, ~460 m/pixel horizontal and ~1 m vertical resolution) digital elevation models. In some areas, contacts are buried underneath ejecta blanket materials or are locally resurfaced; we mapped these sections as uncertain contacts (Fig. 2A).

ADDITIONAL INFORMATION

How to cite this article: Rodriguez, J. A. P. *et al.* Tsunami waves extensively resurfaced the shorelines of an early Martian ocean. *Sci. Rep.* **6**, 25106;

ACKNOWLEDGEMENTS

Funding for JAPR was provided by NASA's Planetary Geologic and Geophysics Program, NASA NPP and KAKENHI 25120006. KLT was also funded by NASA's Planetary Geologic and Geophysics Program. AGF was supported by the Project "icyMARS", funded by the European Research Council, Starting Grant No. 307496. TP was supported by a DFG Grant (PL613/2-1). VCG was funded by MRO HiRISE Co-Investigator funds. HM was funded by KAKENHI 25120006. Publications costs were covered by the Project "icyMARS", funded by the European Research Council, Starting Grant No. 307496. We are thankful to Alexander Cox for his valuable editing.

REFERENCES

1. Parker, T. J., Gorsline, D. S., Saunders, R. S., Pieri, D. C. & Schneeberger, D. M. Coastal geomorphology of the Martian northern plains. J. Geophys. Res. 98, 11061–11078 (1993).
2. Fairén, A. G. et al. Episodic flood inundations of the northern plains of Mars. Icarus 165, 53–67 (2003).
3. Head, J. W. et al. Possible ancient oceans on Mars: Evidence from Mars Orbiter Laser Altimeter data. Science 286, 2134–2137 (1999).
4. Parker, T. J., Saunders, R. S. & Schneeberger, D. M. Transitional morphology in west Deuteronilus Mensae, Mars: Implications for modification of the lowland/upland boundary. Icarus 82, 111–145 (1989).
5. Carr, M. H. & Head, J. W. Oceans on Mars: An assessment of the observational evidence and possible fate. J. Geophys. Res. 108, 5042, 10.1029/2002JE001963 (2003).
6. Kreslavsky, M. A. & Head, J. W. Fate of outflow channel effluent in the northern lowlands of Mars: The Vastitas Borealis Formation as a sublimation residue from frozen ponded bodies of water. J. Geophys. Res. 107, 5121, 10.1029/2001JE001831 (2002).
7. Clifford, S. M. & Parker, T. J. The evolution of the martian hydrosphere: Implications for the fate of a primordial ocean and the current state of the northern plains. Icarus 154, 40–79 (2001).
8. Tanaka, K. L. Sedimentary history and mass flow structures of Chryse and Acidalia Planitiae, Mars. J. Geophys. Res. 102, 4131–4149 (1997).
9. Malin, M. C. & Edgett, K. S. Oceans or seas in the Martian northern lowlands: High resolution imaging tests of proposed coastlines. Geophys. Res. Lett. 26, 3049–3052 (1999).

10. Michael, G. G. Planetary surface dating from crater size-frequency distribution measurements: Multiple resurfacing episodes and differential isochron fitting. Icarus 226, 885–890 (2013).
11. Baker, V. R. Water and the martian landscape. Nature 412, 228–236 (2001).
12. Tanaka, K. L., Skinner, J. A. & Hare, T. M. Geologic map of the northern plains of Mars, (2005) (Date of access: (24/11/2015). U.S. Geological Survey Scientific Investigations Map 2888, scale 1:15,000,000 (1 mm = 15 km) at 90°N and 1:7,500,000 at 0°N, http://pubs.usgs.gov/sim/2005/2888/.
13. Mouginot, J., Pommerol, A., Beck, P., Kofman, W. & Clifford, S. M. Dielectric map of the Martian northern hemisphere and the nature of plain filling materials. Geophys. Res. Lett. 39, L02202, 10.1029/2011GL050286 (2012).
14. Tanaka, K. L. et al. Geologic map of Mars, (2014) (Date of access: (24/11/2015). U.S. Geological Survey Scientific Investigations Map 3292, scale 1:20,000,000, http://pubs.usgs.gov/sim/3292/.
15. Christensen, P. R. et al. Morphology and composition of the surface of Mars: Mars Odyssey THEMIS results. Science 300, 2056–2061 (2003).
16. Rodriguez, J. A. P. et al. The sedimentology and dynamics of crater-affiliated wind streaks in western Arabia Terra, Mars and Patagonia, Argentina. Geomorphology 121, 30–54 (2010).
17. Davila, A. F. et al. Evidence for Hesperian glaciation along the Martian dichotomy boundary. Geology 41, 755–758 (2013).
18. Goto, K., Miyagi, K., Kawamata, H. & Imamura, F. Discrimination of boulders deposited by tsunamis and storm waves at Ishigaki Island, Japan. Marine Geology 269, 34–45 (2010).
19. Fagherazzi, S. & Du, X. Tsunamigenic incisions produced by the December 2004 earthquake along the coasts of Thailand, Indonesia and Sri Lanka. Geomorphology 99, 120–129 (2008).
20. Eaton, J. P., Richter, D. H. & Ault, W. U. The tsunami of May 23, 1960, on the Island of Hawaii. Seis. Soc. Am. Bull. 51, 135–157 (1961).
21. Sugawara, D., Minoura, K. & Imamura, F. In Tsunamiites – Features and Implications (eds Shiki, T., Y. Tsuji, T. Yamazaki & K. Minoura) Ch. 3, 9–49 (Elsevier, 2008).
22. Goto, K., Sugawara, D., Abe, T., Haraguchi, T. & Fujino, S. Liquefaction as an important source of the A.D. 2011 Tohoku-Oki tsunami deposits at Sendai Plain, Japan. Geology 40, 887–890 (2012).
23. Parker, T. J. East Acidalia shoreline morphology at MRO and CTX image scales. Lunar and Planetary Science Conference, 2551 (2009).
24. Iijima, Y., Goto, K., Minoura, K., Komatsu, G. & Imamura, F. Hydrodynamics of impact-induced tsunami over the Martian ocean. Planet. Space Sci. 95, 33–44 (2014).
25. Schulte, P. et al. Tsunami backwash deposits with Chicxulub impact ejecta and dinosaur remains from the Cretaceous–Palaeogene boundary in the La Popa Basin, Mexico, Sedimentology 59, 3, 737–765 (2012).
26. Ivanov, B. A. Mars/Moon cratering rate ratio estimates. Space Sci. Rev. 96, 87–104 (2001).
27. Werner, S. C. & Tanaka, K. L. Redefinition of the crater-density and absolute-age boundaries for the chronostratigraphic system of Mars. Icarus 215, 603–607 (2011).
28. Fairén, A. G. A cold and wet Mars. Icarus 208, 165–175 (2010).
29. Goto, K., Hashimoto, K. D. S., Yanagisawa, H. & Abe, T. Spatial thickness variability of the 2011 Tohoku-oki tsunami deposits along the coastline of Sendai Bay. Mar. Geol. 358, 38–48 (2014).
30. Paris, R. et al. Tsunamis as geomorphic crises: Lessons from the December 26, 2004 tsunami in Lhok Nga, West Banda Aceh (Sumatra, Indonesia). Geomorphology 104, 59–72 (2009).
31. Ice Surge - Saskatchewan Water Security Agency, (2013) (Date of access: (24/03/2016)). YouTube, https://www.youtube.com/watch?v=OgMBQFf64JM.

32. Christensen, P. R. et al. THEMIS Public Data Releases, Image Explorer, (2006), (Date of access: 11/24/2015), Planetary Data System node, Arizona State University, http://themis-data.asu.edu.

CHAPTER 10

Time Evolution of Man-Made Harbor Modifications in San Diego: Effects on Tsunamis

Aggeliki Barberopoulou [1,2,*], Mark R. Legg [3,†] and Edison Gica [4,5,†]

[1]*National Observatory of Athens, Institute of Geodynamics, Lofos Nymphon, 11810 Athens, Greece*

[2]*AIR Worldwide, 131 Dartmouth Street, Boston, MA 02116, USA*

[3]*Legg Geophysical, 16541 Gothard St # 107, Huntington Beach, CA 92647, USA*

[4]*Pacific Marine Environmental Laboratory, Center for Tsunami Research, National Oceanic and Atmospheric Administration, 7600 Sand Point Way NE, Seattle, WA 98115, USA*

[5]*Joint Institute for the Study of the Atmosphere and Ocean, University of Washington, 3737 Brooklyn Ave NE, Box 355672, Seattle, WA 98195-5672, USA*

ABSTRACT

San Diego, one of the largest ports on the U.S. West Coast and home to the largest U.S. Navy base, is exposed to various local and distant tsunami sources. During the first half of the twentieth century, extensive modifications to the port included but were not limited to dredging, expansion of land near the airport and previous tidal flats, as well as creation of jetties. Using historical nautical charts and available Digital Elevation Models, this study gives an overview of changes to San Diego harbor in the last 150+ years due to human intervention and examines the effects of these changes on tsunamis. Two distant and two local scenarios were selected to demonstrate the impact of modified nearshore topography and bathymetry to incoming tsunamis. Inundation pattern, flow depths, and flooded localities vary greatly from year to year in the four scenarios. Specifically, flooded areas shift from the inner harbor to outer locations. Currents induced by the distant tsunamis intensify with modifications and shift from locations primarily outside the harbor to locations inside. A new characteristic in tsunami dynamics associated with port modifications is the introduction of high current spots. Numerical results also show that the introduction of high currents could threaten navigation, vessels, and facilities at narrow openings and also along the harbor "throat"—therefore, at an increased number of locations. Modifications in the port show that changes could have a negative but also a positive impact through constraint of flooding outside of the harbor and shifting of high currents to locations of minimal impact. The results of this study may be used as a first step toward future harbor design plans to reduce tsunami damages.

Keywords: tsunami; San Diego; port modifications

1. INTRODUCTION

Harbor modifications and redevelopment plans (see current plans for the Port of San Diego in [1]) are quite common to enhance or increase space in ports and, thus, to improve operations. In fact, harbor modifications at San Diego Bay decreased the water area by filling shoals around the main channels to add more land, including the area where Lindbergh Field—now known as San Diego Airport—was built. Space was developed for mooring boats, including private yachts and Navy vessels, but the reduction of the water surface area increased tsunami amplitudes and currents by constricting channels between Bay margins. Deepening of channels by dredging to improve navigability also enables greater penetration of tsunami energy into the harbor. Similar problems occurred along major rivers, e.g., the Mississippi River, when construction of levees restricted the channel width, negating the purpose of natural flood plains to distribute and absorb the excess water volume during major floods (e.g., severe flooding in the Midwest and South in 1973 [2]).

Ports are designed to be protected against vertical water fluctuations, but the subject of tsunami currents—a newly recognized problem—is not typically considered in the design of harbor modifications. Tsunami currents form a common threat to ports, as seen in previous tsunamis (22 May 1960, Chile; 27 March 1964, Alaska; 27 February 2010, Chile; 11

March 2011, Japan; e.g., [3,4]); usually, they are not accompanied by significant (>1 m) tsunami amplitudes. Tsunami amplitudes are related to the tsunamigenic potential of a source, but tsunami duration and tsunami currents—even if they relate to relatively small amplitude waves—are also important. The long duration of major trans-Pacific tsunamis (from Mw8+ subduction events) creates a persistent hazard that may produce the greatest damage many hours after the initial tsunami arrival.

In order to understand how anthropogenic changes to ports affect tsunamis, we studied the time evolution of man-made modifications in San Diego Bay since the late nineteenth century and its influence on tsunami amplitudes and currents. San Diego Bay is one of the largest harbors on the U.S. West Coast, and it has been significantly transformed in the twentieth century.

Most of the changes in San Diego Bay have included dredging, enlargement of the North Island/Coronado land area, and creation of new marinas by enhancing already existing sand bars and shoals with fill and creating breakwaters. As done along major rivers, dredging increases the channel depth, but addition of levees, breakwaters, and other shore protection structures combined with narrowing of the channel results in the increase of flow depths and current speeds during inundation (flooding or surge). Tsunami currents can cause damage to structures (e.g., by erosion or scouring; [5]).

We examined the effects of harbor modifications on tsunami hazard by simulating major tsunami events from both local and distant sources. The changes primarily occurred during the first half of the twentieth century. Four specific years were selected as representative of major changes in San Diego Bay: 1892, 1938, 1945 and present day. Post-1945, the bay continued to undergo changes but maintained a similar appearance and, therefore, the bathymetry/topography in and around the bay is represented by a Digital Elevation Model (DEM) showing the harbor shape as we know it today (dated 2014). To reflect these changes, historical nautical charts were used to show the state of the bay during that period.

2. SAN DIEGO BAY

San Diego Harbor is located on the southwestern corner of the United States. It has a long crescent-like shape of approximately 24 km in length and 1–3 km width. The harbor is the fourth largest in California and is home to a U.S. Navy base offering shelter to several aircraft carriers and other naval vessels1. Naval Air Station North Island (NASNI; [6]) is located at the north end of the Coronado peninsula in San Diego Bay. It is the home port of several aircraft carriers of the U.S. Navy, and is part of the largest aerospace-industrial complex in the U.S. Navy, Naval Base Coronado, California ([6]). A cruise ship terminal is located near downtown San Diego, a regular ferry service runs from Coronado to San Diego, and several marinas, for both private and commercial boats, exist throughout the harbor. Although it is naturally well protected from large wave amplitudes, the same does not apply for tsunami currents.

3. TIME EVOLUTION OF SAN DIEGO BAY

San Diego Bay is one of the best natural harbors in the world, protected by the Point Loma Peninsula at the western harbor entrance, and by North Island, Coronado, and the Silver Strand enclosing the inner harbor. Coronado is an affluent resort town located in San Diego County, California, just over 8 km (5.2 mi) SSW of downtown San Diego, with a population of 24,697 ([7]). Coronado Island lies on a peninsula connected to the mainland by a 16-km (10 mi) isthmus called the Silver Strand (locally, The Strand; [8]). North Island was an uninhabited sand flat in the nineteenth century (Figure 1A). It was also referred to as North Coronado Island, because it was separated from South Coronado (now the city of Coronado) by a shallow bay, known as the Spanish Bight, which was later filled in during World War II (1945; see Figure 1and Figure 2). South Coronado, which is not an island but the terminus of a peninsula known as the Silver Strand, became the town of Coronado.

We present nautical charts that portray the changes that have occurred in San Diego Bay since the nineteenth century. Because modifications to the bay happened quite frequently, especially during the first half of the twentieth century, we summarized major changes over selected years. A small set of original nautical charts are presented next. The major changes primarily included modifications to Coronado Island (also referred to as The Island, The North Island, and Coronado, or North Beach Island and Coronado separated by Spanish Bight2 in older charts) and the area around the Dutch Flat (see Figure 1). We use Coronado Island to refer to the entire island of Coronado and North Island after the Spanish Bight was filled in at the post-1945 configuration, and North Island to the pre-1945 configuration to distinguish the northern part of Coronado island separated by Spanish Bight in older charts (see Figure 1).

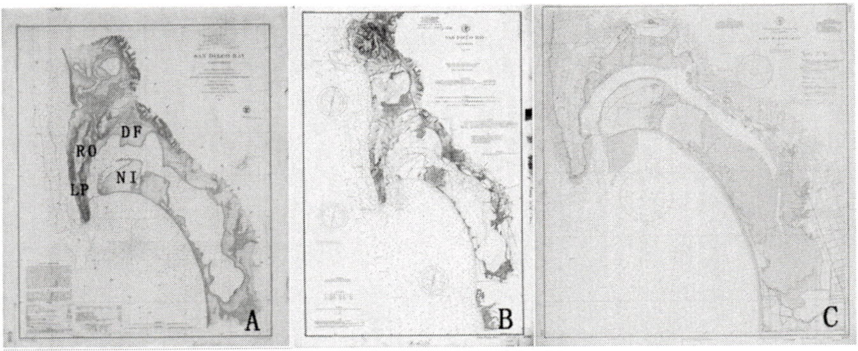

Figure 1. Maps of San Diego Bay. Depths are in feet. (**A**) Nautical chart of 1857. Approximate locations of Dutch Flat (DF), North Island (NI), Roseville (RO), and La Playa (LP) are shown in bold capitals; (**B**) Nautical chart of 1892. The major difference in this chart compared to (A) is the increase of depth at the entrance of the bay, east and northeast of Point Loma (between Roseville and LaPlaya) and between Ballast Point and North Island as a result of dredging. The change appears at least five-fold when compared to the earlier map of 1857; (**C**) Nautical chart of 1938. The major modifications include filling west and north of the North Island (compared to maps A and B) to unify with existing sand dunes and sand spits into North Island and Coronado, with a more regular shape and increased land space. The maps were downloaded from the historical map and chart collection of the Office of Coast Survey, NOAA [9].

3.1. Pre-Nineteenth Century to 1895

During this period, as shown in Figure 1B, the bay remained largely unmodified, with some dredging performed in the late nineteenth century. The depths inside the bay were fairly shallow at the time (characteristic depths 15–54 ft or 4.5–16.5 m near the entrance; Figure 1; [9]). The major characteristics of the bay were the irregularly shaped North Island3 and the width of the bay around the Dutch Flat (North East of the Bay entrance; see Figure 1) to the northeast. The first tide gauge for San Diego was installed in 1854 at La Playa (near Roseville). It was removed in 1872 and in 1906, it was relocated to the south side of Navy Pier in downtown San Diego. The Dutch Flat area was part of the fan delta complex (depositional area) of the San Diego River, which was diverted to the west through Mission Bay (originally called False Bay) via levee and jetty construction after major floods in the mid-nineteenth century (*ca.* 1862; [12,13]).

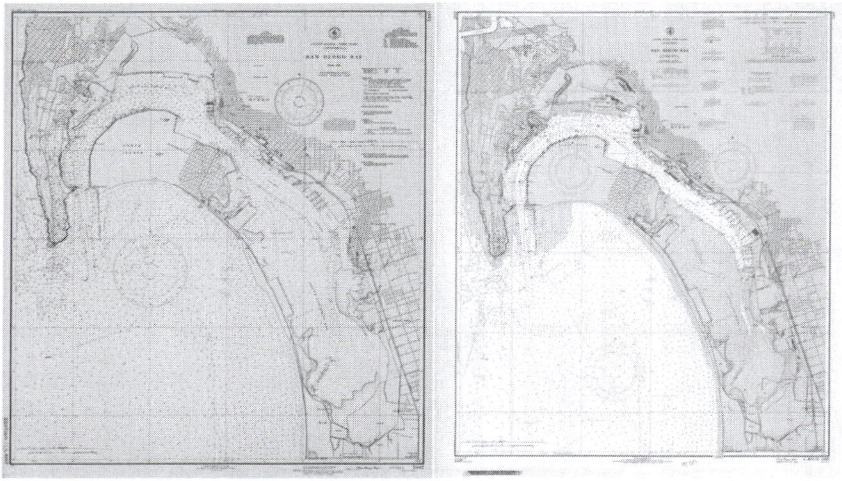

Figure 2. (**Left**) Nautical chart of San Diego Bay of 1945; (**Right**) Nautical chart of San Diego Bay of 1966. The maps were downloaded from the historical map and chart collection of the Office of Coast Survey, National Oceanic and Atmospheric Administration (NOAA; [9]).

3.2. Post-1895 and Early Twentieth Century

Most of the modifications for the enhancement of space in the bay happened during this period and prior to the 1950s. The modifications included dredging at and around the entrance to the bay and expansion around the Dutch Flat, probably to enable better use of the Naval Training Station and the U.S. Marine Corps base near the last location[4]. In addition to the previously noted modifications, sand spits and sand dunes close to North Island were connected through filling, for the creation of the more regularly shaped North Island[5] in the 1930s (Figure 1C).

3.3 Circa 1945 (1945 to 1960s)

By 1945, San Diego Bay had changed considerably, looking very close to how we know it today. The major changes to the harbor configuration resulted in narrow channels between islands and mainland shorelines, specifically from the harbor entrance to New San Diego (downtown San Diego). The area between San Diego and Coronado has always been narrow and the bathymetry fairly shallow, although it has increased over the years from 2.5 m (8 ft) in 1857 to about 13 m (40 ft) today; North Island and Coronado Island may also be uplifts flanking the Spanish Bight and Coronado fault segments of the Rose Canyon fault zone [10]. In fact, uplift exists along the west side of the north-trending Coronado Fault, which passes from the east side of Coronado Island to the mainland shore along the Pacific Highway (note how straight this section of coast was in the nineteenth century maps; Figure 1; west of New San Diego).

The Spanish Bight was a narrow channel that separated Coronado from North Island up to the early 1940s (Figure 1). The development of North Island by the U.S. Navy prior to and during World War II led to the filling of the Bight in 1943, combining the land areas into a single body. The Navy still operates NASNI on Coronado. On the southern side of the town is Naval Amphibious Base Coronado, a training center for Navy SEALs. Both facilities are part of the larger Naval Base Coronado complex [14].

Sand bars (shoals) to the north of the bay were further enhanced to create new marina locations (between Quarantine Station, now known as La Playa6, and Roseville) in North San Diego Bay (Figure 1 and Figure 2). By 1966, Shelter Island and Harbor Island were created by dredging and filling the area east of Roseville and Dutch Flat, providing more space for marinas (Figure 3). These islands are actually connected to the mainland by causeways on landfill. The depths in the bay are more uniform and larger than before. The area around Dutch Flat, Fisherman's Point and the municipal airport were filled, with the exception of a channel east of Roseville (Figure 2A). This channel provided boat access to the Marine Corps base. Small, elongated boat basins now exist between Shelter Island, Harbor Island and the mainland.

Figure 3. View of Shelter Island in 1937, the 1950s and the 1990s, left to right. Photos courtesy of Unified Port of San Diego [1]. An example of extreme transformation showing how the sand spit, or "mud island" as it was referred to, now offers home to numerous vessels. View is southwest to northeast. The entrance to Shelter Island has also been subject to high currents in past tsunamis (e.g., 22 May 1960, Chile and 11 March 2011, Japan; [4]).

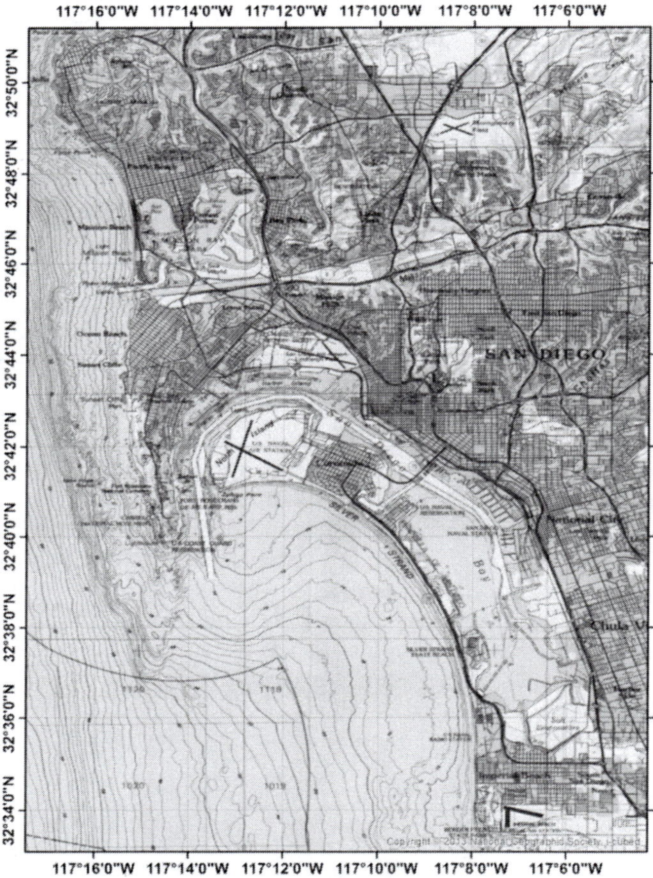

Figure 4. Map of San Diego Bay, showing the state of the harbor today with locations that are referenced frequently throughout the text.

4. NUMERICAL GRIDS OF SAN DIEGO

Four different DEMs for San Diego were used to depict the evolution of the port due to man-made modifications. Three were digitized from historical charts of the historical map and chart collection of the Office of Coast Survey, NOAA [9] for the purpose of our numerical simulations. The present day (2014) DEM originated from the National Geophysical Data Center (now National Centers for Environmental Information, or NCEI) [15]. The quality of the paper chart (*i.e.*, detail) and the years representing the major changes in the port were the primary factors that affected our choice of charts. The years representing the major changes to the port are: 1892, 1938, 1945, and present day for the current harbor configuration (Figure 4).

All numerical grids (1892, 1938, 1945, and present day) share two outer grids (usually referred to as A and B grids) that form a nested system. The A grid, or outermost grid, was created from interpolated ETOPO1 data to 30 arc-sec (Figure 5; ~ 900 m [16]). The B grid was also created from interpolated ETOPO1 data to 12 arc-sec [16], while the C grid uses a 3 arc-sec resolution, subsampled from a 1/3 arc-sec resolution dataset (~ 90 m; Figure 5; [17]).

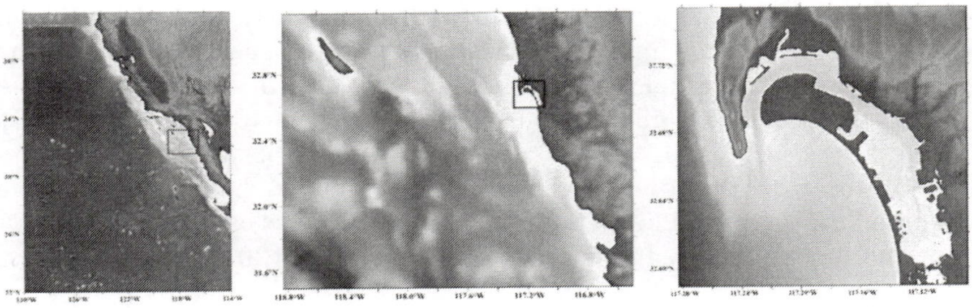

(A) (B) (C)

Figure 5. Extent of numerical grids **A (left)**, **B (middle**; shown as red rectangle within the A grid) and **C (right**; shown in blue colour within the B grid). C grid here only shows the current configuration but the extent of all four C grids (1892, 1938, 1945 and 2014) is the same.

Digitization of old nautical charts is challenging, especially in cases where not only the coastline and bathymetry around a port have changed, but also the topography has changed due to urbanization. Nautical charts are meant for navigation, and therefore, elevation information provided in them is limited. Elevation data for past years are also generally difficult to find so a number of assumptions were made for the creation of the DEMs. Elevation data of San Diego for the high elevation areas to the north of San Diego toward Mission Bay were assumed to have changed little in the last 100 years. This may be a reasonable assumption since construction of the restraining wall (levee) to redirect the San Diego River helped to maintain topography in the low area between Mission Bay and San Diego Bay. This allowed us to use topography from the United States Geological Survey (USGS) with datum NAD837 (North American Vertical Datum 1988, or NAVD88; [18]). Flat areas close to shores usually have large uncertainties and were manually digitized to match available contour data (Figure 6). The coastline was digitized from the charts in addition to bathymetric points (Figure 6). All these data were combined to create the final DEMs at 90 m resolution and mean high water (MHW) depth.

Topographic maps (e.g., from USGS) generally have elevations referenced to orthometric datums, either the NAVD 88 or the older National Geodetic Vertical Datum 1929 (NGVD 29). All GPS positioning data are referenced to one of many 3-D/ellipsoid datums. NOAA's nautical charts have depths referenced to mean lower low water (MLLW), and bridge clearances are referenced to MHW. The legal shoreline in the U.S., which is the shoreline represented on NOAA's nautical charts, is the MHW shoreline; that is, the land-water interface when the water level is at an elevation equal to the MHW datum. For the construction of DEMs a tidal datum—representing a reference plane—is necessary. A tidal datum refers to an average height of the water level at different phases of the tidal cycle ([19]). The MLLW line is also depicted on NOAA's charts8 [19]. Mean Sea Level (MSL) is usually the average between MLLW and MHHW9 as provided on the charts. MHW is more challenging, but we assumed the value of 1.52 m (4.98 ft) above MLLW as estimated by NOAA for San Diego for the years between 1993 and 2003 [20].

We also assumed that the highest elevations on Coronado Island have not changed considerably over time (a 6 m (20 ft) contour is the highest elevation contour that appears in early charts; Figure 1, [9]). The plane of reference used in the charts is the average of MLLW (Figure 1 and Figure 2).

5. NUMERICAL MODELING

For the numerical modeling of tsunamis, we use the well-known and tested MOST (Method of Splitting Tsunami) code [21], which solves the depth-averaged, nonlinear shallow-water wave equations that allow wave evolution over variable bathymetry and topography [21,22]. This is a suite of numerical codes that model all phases of the tsunami simulation (generation, propagation, and runup). The initial displacement on the surface of the water is obtained directly from the terminal coseismic deformation of the ocean floor using Okada's formulation for the surface expression of shear and tensile faults in a homogeneous, elastic half space [23]. The equations of motion are then solved numerically to propagate the water surface disturbance across the computational domain [21,22,24]. A 0.0009 Manning's roughness coefficient was used in all simulations.

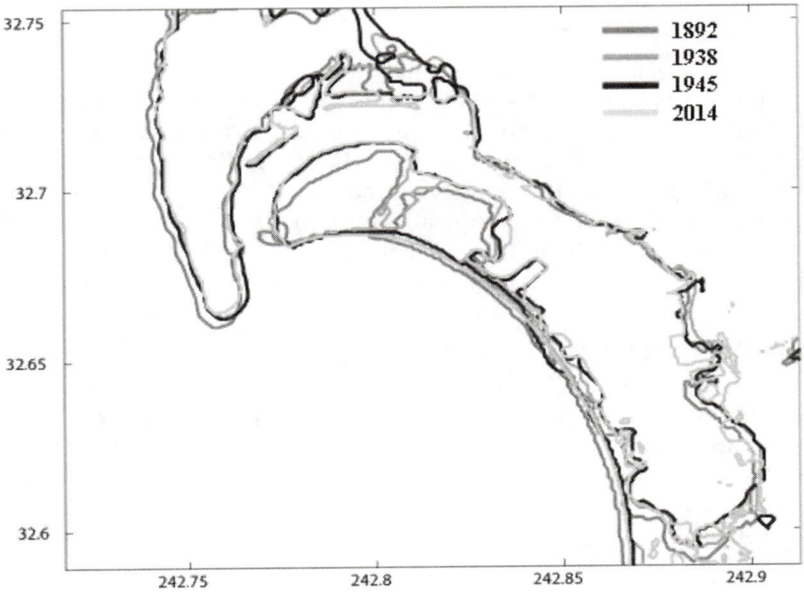

Figure 6. Coastlines of the four digital elevation models (DEMs) used in the numerical simulations.

MOST has been tested extensively, complies with the standards and procedures outlined in [25], and has shown reliability in forecasting inundation. The Tohoku 2011 source used in this article was also selected due to a really good fit to DART® buoy data from the Pacific Ocean [26]. More recently it was also shown to compare fairly well against tsunami currents measured by recently installed acoustic Doppler current profilers (ADCP; [27]).

A set of four scenarios were considered on all numerical grids. A common A and B grid was used for all four DEMs; those are the outermost and intermediate grids used in the simulations for computational efficiency, respectively. We applied two distant scenarios and two local/regional scenarios to investigate the role of modified bathymetry and nearshore topography on our results. Distant scenarios were run for approximately 24 h while local scenarios for 12 h each.

6. TSUNAMI SOURCES

San Diego is exposed to tsunamis from both local and distant sources. A small but balanced set of sources was selected (two distant, two local) to allow for the arrival of waves from different directions in order to investigate, among other factors, the contribution of source location to tsunami impact in San Diego. Specifically, two distant tsunamis generated in the south and in the northwest Pacific were considered. With respect to near-field sources, one local tectonic source and one landslide source were included for locations to the southwest, where the waves' approach is directly into the bay.

The distant sources are major historical subduction zone earthquakes (in opposite hemispheres): the 22 May 1960 Chile earthquake (M_w 9.5) and the 11 March 2011 Tohoku, Japan (M_w 9.0) earthquake. The local sources are theoretical and include a large transpressional earthquake (M_w 7.4) on the San Clemente fault (SCF), located about 70 km southwest of San Diego Bay, and a large submarine landslide located in Coronado Canyon about 25 km southwest of San Diego. For a comprehensive tsunami hazard assessment for San Diego harbor, considering all types of local, regional, and distant sources capable of generating tsunamis that may be damaging, see [28]. A more detailed discussion of the sources with impact to San Diego, as well as the most significant tsunami records in San Diego, can also be found in the same article.

6.1. Chile 1960

A simple rectangular source (800 km × 200 km) was used to represent the fault surface that produced the 22 May 1960 M_w9.5 earthquake and transpacific tsunami [29]. The collision zone of Nazca and South American plates (NazSA10) is an important source of tsunamis and hazard studies for the U.S. West Coast, and California in particular. The 22 May 1960 Chile earthquake is known as the largest recorded earthquake and was preceded and followed by large earthquakes within a few minutes (~Mw 8.0 or greater; see [30,31]). Rupture of the NazSA zone is likely more complex and possibly longer, but far-field studies are not affected greatly by such variation. We consider a Chile 1960 type of source to represent a source similar to the 22 May 1960 event, represented by a single fault plane [29] with a slip of 24 m. More details are provided in Table 1.

Table 1. Table of sources.

Source Name (Far-field)	Fault Length (km)	Fault Width (km)	Slip (m)	Dip (degrees)	Rake (degrees)	Strike (degrees)	Depth[11] (km)
Chile 1960[12]	800	200	24	10	90	10	87.73
			Tohoku 2011[13]				
Segment 1	100	50	4.66	19	90	185	5
Segment 2	100	50	12.23	19	90	185	5
Segment 3	100	50	26.31	21	90	188	21.28
Segment 4	100	50	21.27	19	90	188	5
Segment 5	100	50	22.75	21	90	198	21.28
Segment 6	100	50	4.98	19	90	198	5
Source Name (Near-field)	**Fault Length (km)**	**Fault Width (km)**	**Slip (m)**	**Dip (degrees)**	**Rake (degrees)**	**Strike (degrees)**	**Depth (km)**
San Clemente Fault Bend Region							
Segment 1	18.52	10	1.6	89	−161.6	330.4	0.5
Segment 2	8.86	10	2.7	89	158.2	310.2	0.5
Segment 3	6.96	10	7.0	80	135.0	307.8	0.5
Segment 4	23.65	10	4.2	89	166.0	304.9	0.5
Segment 5	12.92	10	5.8	85	149.0	314.5	0.5
Segment 6	6.9	10	3.2	89	161.6	312.4	0.5
Segment 7	19.5	10	1.6	70	−161.6	311.8	0.5
Landslide Case	**Lat**	**Lon**	**Depth (m)**	**Uplift (m)**	**Subsidence (m)**	**Direction (deg)**	
Coronado Canyon	32.519	−117.338	370	5	5	260	

6.2. Tohoku 2011

The 11 March 2011 M_w 9.0 Tohoku Japan tsunami caused wide devastation on the Japanese coast, damaged facilities and vessels on the U.S. West Coast [4] and was recorded by multiple sea level stations across the Pacific. As a result of the tsunami, at least 20,000 perished [24] and the nearshore topography was completely redefined [32,33]. The tsunami source used in this study is based on real-time forecasting done by NOAA Center for Tsunami Research. Tsunami wave data from two DART® stations near Japan (21418 and 21401; [34]) were used by an inversion algorithm [26] that selected six unit sources from a pre-computed propagation database and its corresponding coefficients [35]. Parameters of each unit source and their corresponding coefficients are listed in Table 1.

6.3. San Clemente Fault (SCF)—Bend Region

We modeled a major left bend in the SCF offshore northern Baja California, Mexico (Table 1). This is a zone of oblique shortening, where the fault impedes the general northwest movement of the Pacific plate. Seafloor uplift resulting from this restraining bend covers a region of about 875 km^2 and represents as much as 720 m of tectonic uplift [36,37,38]. Recent submersible investigations of the SCF revealed vertical seafloor scarps, representative of seafloor uplift during the most recent large earthquakes on the SCF [39]. The largest scarps are 1–3 m high and would be associated with earthquakes of magnitude 6.5 or greater. Three moderate earthquakes (*M*6) along the SCF have been recorded by modern seismographs in the past 83 years.

6.4. Coronado Canyon Landslide

Sediment and sedimentary rock slope failures may occur along the Coronado Escarpment within Coronado and La Jolla submarine canyons. A large failure in Coronado Canyon was modeled to represent a submarine landslide close to San Diego Bay. Our landslide case was represented as a dipole (uplift pattern; see Table 1; [40,41]).

7. RESULTS

7.1. Chile 1960

The inundation extent and corresponding inundation pattern varies greatly from year to year in this scenario for San Diego (Figure 7). Inundation for the present-day (2014) configuration of San Diego is primarily constrained to the south. For earlier years we see large inundation on the inner side of the harbor and NNE in the area of what used to be large tidal lands. Large inundation is evident for the 1945 DEM, especially north of the San Diego region and close to Lindbergh Field (San Diego airport), but that was reversed with the expansion of the Dutch Flat. The early bay, mostly unmodified by human activity, suffered relatively minor wave amplitudes beyond the harbor entrance. This is probably a result of shallow water at the entrance, which reduces the surge into the harbor. Inundation was increased locally at narrow embayments north and west of San Diego where dredging created deeper channels for boats, and fill areas further restricted the channel width. In contrast, the low-lying Silver Strand was overtopped in earlier configurations, but widening and increased elevation reduced the area of potential inundation in more recent years.

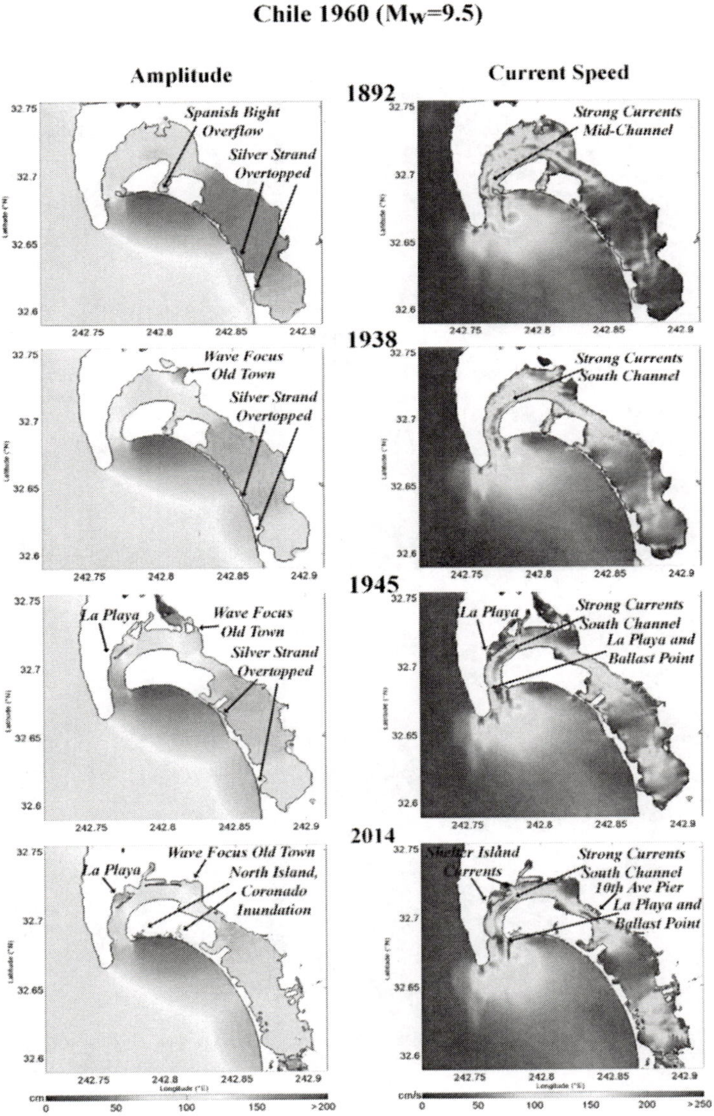

Figure 7. Maximum wave amplitude and maximum current distribution in San Diego harbor for four different configurations (1892, 1938, 1945, and 2014). A Chile 1960 type of scenario was used to investigate changes in wave dynamics as a result of anthropogenic changes. Years indicate time tagging of large changes in the harbor. **Left column** shows wave heights in cm; **right column** tsunami currents in cm/s.

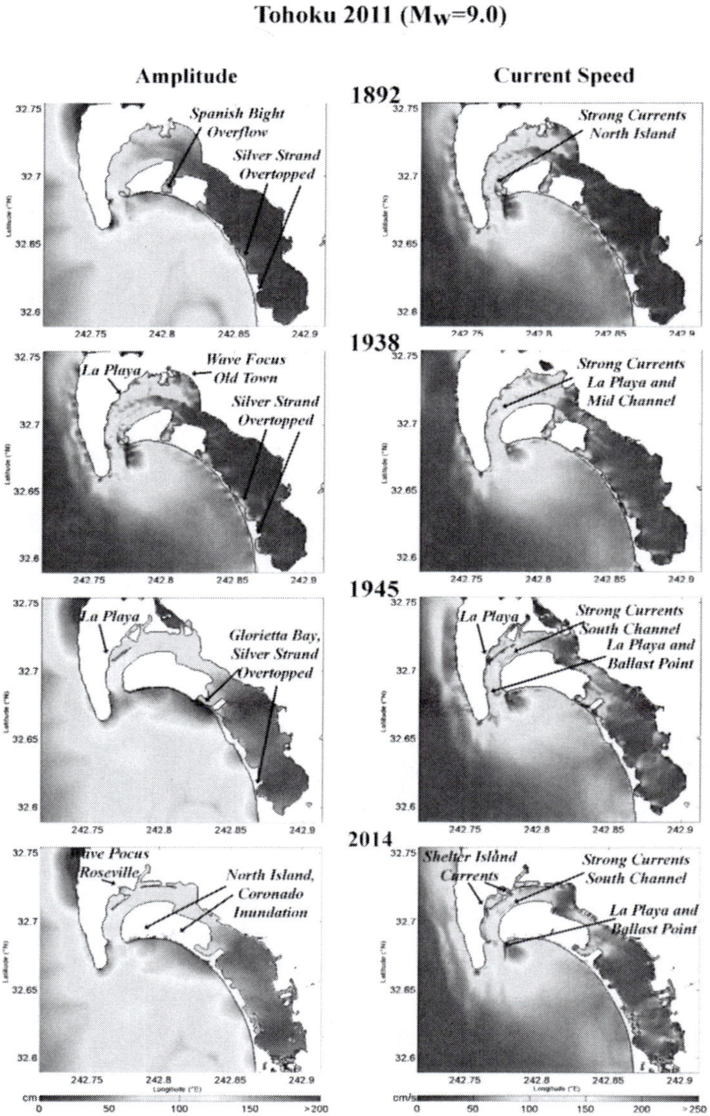

Figure 8. Maximum wave amplitude and maximum current distribution in San Diego harbor for four different configurations (1892, 1938, 1945, and 2014). A tsunami scenario resembling the 11 March 2011 Tohoku tsunami was used to investigate changes in wave dynamics as a result of anthropogenic changes. Years indicate time tagging of large changes in the harbor. **Left column** shows wave heights in cm; **right column** tsunami currents in cm/s.

Generally, the pattern of wave amplitude distribution away from (or outside) the harbor has remained the same over the years but wave amplitudes inside the harbor show a gradual increase as a result of modifications. Currents also show a gradual increase as a function of further modifications for this event. A high current curve following the shape of the harbor is evident for all DEMs in this South American scenario. Additional spots of high current activity also appear in later years.

7.2. 11 March 2011 Japan Tsunami

Similar to the previous distant (Chile 1960) scenario, smaller wave amplitudes appear in older configurations. More specifically, at the earliest year (1892), numerical modeling shows energy spots constrained outside the harbor along the coastline and at the entrance of the harbor. Further development in the bay introduced larger waves inside the harbor around La Playa (Figure 8, 1938). In later years, the creation of the jetty near La Playa changed this pattern and redistributed waves away from the original coastline and along the jetty. Currents also show an increase, but the overall trend is not so clear. Generally, in earlier years, currents were more widely distributed compared to more recent years, while spots of high currents were introduced. The high current spots within the harbor correspond to a new characteristic introduced in the second half of the twentieth century. This is common in both distant scenarios.

The tsunami current distribution from both distant scenarios is worth comparing. In the case of Chile 1960, high currents appear to be more widely distributed along the harbor, while for the Japan tsunami, high currents appear more concentrated near the entrance. This would likely be attributed to the location of the tsunami source and directivity of waves. The maximum amplitudes occur on the south side of Coronado/North Island for the Chile event, showing the effect of the southern source (directivity). In contrast, high amplitudes occur west of Point Loma for the Tohoku event, which arrives from the north and west. Multiple wave arrivals due to refraction along Pacific Ocean waveguides, e.g., Emperor–Hawaiian Seamounts produce strong energy many hours after the first wave arrivals ([42]). Later wave arrivals may involve local resonant effects, so large amplitudes and strong currents were observed during both events. The long period character of these distant subduction megathrust events is very important for emergency planning. Major surges arriving several hours after the first waves caused serious and unexpected damage in 2011 (the surges also persisted for several hours in 1960).

We notice high currents in all plots, but it is clear that the creation of narrow islands and boat basins changes the distribution pattern in the harbor. Although there is an increasing trend with the modifications, enlargement of land space and dredging appear also to have contributed to reduction of currents in some cases while introducing high currents elsewhere (see 1892 and 1938 results; Figure 5 and Figure 6). Currents are high in narrow openings as one might expect, and more such spots of high currents appear in later years as San Diego land expansion led to a narrower opening in the port. The development of a narrow opening combined with channel deepening to the harbor appears also to be responsible for the intensification of the currents that follow a curved wider line along the harbor in 1938 as opposed to 1892. This implies increased danger to vessels not only at the entrances of narrow openings, but also along the "throat14" of the San Diego harbor. Strong currents were indeed observed in Harbor Island and Shelter Island during the Tohoku tsunami and in other harbors in California where narrow channels exist (e.g., Catalina harbor, King Harbor in Redondo beach *etc.*).

7.3. San Clemente Fault—Bend Region

Wave amplitudes for this scenario appear relatively low (predominantly 1–2 m). The pattern does not change with modifications. Tsunami currents also appear low for all years, but isolated spots of high energy are introduced after the extended modifications at the entrance of boat basins (e.g., La Playa; see Figure 9). The small changes for this event likely are attributed to waves of shorter wavelength.

7.4. Coronado Canyon Landslide

Wave amplitude pattern does not appear vastly changed during the years for this scenario. Currents also generally maintain their distribution pattern while some spots of high activity get introduced. High currents appear constrained mainly along Coronado Island and Silver Strand outside the harbor (Figure 10). Specifically, we notice similar patterns to the other scenarios with increased currents where channels narrow. Most importantly, for these scenarios, amplitudes are similar to distant events or greater (2 m or higher) due to proximity of the source and the landslide mechanism (larger displacement). The Coronado Canyon landslide source has a nearly direct path into San Diego Bay coming from the south.

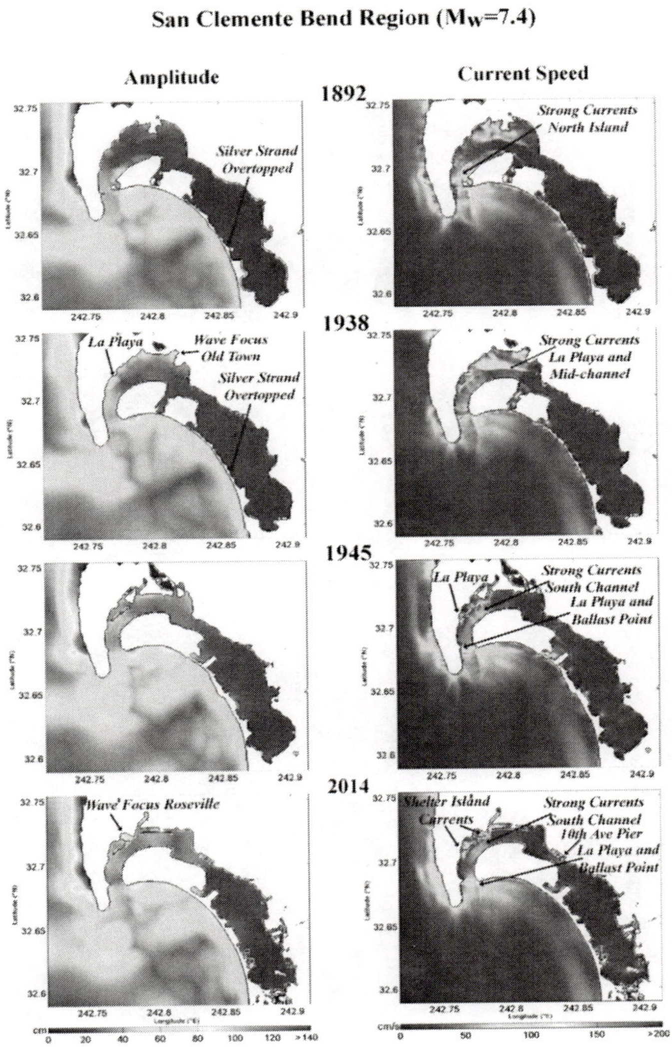

Figure 9. Maximum wave amplitude and maximum current distribution in San Diego harbor for four different configurations (1892, 1938, 1945, and 2014). A San Clemente fault rupture type of tsunami scenario was used to investigate changes in wave dynamics as a result of anthropogenic changes. Years indicate time tagging of large changes in the harbor. **Left column** shows wave heights in cm; **right column** tsunami currents in cm/s.

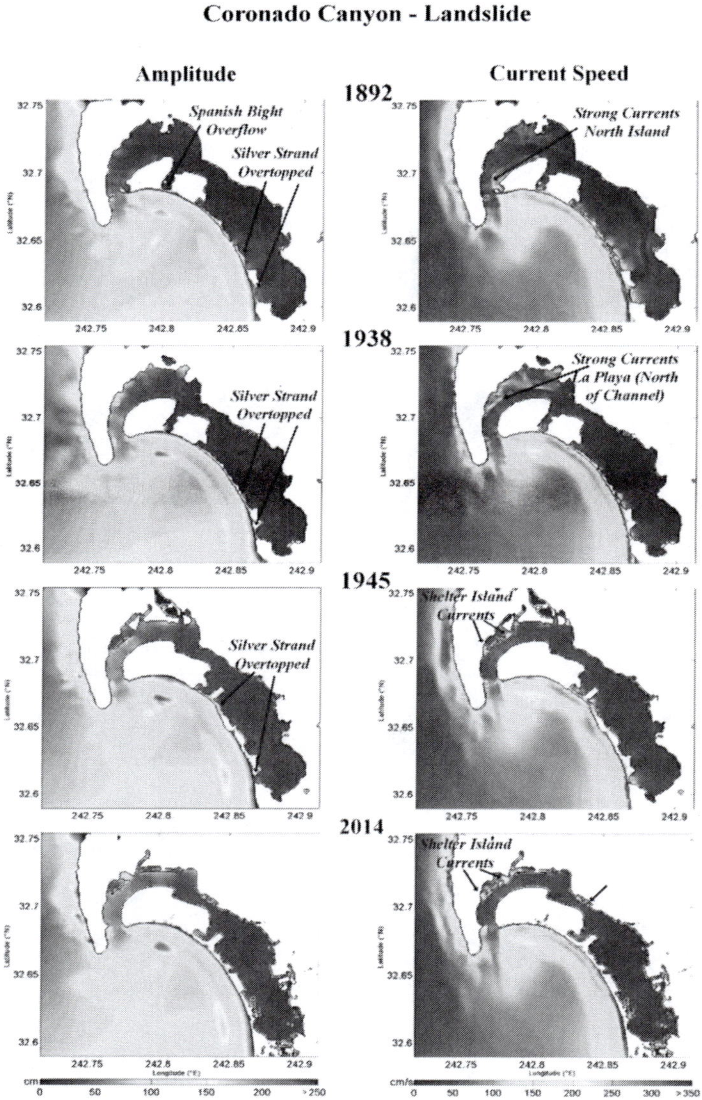

Figure 10. Maximum wave amplitude and maximum current distribution in San Diego harbor for four different configurations (1892, 1938, 1945, and 2014). A hypothetical tsunami scenario generated by a landslide on Coronado Canyon was used to investigate changes in wave dynamics as a result of anthropogenic changes. Years indicate time tagging of large changes in the harbor. **Left column** shows wave heights in cm; **right column** tsunami currents in cm/s.

8. DISCUSSION/CONCLUSIONS

San Diego harbor has undergone extensive modifications and development. Four scenarios were run on four different DEMs to investigate the effects of the major harbor changes to tsunamis. The set of scenarios was selected to ensure a balanced set of scenarios that included both distant and local sources with varying geographical locations. The numerical code MOST was used to model all phases of the tsunami (generation to inundation).

Results show varying inundation patterns for the scenarios with some common characteristics for the distant events when compared to the local events. Inundation within the bay generally was reduced and constrained to locations outside the bay mainly due to expansion of land (channel narrowing) and further filling to elevate tidal flats. Introduction of jetties also caused redistribution of wave heights.

With respect to tsunami currents, it appears that spots of high energy increased in later years. This appears to be associated with the narrowing and deepening of the channel combined with the construction of jetties. In some occasions, changes had a positive effect (reduction of currents), but these positive effects appear to have been counteracted by creation of new high energy spots, due to the creation of narrow entrances to boat basins, in order to accommodate more vessels.

Altering the location of entrances to the boat basins or changing the shape of the boat basins might lead to alternative designs that attenuate currents. While the harbor modifications improve navigation, anchorage and mooring areas, the narrowed channels, especially where elongate islands were created near the harbor entrance, create intense currents during long period wave activity from major distant tsunamis. These currents have created havoc, damaging boats and docking facilities during large events in 1960 and 1988 (Chile), 1964 (Alaska), and 2011 (Japan) [4]. The short period character and shorter duration of tsunamis from local sources reduce these effects. Some narrow low-lying areas along the Silver Strand may be susceptible to overtopping and inundation during large amplitude events, depending also on tidal conditions during the tsunami arrival. In addition, strong currents along beaches may cause severe erosion and scour, and pose serious threat to swimmers and small boats.

The effects of harbor modifications on tsunami impact is a subject that merits further investigation. However, the work presented here leads to the following more general conclusions:

- Enhancement of port space increases exposure to assets within the harbor compared to previous years as more vessels are now accommodated within the port facilities.
- Numerical results also show that the introduction of high currents could threaten navigation, vessels and facilities at narrow openings and also along the harbor "throat".
- Harbor modifications can have both positive and negative effects on tsunami impact. For example, currents can be constrained to areas of minimal impact (e.g., outside of the harbor), thereby exposing less assets to dangerous wave action.
- The long duration of distant subduction megathrust tsunamis (Mw >8 source magnitude) spans several tidal cycles so that serious currents occur for several hours, or even days, after the first wave arrivals.

ACKNOWLEDGMENTS

We would like to thank Finn Scheele for help with some of the maps in this manuscript. We would like to thank Cynthia Holder, Deputy District clerk/records manager of the Unified Port of San Diego for giving us access to the historical photographs of San Diego Bay.

This publication is partially funded by the Joint Institute for the Study of Atmosphere and Ocean (JISAO) at the University of Washington under NOAA Cooperative Agreement No. NA17RJ1232, Contribution No. 2445. This is Contribution No. 4364 from the NOAA/Pacific Marine Environmental Laboratory.

AUTHOR CONTRIBUTIONS

A.B. wrote the majority of the paper, digitized the nautical charts for the simulations and made about half of the figures appearing in the manuscript. M.R.L. had the original idea for this paper, provided information on the local sources and wrote sections of the paper relative to the sources and San Diego Bay history. E.G. ran the simulations, helped with the decision-making of the distant sources and contributed to the editing of this paper.

CONFLICTS OF INTEREST

The authors declare no conflict of interest.

REFERENCES

1. Unified Port of San Diego. Available online: http://www.portofsandiego.org (accessed on 7 July 2015).
2. Kemp, K. The Mississippi Levee System and the Old River Control Structure. Available online: http://www.tulane.edu/~bfleury/envirobio/enviroweb /Flood Control.htm (accessed on 7 July 2015).
3. Lander, J.F.; Lockridge, P.A. United States Tsunami 1890–1988; National Geophysical Data Center, National Environment Satellite, Data: Boulder, CO, USA, 1989.
4. Wilson, R.I.; Admire, A.R.; Borrero, J.C.; Dengler, L.A.; Legg, M.R.; Lynett, P.; McCrink, T.P.; Miller, K.M.; Ritchie, A.; Sterling, K.; et al. Observations and impacts from the 2010 Chilean and 2011 Japanese tsunamis in California (USA). Pure Appl. Geophys. 2012, 170, 22.
5. Shuto, N. Damages to coastal structures by tsunami-induced currents in the past. J. Disaster Res. 2009, 4, 462–468.
6. Military.com. Naval Air Station North Island. Available online: http://www. military.com/base-guide/naval-air-station-north-island (accessed on 13 July 2015).
7. 2010 Census—Census.gov, 2010. Available online: http://www.census.gov /2010census/ (accessed on 14 September 2015).
8. California Department of Parks and Recreation. About the Beach. Available online: http://www.parks.ca.gov/?page_id=984 (accessed on 13 July 2015).
9. Office of Coast Survey, NOAA. Historical Charts. Available online: http://historicalcharts.noaa.gov/ (accessed on 8 July 2015).
10. Fischer, P.J.; M ills, G.I. The Offshore Newport-Inglewood-Rose Canyon Fault Zone, California: Structure, Segmentation and Tectonics; Environmental Perils of the San Diego Region: San Diego, CA, USA, 1991; pp. 17–36.
11. Harper & Row. Shepard, Submarine Geology, 3rd ed.; Harper & Row: New York, NY, USA, 1973.
12. Gohres, S.D.H.C.-D. Tidal Marigrams. Available online: https://www.sandiegohistory.org/journal/64october/marigrams.htm#Gohres (accessed on 10 July 2015).
13. Gohres, H. Tidal Marigrams. J. San Diego Hist. 1964, 10, 4.
14. Military.com. CNIC|Naval Base Coronado Complex. Available online: http://www.cnic.navy.mil/regions/cnrsw/installations/navbase_coronado.html (accessed on 13 July 2015).
15. NOAA, National Centers for Environmental Information. NOAA Tsunami Inundation Digital Elevation Models (DEMs). Available online: http://www.ngdc.noaa.gov/mgg/inundation/ (accessed on 8 July 2015).
16. NOAA, Natio nal Centers for Environmental Information. ETOPO1 Global Relief Model. Available online: https://www.ngdc.noaa.gov/mgg/global/global.html (accessed on 8 July 2015).
17. Carignan, K.S.; Taylor, L.A.; Eakins, B.W.; Friday, D.Z.; Grothe, P.R.; Love, M. Digital Elevation Models of San Diego, California: Procedures, Data Sources and Analysis; National Geophysical Data Center, National Oceanic Atmospheric Administration: Silver Spring, MD, USA, 2012.
18. National Geodetic Survey, NOAA. Vertical Datums. Available online: http://www. ngs.noaa.gov/datums/vertical/(accessed on 8 July 2015).

19. NOAA, Center for Operational Oceanographic Products and Services. Tidal Datums and Their Applications; US Department of Commerce: Silver Spring, MD, USA, 2000.
20. NOAA, Tides and Currents-Superseded Bench Mark Sheet for 9410170, San Diego CA. Available online: http://beta.tidesandcurrents.noaa.gov/benchmarks.html?id=9410170&type=superseded (accessed on 10 July 2015).
21. Titov, V.V.; Synolakis, C.E. Numerical Modeling of tidal waver runup. J. Waterw. Port Coast. Ocean Eng. 1998, 124, 15.
22. Titov, V.V.; Gonzalez, F.I. Implementation and Testing of the Method of Splitting Tsunami (MOST) Model; NOAA Technical Memorandum ERL PMEL-112; NOAA: Seattle, WA, USA, 1997.
23. Okada, Y. Surface deformation due to shear and tensile faults in a half-space. Bull. Seismol. Soc. Am. 1985, 75, 20.
24. International Tsunami Information Center (ITIC). m.yamamoto@unesco.org (accessed on 9 October 2012).
25. Synolakis, C.E.; Bernard, E.N.; Titov, V.V.; Kanoglu, U.; Gonzalez, F.I. Validation and Verification of Tsunami Numerical Models. Pure Appl. Geophys. 2008, 165, 38.
26. Percival, D.B.; Denbo, D.W.; Eble, M.C.; Gica, E.; Mofjeld, H.O.; Spillane, M.C.; Tang, L.; Titov, V.V. Extraction of tsunami source coefficients via inversion of DART® buoy data. J. Int. Soc. Prev. Mitig. Nat. Hazards 2011, 58, 567–590.
27. Admire, A.R.; Dengler, L.A.; Crawford, G.B.; Uslu, B.U.; Borrero, J.C.; Greer, S.D.; Wilson, R.I. Observed and modeled currents from the Tohoku-oki Japan and other recent tsunamis in Northern California. Pure Appl. Geophys.2014, 171, 18.
28. Barberopoulou, A.; Legg, M.R.; Uslu, B.; Synolakis, C.E. Reassessing the tsunami risk in major ports and harbors of California I: San Diego. Nat. Hazards 2011, 58, 18.
29. Kanamori, H.; Cipar, J.J. Focal Process of the great Chilean earthquake 22 May 1960. Phys. Earth Planet. Inter. 1974, 9, 9.
30. Kanamori, H. The energy release in great earthquakes. J. Geophys. Res. 1977, 82, 7.
31. Barrientos, S.E.; Ward, S.N. The 1960 Chile earthquake: Inversion for slip distribution from surface deformation. Geophys. J. Int. 1990, 103, 9.
32. Tanaka, H.; Tinh, N.X.; Umeda, M.; Hirao, R.; Pradjoko, E.; Mano, A.; Udo, K. Coastal and estuarine morphology changes induced by the 2011 Great East Japan Earthquake Tsunami. Coast. Eng. J. 2012, 54, 25.
33. Taylor, A. Japan Earthquake, 2 Years Later: Before and After. Available online: http://www.theatlantic.com/photo/2013/03/japan-earthquake-2-years-later-before-and-after/100469/#img18 (accessed on 13 July 2015).
34. NOAA, National Data Buoy Center. Available online: http://www.ndbc. noaa.gov /dart.shtml (accessed on 10 July 2015).
35. Gica, E.; Spillane, M.C.; Titov, V.V.; Chamberlin, C.D.; Newman, J.C. Development of the forecast propagation database for NOAA's Short-Term Inundation Forecast for Tsunamis (SIFT); NOAA Technical Memorandum, OAR PMEL-139; NOAA: Seattle, WA, USA, 2008.
36. Legg, M.R. Geologic Structure and Tectonics of the Inner Continental Borderland Offshore Northern Baja California, Mexico, Santa Barbara, CA. Ph.D. Dissertation, University of California, Santa Barbara, CA, USA, 1985.
37. Legg, M.R.; Borrero, J.C. Tsunami potential of major restraining bends along submarine strike-slip faults. In Proceedings of the International Tsunami Symposium 2001, NOAA/PMEL, Seattle, WA, USA, 7–10 August 2001.
38. Legg, M.R.; Goldfinger, C.; Kamerling, M.J.; Chaytor, J.D.; Einstein, D.E. Morphology, structure and evolution of California Continental Borderland restraining bends. Tecton. Strike-Slip Restraining Releas. Bends Cont. Oceanic Settings2007, 290, 143–168.
39. Goldfinger, C.; Legg, M.R.; Torres, M.E. New Mapping and Submersible Observations of Recent Activity on the San Clemente Fault; EOS, Transactions of the American Geophysical Union: Washington, DC, USA, 2000.

40. Synolakis, C.E.; Borrero, J.C.; Eisner, R. Developing inundation maps for Southern California. In Proceedings of the 2002 Coastal Disasters Conference, San Diego, CA, USA, 24–27 February 2002.
41. Synolakis, C.E.; Bardet, J.P.; Borrero, J.; Davies, H.; Okal, E.; Silver, E.; Sweet, J.; Tappin, D. Slump origin of the 1998 Papua New Guinea tsunami. Proc. R. Soc. Lond. A 2002, 458, 763–789.
42. Barberopoulou, A.; Legg, M.R.; Gica, E. Multiple Wave Arrivals Contribute to Damage and Tsunami Duration on the US West Coast; Springer: Berlin, Germany; Heidelberg, Germany, 2014.

NOTES

[1]There are also some submarines, an explosives loading area, and mooring facilities and docks for many other naval vessels throughout the bay ([6]).

[2]Spanish Bight is a graben associated with the Rose Canyon fault zone, e.g., [10]. San Diego Bay lies within the releasing stepover of the Rose Canyon fault zone, which created the bay and uplift of Point Loma. The Silver Strand is a narrow sand spit that connects North Island/Coronado to the mainland at Imperial Beach to the south. San Diego Bay was a natural lagoon formed behind this barrier beach ([11]).

[3]North Island was the source of fresh water for the lighthouse keeper on Point Loma during the nineteenth century.

[4]Originally, this area was the fan delta for the San Diego River, which was flooded in the 1861/62 "Noachian Deluge"—a severe flood event. Levees (restraining wall on 1892, no 5106 chart) were built to redirect the river flow to False Bay (Mission Bay), which finally succeeded after a couple more floods broke through to San Diego Bay. Dutch Flat was expanded, and adjoining basins were subsequently filled as Lindbergh Field (the municipal airport) grew and adjacent land was developed. Finally, Shelter Island and Harbor Island were constructed to provide additional protection for marinas and land for hotels, restaurants, *etc.* (e.g., Figure 3).

[5]Spanish Bight was mostly filled to connect North Island to Coronado, and additional dredging created local bays, such as Glorietta Bay, for marinas and naval vessels. Piers were constructed at San Diego for commercial boats and naval shipyards to the south at National City where major shipyards were built. The southern end of the bay had salt evaporation ponds, although a marina was also built for the City of Chula Vista. The Silver Strand was enhanced to reduce flooding during storms, and marinas were constructed in some locations (other locations were restricted to military operations including an amphibious base and radio station).

[6]La Playa is the location of the original tide gauge for San Diego. It was first installed in 1854. A strong local earthquake, 27 May 1862 M~6, created a wave in the bay that was observed by the tide gauge engineer (Andrew Cassidy) who was repairing the pier for the tide gauge at the time of the earthquake. He observed a runup of about 1–1.2 m (3–4 ft) on the beach at La Playa but the water returned to its normal level. We do not know if this runup was a vertical or horizontal distance [12,13]. The Rose Canyon fault was likely the source of the earthquake but data are inadequate to determine the epicenter with certainty.

[7]Although contours were matching fairly well with old charts an additional small shift ensured a better fit to the contours in the nautical charts and the coastline.

[8]MLLW is the lower low water height average of each tidal day calculated over a tidal datum epoch (19 year period adopted by the National Ocean Service).

[9]MHHW is the higher high water height average of each tidal day calculated over a tidal datum epoch (19 year period adopted by the National Ocean Service).

[10]We abbreviated the source simply for easy reference in this article.
[11]Depth is measured from ground surface to top of the fault.
[12][29].
[13]Source from NOAA Center for Tsunami Research.
[14]Numerous piers and anchorages exist in the area between San Diego and Coronado. Strong currents in these areas were damaging to Navy ships and other boats, while the ferry service was disrupted during the 1960 event. Docks were damaged and some boats dragged from their moorings [3].

CHAPTER 11

Possible Failure Mechanism of Buildings Overturned during the 2011 Great East Japan Tsunami in the Town of Onagawa

Panon Latcharote[1*], Anawat Suppasri[1], Akane Yamashita[2], Bruno Adriano[1], Shunichi Koshimura[1], Yoshiro Kai[3] and Fumihiko Imamura[1]

[1]*International Research Institute of Disaster Science, Tohoku University, Sendai, Japan*

[2]*Earthquake Research Institute, University of Tokyo, Tokyo, Japan*

[3]*Department of Infrastructure Systems Engineering, Kochi University of Technology, Kochi, Japan*

ABSTRACT

Six buildings were overturned in the town of Onagawa during the 2011 Great East Japan tsunami. This study investigates the possible failure mechanisms of building overturning during tsunami flow. The tsunami inundation depth and flow velocity at each overturned building were recalculated by using a tsunami numerical simulation and verified using a recorded video. The overturning moment is a result of hydrodynamic and buoyancy forces, whereas the resisting moment is a result of building self-weight and pile resistance force. This study aimed to demonstrate that the building foundation design is critical for preventing buildings from overturning. The analysis results suggest that buoyancy force can generate a larger overturning moment than hydrodynamic force, and the failure of a pile foundation could occur during both ground shaking and tsunami flow. For the pile foundation, pile resistance force plays a significant role due to both tension and shear capacities at the pile head and skin friction capacity between the pile and soil, which can be calculated from 18 soil boring data in Onagawa using a conventional method in the AIJ standards. In addition, soil liquefaction can reduce skin friction capacity between the pile and soil resulting in a decrease of the resisting moment from pile resistance force.

Keywords: building overturning, Onagawa, tsunami flow, pile foundation, soil liquefaction

INTRODUCTION

During the 2011 Great East Japan earthquake and tsunami, many buildings were seriously damaged by a combination of ground shaking, tsunami flow, debris impact, and soil liquefaction. After the 2011 Great East Japan tsunami, overturned buildings were found unexpectedly in two locations: (1) six buildings [i.e., five reinforced concrete (RC) buildings and one steel-frame building] in the town of Onagawa in Miyagi Prefecture and (2) two buildings in the city of Miyako in Iwate Prefecture. This study focused on the overturned buildings in Onagawa because of the comprehensiveness of the building-related information, soil information, and tsunami simulation results. These overturned buildings were built more than 30 years ago over filled soil foundations. A field survey revealed that one of the six overturned buildings in Onagawa was built on a shallow foundation, and the other buildings had a pile foundation; one of the buildings was overturned and moved 70 m from its original position (Suppasri et al., 2012; Latcharote et al., 2014). Based on inundation data, these overturned buildings were fully (or at least nearly) submerged and overturned by the following possible causes: (a) hydrodynamic force, including debris effects; (b) buoyancy force; and (c) weakened foundation associated with soil instability (Yeh et al., 2013). Therefore, the maximum inundation depth exceeded the height of all overturned buildings in Onagawa, and it was assumed that those buildings were overturned during the overtopping tsunami flow.

The one famous instance of an overturned RC building occurred in 1946 following an Aleutian tsunami, in which an 18-m-tall lighthouse at a ground elevation of 10 m was overturned by a 30-m tsunami. In Japan, building overturning had not been reported in previous earthquakes and subsequent tsunamis and thus was not considered in building foundation design. However, building overturning is now considered in the design guidelines for building foundations (Architectural Institute of Japan, 2001), particularly for tsunami evacuation buildings. In recent years, the seismic performance design of pile foundations has considered the rocking of pile caps and the negative friction of piles to resist the uplift of buildings. During tsunami flow, building overturning can occur as a result of lateral force (hydrodynamic force) and uplift force (buoyancy force), the latter of which depends on the dimensions from the top of the window opening to the ceiling in buildings. Based on the surveyed data, most of the piles were likely broken by tension and shear failures at the pile head and insufficient friction between the pile and soil, including the effect of soil liquefaction, which caused the piles to be easily pulled out of the ground. During soil liquefaction, the soil shear strength is decreased, thereby decreasing the shaft resistance (skin friction) between the pile surface and soil around the pile (Fraser et al., 2013). This decreased shaft resistance would allow the greater vertical movement of the piles while in the ground. The piles would then be pulled from the ground more easily when the building is subjected to uplift and lateral forces from tsunami flow and debris impact, which were significant in Onagawa due to the extreme inundation depth (Fraser et al., 2013). Soil liquefaction changed the soil properties and caused a loss of skin friction capacity between the pile and soil because of the loosening of soil around the pile.

Therefore, the building foundation would be the main consideration of building overturning in Onagawa. The overturning mechanism of these overturned buildings has been thoroughly investigated in previous studies, such as Ishida et al. (2015), Tokimatsu et al. (2016), and Yeh and Sato (2016), and few design considerations have been suggested. This study investigates the overturning mechanism in different ways and also considers the effect of soil liquefaction, which can result in a decrease of skin friction capacity between the pile and soil. Based on experimental studies of geotechnical problems, this study suggests a conventional method to evaluate the approximate skin friction capacity when soil liquefaction occurs. The possible failure mechanism of the overturned buildings can then be investigated by comparing the overturning moment and resisting moment. The results of this study can be used to improve the recommendations for building foundation design in a building design code.

CHARACTERISTICS OF FIVE OVERTURNED BUILDINGS

Our survey team observed that most of six overturned buildings in Onagawa had shifted away from sea and thus appeared to be overturned by striking wave. The tsunami force is estimated to be many tons per square meter with a long-period wave (e.g., 30 min), which led to a prolonged interaction of tsunami flow acting on these overturned buildings. The water released from the uppermost floors of the buildings generated uplift force, which caused a large overturning moment with hydrodynamic force. Small openings were observed in these overturned buildings, which could also generate large uplift force. However, there was sufficient time for water to flow inside the buildings because of the long-period wave. Thus, only the accumulated air between the top of the windows and the ceiling generated buoyancy force (Suppasri et al., 2013). The survey team also stated that most of the piles were probably pulled out and broken at the pile head as a result of ground shaking, hydrodynamic force, buoyancy force, and soil liquefaction. Five overturned buildings (i.e., Buildings A, B, C, D, and E) in Figure 1 were analyzed in this study, and the characteristics of each overturned building are provided below.

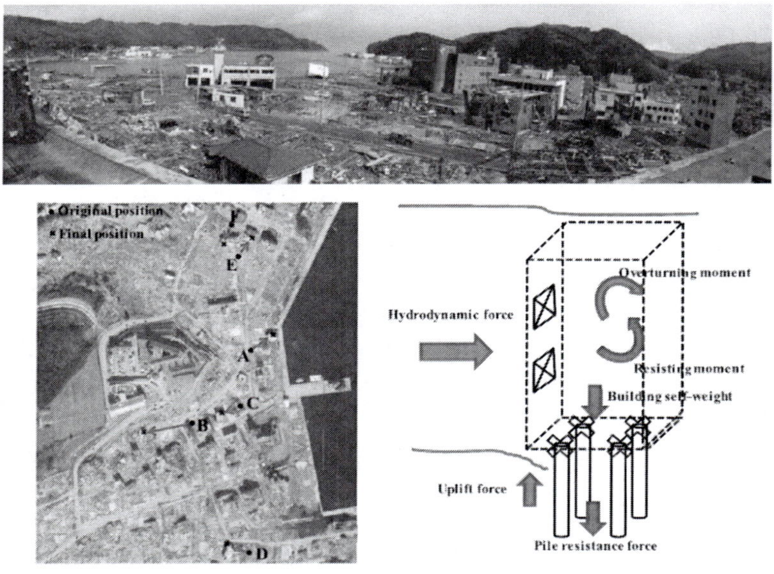

FIGURE 1. Six overturned buildings in the town of onagawa and free-body-diagram of building overturning. Note: taken by our survey team in March 29, 2011 at the town of Onagawa and Google Earth.

Building A

Building A is believed to have been built between 1965 and 1970 adjacent to the shoreline (Onagawa, 2013). It was used as a repair shop of fishing boats in the past but was being used as a commercial store before the tsunami (Onagawa, 2013). The building was submerged by 0.4 m of seawater at high tide for many months after the tsunami due to the residual subsidence from the earthquake. The building was a three-story RC structure with a mat foundation on hard ground, as shown in Figure 2A. Small openings were observed on the face of this building subject to tsunami flow. As shown in Figure 1, this building overturned seaward, but it is expected that the initial failure was landward (consistent with the other buildings) and this building was then moved during tsunami return flow to its final position (Fraser et al., 2013). On the other hand, the building may have been overturned seaward by the receding wave (Onagawa, 2013). Due to the mat foundation, only building self-weight could provide a resisting moment against the overturning moment from hydrodynamic and buoyancy forces.

Building B

Building B is believed to have been built between 1955 and 1975 based on its type of foundation (Nikkei BP Company, 2011). It was an accommodation building with a four-story RC structure and a pile foundation, as shown in Figure 2B. The pile foundation had 32 hollow concrete pipe piles with a pile diameter of 20 cm. This building was moved 70 m from its original position. As shown in Figure 2B, some piles were pulled from the ground, and some were broken under the foundation. Twelve of the 32 piles under the foundation appear to have been effective in resisting the overturning moment (Kabeyasawa et al., 2012). No spiral reinforcing bar was observed inside the piles, but six longitudinal reinforcing bars were observed inside each pile. The settled ground around a neighboring building indicated the occurrence of soil liquefaction (Tokimatsu et al., 2012).

Building C

Building C is believed to have been built between 1980 and 1985 (Onagawa, 2013). The building was used as an accommodation facility in the past but was being used as offices for private business and accommodations for sailors before the tsunami (Onagawa, 2013). The building was moved approximately 10–16 m from its original location by the tsunami (Onagawa, 2013). The building was submerged by 0.2 m of seawater at high tide for a short period after the tsunami. It was a four-story steel-frame building with a pile foundation, as shown in Figure 2C. The pile foundation had 8 pile caps with 20 piles having a diameter of 25 cm. Most hollow concrete pipe piles failed at the connection to the pile cap under the foundation unless one pile on the upper right was pulled out. A spiral reinforcing bar and six longitudinal reinforcing bars were observed inside each pile, and the six reinforcing bars had ruptured. This building was constructed from steel

FIGURE 2. THE CHARACTERISTICS OF FIVE OVERTURNED BUILDINGS IN THE TOWN **OF ONAGAWA. (A)** Building A, **(B)** Building B, **(C)** Building C, **(D)** Building D, and **(E)** Building E. Note: taken by our survey team in March 29, 2011 and July 9, 2011 at the town of Onagawa.

frames and ALC walls; thus, building self-weight was less than that of RC buildings. This building was floated, carried away and then overturned by tsunami flow, and most of the piles were broken at their joints with the pile caps (Tokimatsu et al., 2012).

Building D

Building D is believed to have been built between 1965 and 1975 (Nikkei BP Company, 2011). It was used as a refrigerated warehouse with a two-story RC structure and a pile foundation, as shown in Figure 2D. The pile foundation had six pile caps with four piles with a diameter of 20 cm in each pile cap. All piles were broken at the pile caps. No spiral reinforcing bar was observed inside the piles, but six longitudinal reinforcing bars were observed inside each pile. This building was floated more than 1 m and moved approximately 7 m, and all piles were ruptured at or near the joints (Tokimatsu et al., 2012). No pile remained connected to the pile caps, suggesting a higher level of shear in the overturning motion than was experienced in the other overturned building with a pile foundation (Fraser et al., 2013). This building was lifted by the hydrostatic buoyancy off of its pile foundation, which did not have tension capacity due to the minimal reinforcing steel (Chock et al., 2013). In addition, it was lifted off its original site and carried over a low wall before being deposited approximately 15 m inland from its original location (Chock et al., 2013).

Building E

Building E is believed to have been built in 1980 (Onagawa, 2013). It was a police box on the first floor and a rest area on the second floor (Onagawa, 2013). The building was overturned near its present location and then moved to its present location by the receding wave (Onagawa, 2013). Some damage from floating debris can be observed on the upper part of the building. The building was submerged by 0.3 m of seawater at high tide for a short period after the tsunami. The building was a two-story RC building with a pile foundation, as shown in Figure 2E. The pile foundation had 6 pile caps with 14 piles having a diameter of 25 cm, and most of the piles were pulled out of the ground. No spiral reinforcing bar was observed inside the piles, but six longitudinal reinforcing bars were observed inside each pile. The settled ground near a neighboring building indicates that liquefaction occurred (Tokimatsu et al., 2012).

FACTORS INFLUENCING BUILDING OVERTURNING

Tsunami Inundation

Based on a video recorded from the rooftop of a building in Onagawa, a thorough analysis of the tsunami inundation was conducted using numerical modeling and measurements (Adriano et al., 2016). Numerical tsunami simulations were performed to reproduce the calculated time series of tsunami flow using the tsunami source model proposed by the Cabinet Office, Government of Japan. The simulations demonstrate that the maximum inundation depth due to the first incoming wave was over 16 m, and more than 500 buildings were washed away by this first wave, which is consistent with the video data (Adriano et al., 2016). The source model was verified with the observed tsunami inundation depth and flow velocity interpreted from the video record including the measured depth of maximum tsunami inundation of 30 points from tsunami watermark and the inundation area measured by field survey and satellite image analysis (Adriano et al., 2016). This study extended this reliable source model of tsunami numerical simulations to reproduce the waveforms of tsunami inundation depth and flow velocity at each overturned building. Then, these waveforms were used to estimate hydrodynamic and buoyancy forces in time series in order to investigate the possible mechanism of building overturning.

The tsunami inundation depth and flow velocity in the time series at each overturned building validated by the interpretation of the video data are shown in Figure 3. The peak inundation depth and peak flow velocity occurred at different times, which could result in the induction of large hydrodynamic force at any time during the striking or receding wave. Figure 3A shows the calculated time series of the tsunami inundation depth and flow velocity in 2014 (Adriano et al., 2014). In this 2014 simulation, the maximum inundation depth was not consistent with the video analysis, although the maximum flow velocity was somewhat consistent. Figure 3B shows the calculated time series of the tsunami inundation depth and flow velocity in 2016, including the crustal deformation (Adriano et al., 2016). In this 2016 simulation, the maximum inundation depth was consistent with the video analysis, whereas the maximum flow velocity was lower than that of the video analysis. However, these numerical tsunami simulations were generated by the tsunami source model, which is typically used to reproduce tsunami propagation for all affected areas in Japan after the 2011 Great East Japan tsunami. This source model is not specific to only the studied area in Onagawa, which makes it different from other reverse models that attempt to be consistent with the video evidence. In this study, sets of inundation depth and flow velocity from both 2014 and 2016 simulations were used to estimate hydrodynamic and buoyancy forces.

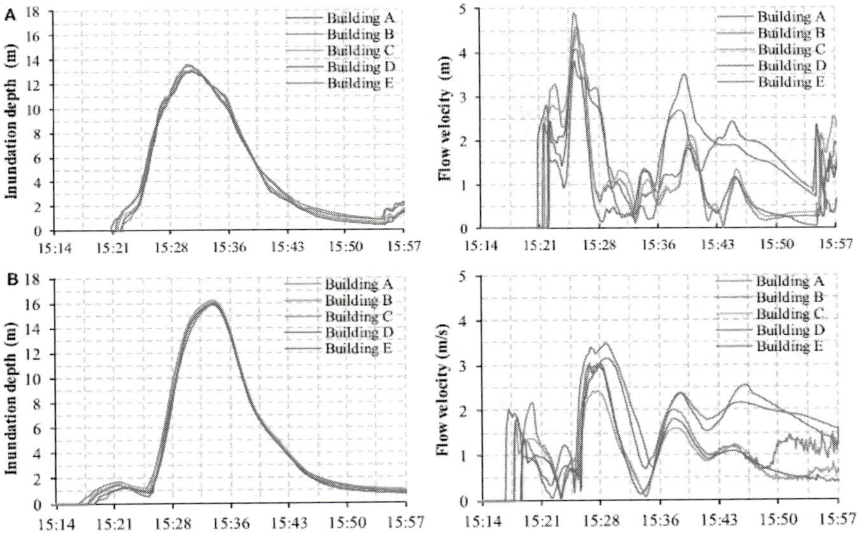

FIGURE 3. TSUNAMI INUNDATION DEPTH AND FLOW VELOCITY FROM 2014 AND 2016 SIMULATIONS. (A) 2014 simulation and **(B)** 2016 simulation.

Hydrodynamic Force

The overturning moment is partially the result of hydrodynamic force, which can be calculated from the inundation depth and flow velocity, as shown in Figure 3. This study assumed that these overturned buildings were surrounded by water and had a minimum unbalanced hydrostatic force, i.e., a minimum tsunami load. For each overturned building, hydrodynamic force (F_d) was applied as a uniform load over the depth of tsunami flow (Federal Emergency Management Agency, 2011) as

$$F_{\mathrm{d}} = \frac{1}{2}\rho_{\mathrm{s}} C_{\mathrm{d}} B\left(hu^2\right),$$

where ρ_s is the density of salt water with sediment (1,200 kg/m^3), C_d is the drag coefficient (2.0), B is the building width in the plane normal to the direction of flow, h is the inundation depth or building height, and u is the flow velocity.

Buoyancy Force

The overturning moment is also partially the result of buoyancy force, which can be calculated from the inundation depth shown in Figure 3. For each overturned building, buoyancy force (F_b) was set equal to the water weight of the residual air space inside it, which also depends on the opening ratio. A residual air space ratio (C_b) of 0 indicates that the entire building was filled with water, whereas a residual air space ratio of 1.0 indicates that no water entered the building, expressed as

$$F_b = \rho_s C_b g B D h,$$

where ρ_s is the density of salt water with sediment (1,200 kg/m^3), C_b is the residual air space ratio varying with the relative volume of entrapped air inside the building (0.0–1.0), g is the gravitational acceleration (9.81 m/s^2), B is the building width, D is the building depth, and h is the inundation depth or building height.

Building Self-Weight

The resisting moment is also partially the result of building self-weight, which can be calculated from the weight per unit area of the RC and steel buildings. The concrete density is typically 2,400 kg/m^3. In this study, the weight per unit area of the RC buildings is assumed to be 14 kPa, whereas the weight per unit area of the steel buildings is assumed to be 8 kPa.

Pile Resistance Force

The resisting moment is also partially the result of pile resistance force, which can be calculated from the pile and soil. From the damage observed in the overturned buildings with a pile foundation, two possible cases of pile damage were identified: tension or shear failure at the pile heads and pulling of the pile from the ground.

Tension and Shear Capacities

In the case of tension failure, pile resistance force (R_{TC}) can be calculated from the fracture strength of the PC steel wire (F_u) inside the pile and from the shear strength of the pile section (Q_u) for the case of shear failure. Tokimatsu et al. (2016) suggested the tension and shear capacities of piles in a pile foundation based on the catalog specifications.

Skin Friction Capacity

Eighteen soil boring data were obtained from the Onagawa office to represent the soil profiles of overturned buildings with pile foundations, as shown in Figure 4. The coastal area in Figure 4A (covered by the red-dashed line) was largely filled by soil (Onagawa, 1960). A soil boring data contains soil layers, such sand, gravel, silt, and clay, and *N* value, as shown in Figure 4B. In the case of pulling a pile out of the ground, pile resistance force (R_{TC}) is calculated from skin friction capacity (Q_s) between the pile and soil based on the recommendations for the design of building foundations (Architectural Institute of Japan, 2001) as

$$Q_s = \left(\sum \tau_{st} L_s + \sum \tau_{ct} L_c\right) \varphi_p,$$

where τ_{st} is the friction stress in a sand layer, L_s is the length of a sand layer, τ_{ct} is the friction stress in a clay layer, L_c is the length of a clay layer, and φ_p is the peripheral length of a pile. In this equation,

$$\tau_{st} = 2.0\ \mathrm{N},$$

$$\tau_{ct} = \beta C_u,\ \beta = \alpha_p \mathrm{L_f},\ C_u = q_u / 2,\ q_u = \max\left(25\ \mathrm{N}, 60\right),$$

where α_p is the adhesive factor (0.5–1.0), L_f is the length index (0.7–1.0), and C_u is the average undrained shear strength.

The adhesive factor (α_p) varies with the ratio of undrained shear strength and effective overburden pressure of silt and clay, whereas the length index (L_f) varies with the ratio of the layer thickness and the pile diameter.

Soil Liquefaction

In general, the evaluation of soil liquefaction during ground shaking and its effects on a pile foundation are highly complex because the seismic ground motion at the site must be considered. In addition, the dissipation of pore-water pressure and lateral ground spreading must be considered to investigate their effects on the soil–pile interaction, which could be caused by the main shock before the tsunami arrival. Therefore, the liquefaction process might be completed before the tsunami arrival and the shear strength of the soil might have recovered by the dissipation of pore-water pressure. However, it may take longer for the soil to regain its shear strength so it might occur within the several tens of minutes between the main shock and tsunami arrival, such as with the repeated occurrence of soil liquefaction in New Zealand within a year due to the September 2010 Canterbury earthquake and the February 2011 Christchurch earthquake.

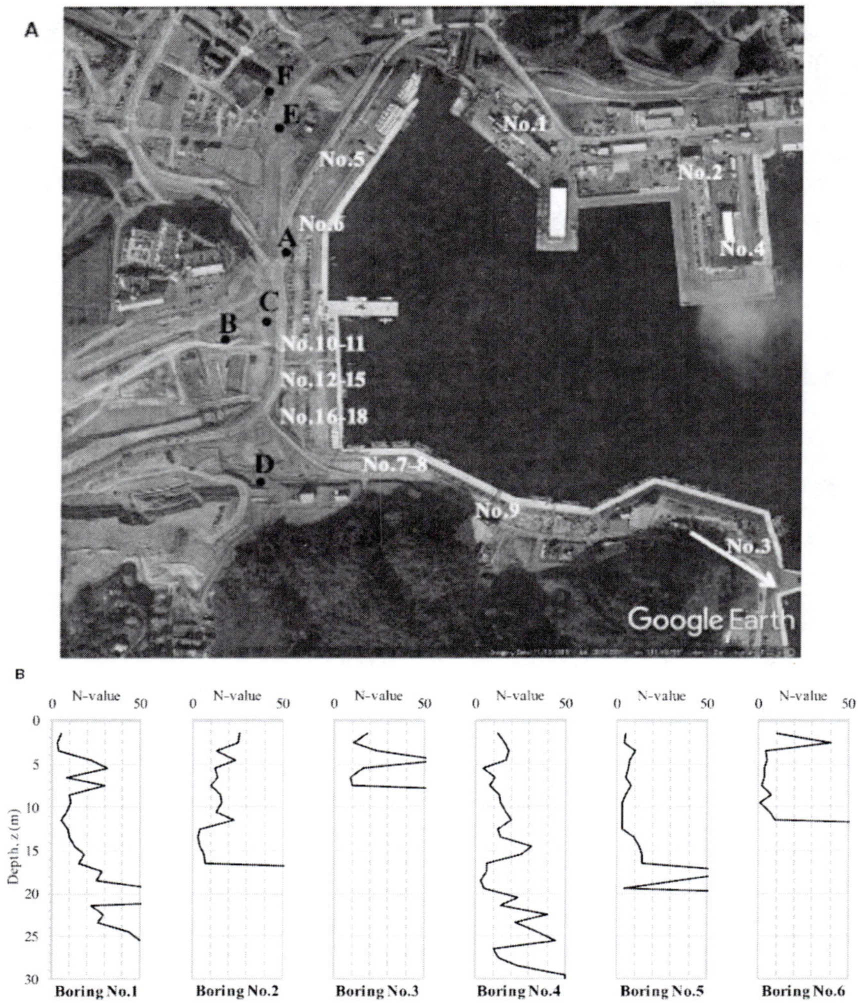

FIGURE 4. 18 SOIL BORING DATA IN THE TOWN OF ONAGAWA. **(A)** 18 boring locations (Google Earth). **(B)**Standard penetration test (SPT)-*N* value.

Because soil liquefaction is an interaction between soil and water, it takes time for the soil and water to slide out. Based on field surveys after two earthquakes in Japan, Mizutani (2008) reported that the soil liquefaction process took considerably longer than the ground shaking process. In the case of the 1964 Niigata earthquake, the sand boiling started after the ground shaking ended, and a large amount of soil boiling was observed for more than 10 to several tens of minutes. In the case of the 1983 Japan Sea earthquake, there was a report of a large sand boil hole with a diameter of 8 m and a depth of 1.5 m that sprayed soil with to a height of 10 m (Japanese Geotechnical Society, Tohoku Branch, 1986), and the outflow of water lasted for more than half a day. Therefore, the direct damage caused by strong ground shaking occurs over several tens of seconds, whereas the indirect damage caused by soil liquefaction requires much more time, and thus is responsible for minimal or no fatalities.

Yeh et al. (2013) suggested that soil liquefaction due to strong ground shaking of the earthquake that had occurred approximately 40 min prior to the tsunami arrival may have further promoted overturning failure. Fraser et al. (2013) also suggested that although any evidence of liquefaction was washed away in the tsunami, it may have contributed by loosening the soil around the piles prior to the overturning motion. Although, the liquefaction process in Onagawa is still not fully understood, it is worth considering the effect of soil liquefaction based on a building design standard because the 2011 tsunami arrived at Onagawa within approximately 40 min of its initiation, when soil liquefaction could not have been negligible. In this study, cases with and without soil liquefaction will be analyzed and compared.

EFFECT OF SOIL LIQUEFACTION ON SKIN FRICTION CAPACITY

The possibility of soil liquefaction might not have been considered in the design of these overturned buildings because the effect of soil liquefaction on pile foundation had likely not been clearly described. Many factors influence the occurrence of soil liquefaction, such as the earthquake magnitude, peak ground acceleration, and soil condition. The Technical Standard Manual for Building [Building Center of Japan (BCJ), 2007] proposed the determination of soil conditions, such as liquefaction hazards. It suggested that soil liquefaction can occur in sandy soil based on four conditions, including alluvium within 20 m from the ground surface, saturated soil, less fine particles, and a lower N value. In this study, a conventional method based on the AIJ standards was used to evaluate the effect of soil liquefaction on the pile foundations. Based on the recommendations for design of building foundations (Architectural Institute of Japan, 2001), a description of the soil properties in this conventional method can be used to calculate the safety factor of each soil layer against the occurrence of soil liquefaction. This safety factor indicates the potential for soil liquefaction, which can cause a loss of skin friction between the soil and pile and result in a reduction of total pile resistance force.

Because boring data were not available at the exact location of each overturned building, skin friction capacity was evaluated from 18 boring data in an adjacent area. The majority of the filled soil at these 18 boring locations contains sand and gravel, which can cause a loss of skin friction between the pile and soil when soil liquefaction occurs. With this assumption, the conventional method was sufficient to calculate skin friction capacity when soil liquefaction occurs. Based on the report from field surveys after the 2011 Great East Japan tsunami (National Institute for Land and Infrastructure Management (NILIM) and Building Research Institute (BRI) in Japan, 2012), a sample calculation of skin friction capacity from this conventional method for boring No. 1 in Figure 4 is shown in Table 1. Skin friction capacity (Q_s) can be calculated from the friction stress of the pile in sand and clay layers. The objective of this conventional method is to evaluate skin friction capacity when soil liquefaction occurs (Q′s)(Qs′), as shown in Table 1. The soil parameters in Table 1 are explained in Table 2. The cyclic resistance ratio (CRR) and cyclic stress ratio (CSR) were used to calculate the safety factor (FS), which can determine skin friction capacity during soil liquefaction as

$$\mathrm{FS} = \frac{\mathrm{CRR}}{\mathrm{CSR}},$$

$$\mathrm{CRR} = 0.041\left[\sqrt{N_\mathrm{a}} + 0.00903\left(N_\mathrm{a} / 10\right)^7\right],$$

$$\mathrm{CSR} = \gamma_\mathrm{n} \frac{\alpha_\mathrm{max}}{g} \frac{\sigma_\mathrm{z}}{\sigma'_\mathrm{z}} \gamma_\mathrm{d},\ \gamma_\mathrm{n} = 0.1\left(M - 1\right),$$

where M is the earthquake magnitude, α_{max} is the peak ground acceleration, and γ_d is the stress reduction factor, which equal to $1 - 0.015z$.

TABLE 1. Evaluation of skin friction capacity for boring no. 1.

z (m)	h (m)	Soil type	N	γ (kg/m³)	σ'_z (kPa)	σ_z (kPa)	F_c (%)	D_{50} (m)	N_a	CRR	CSR (500 gal)	FS (500 gal)	Q_s/φ_p (kN/m)	Q'_s/φ_p (kN/m)
1.00	1.00	Sandy gravel mixed	–	–	–	–	–	–	–	–	–	–	–	–
1.60	0.60	with clay (fill)	5	2,100	6.47	12.36	15	0.03	16	0.17	0.76	0.23	6.00	0.00
2.00	0.40	Silty fine sand (fill)	5	1,800	9.61	19.42	15		23	0.32	0.80	0.39	4.00	0.00
2.70	0.70		3	1,800	15.11	31.78	15		15	0.16	0.82	0.20	4.20	0.00
3.00	0.30	Sandy silt (fill)	3	1,800	17.46	37.08							3.94	3.94
3.25	0.25		4	1,800	19.42	41.50							4.38	4.38
4.00	0.75	Silty sand (fill)	4	1,800	25.31	54.74	15		15	0.16	0.83	0.20	6.00	0.00
4.50	0.50		22	1,800	29.23	63.57	15		47	18.80	0.83	22.72	22.00	22.00
5.00	0.50	Sandy gravel mixed	22	2,100	34.63	73.87		0.005	26	0.48	0.81	0.60	22.00	0.00
6.00	1.00	with pebble (fill)	31	2,100	45.42	94.47		0.005	32	1.39	0.77	1.80	62.00	62.00
7.00	1.00		8	2,100	56.21	115.07		0.005	7	0.11	0.75	0.15	16.00	0.00
7.95	0.95		30	2,100	66.46	134.64		0.005	25	0.45	0.73	0.62	57.00	0.00
8.00	0.05	Sandy silt mixed with	30	1,800	66.86	135.53							6.56	6.56
8.60	0.60	gravel	10	1,800	71.56	146.12							26.25	26.25
9.00	0.40	Sandy gravel	10	2,100	75.88	154.36		0.002	9	0.12	0.72	0.17	8.00	0.00
10.00	1.00		11	2,100	86.67	174.96		0.002	9	0.13	0.70	0.18	22.00	0.00
10.40	0.40		8	2,100	90.99	183.20		0.002	7	0.11	0.69	0.15	6.40	0.00
11.00	0.60	Silty clay	8	1,500	93.93	192.03							21.00	21.00
11.45	0.45		5	1,500	96.14	198.65							9.84	9.84
12.00	0.55	Sandy gravel	5	2,100	102.07	209.98		0.002	4	0.08	0.69	0.12	5.50	0.00
13.00	1.00		9	2,100	112.86	230.58		0.002	7	0.11	0.67	0.16	18.00	0.00
13.90	0.90		10	2,100	122.58	249.12		0.002	7	0.11	0.66	0.17	18.00	0.00
14.00	0.10	Fine sand mixed	10	1,950	123.51	251.04			9	0.12	0.66	0.19	2.00	0.00
14.50	0.50	with silt	13	1,950	128.17	260.60			11	0.14	0.65	0.21	13.00	0.00
14.80	0.30	Clay mixed with gravel	13	1,500	129.64	265.02							17.06	17.06
15.00	0.20	Sandy gravel	13	2,100	131.80	269.14		0.002	9	0.12	0.65	0.19	5.20	0.00
16.00	1.00		18	2,100	142.59	289.74		0.002	12	0.14	0.63	0.23	36.00	0.00
17.00	1.00		15	2,100	153.38	310.34		0.002	10	0.13	0.62	0.21	30.00	0.00
18.00	1.00		28	2,100	164.17	330.94		0.002	17	0.19	0.60	0.31	56.00	0.00
19.00	1.00		25	2,100	174.96	351.54		0.002	15	0.16	0.59	0.28	50.00	0.00
20.00	1.00		50	2,100	185.75	372.14		0.002	29	0.83	0.57	1.45	100.00	100.00
21.00	1.00		50	2,100	196.54	392.74		0.002	28	0.72	0.56	1.28	100.00	100.00
22.00	1.00		22	2,100	207.33	413.34		0.002	12	0.14	0.55	0.26	44.00	0.00
23.00	1.00		29	2,100	218.13	433.95		0.002	15	0.17	0.53	0.32	58.00	0.00
24.00	1.00		26	2,100	228.92	454.55		0.002	13	0.15	0.52	0.30	52.00	0.00
24.35	0.35		44	2,100	232.69	461.76		0.002	23	0.31	0.51	0.60	30.80	0.00
–	–	Bedrock	–	–	–	–	–	–	–	–	–	–	–	–

R'_{fc}: pile resistance force in case of soil liquefaction.
φ_p: peripheral length of a pile.
The red color represents that the safety factor (FS) is less than 1.0, so that Qs become zero.
The bold font represents the pile length of 4.0 m, 6.0 m, and 8.0 m in ***Table 3***.

The CRR can be estimated from the N value of the standard penetration test, and the CSR can be estimated from the peak ground acceleration and earthquake magnitude. As noted above, sand and gravel may cause a loss of skin friction capacity between the pile and soil when soil liquefaction occurs. As shown in Table 1, skin friction capacity becomes 0 if the safety factor is less than 1. Total pile resistance force of each pile was calculated by the summation of all soil layers multiplied by the peripheral length of a pile. To calculate the resisting moment from pile resistance forces, the load distribution on the pile group was considered for the summation of all piles. From these 18 boring data in the adjacent area, skin friction capacity in each pile length was evaluated, resulting in a decrease of total pile resistance force, as shown in Table 3. The overturning moment from hydrodynamic and buoyancy forces and the resisting moment from building self-weight and pile resistance force were considered to investigate possible overturning mechanisms in the next section.

TABLE 2. Description of each parameter (**Architectural Institute of Japan, 2001**).

Symbol	Soil properties	Determination method
N	Standard penetration test (SPT)-N value	Obtained from SPT
γ	Unit weight of soil	Assumed from each soil type
σ'_z	Effective stress	$(\gamma - \gamma_w)h$
σ_z	Absolute stress	γh
F_c	Fine particle ratio	Estimated from each soil type (only sand and gravel)
D_{50}	50% Particle size	Estimated from each soil type (only sand and gravel)
N_a	Corrected N value	$N_a = N_1 + N_f$
N_1	Conversed value of N	$N_1 = C_N N,\ C_N = \sqrt{1/\sigma'_z}$
N_f	Correction factor for fine content	Varied with fine particle ratio

POSSIBLE OVERTURNING MECHANISM

When investigating the possible mechanism of each overturned building, the overturning moment is the result of hydrodynamic and buoyancy forces, whereas the resisting moment is the result of building self-weight and pile resistance force. For a pile foundation, all of the piles could fail during the ground shaking due to large base shear force between the pile heads and pile caps; thus, the overturned buildings could resist tsunami flow using only their building self-weight. These overturned buildings floated and then moved from their original positions, so buoyancy force rapidly exceeded building self-weight after overturning. However, all piles may also have still been in good condition to resist tsunami flow after the ground shaking. Thus, two possible mechanisms of these overturned buildings with pile foundation are tension failure at the pile heads and pulling of the piles out of the ground.

TABLE 3. Skin friction capacity of 18 soil boring data for each pile length.

Boring data	Pile length 4.0 m		Pile length 6.0 m		Pile length 8.0 m	
	Q_s/φ_p	Q'_s/φ_p	Q_s/φ_p	Q'_s/φ_p	Q_s/φ_p	Q'_s/φ_p
No. 1	28.51	8.31	134.51	92.31	214.08	98.88
No. 2	128.00	102.00	200.00	102.00	248.00	102.00
No. 3	106.00	36.00	238.00	136.00	276.00	136.00
No. 4	92.00	0.00	134.00	0.00	172.00	0.00
No. 5	38.00	0.00	64.00	0.00	118.50	52.50
No. 6	117.50	117.50	152.50	152.50	176.13	176.13
No. 7	231.75	231.753	262.38	262.38	303.06	303.06
No. 8	62.00	31.20	90.00	31.20	124.00	31.20
No. 9	66.55	43.75	106.55	43.75	148.55	43.75
No. 10	31.50	31.50	52.50	52.50	79.33	76.13
No. 11	24.00	0.00	38.00	0.00	48.00	0.00
No. 12	18.00	0.00	42.00	0.00	78.00	0.00
No. 13	31.50	31.50	49.90	48.30	165.90	148.30
No. 14	16.00	0.00	32.93	6.13	238.55	211.75
No. 15	106.00	42.00	156.00	42.00	356.00	242.00
No. 16	102.38	102.38	426.13	426.13	736.75	736.75
No. 17	164.00	146.00	262.25	243.25	283.25	264.25
No. 18	166.00	100.00	202.00	100.00	312.00	200.00

Unit (kN/m): force per peripheral length of pile.
Pile length: depth (z) from ground in ***Table 1****.*

Potential Shear Failure of Piles during Ground Shaking

For low-rise regular buildings, equivalent static seismic loads are sufficient to consider base shear force instead of dynamic loads during the ground shaking. The equivalent static seismic loads can be calculated based on a response spectrum analysis using natural period. Based on the AIJ Recommendations for Loads on Buildings (Architectural Institute of Japan, 2004), a simplified method can be used to evaluate the equivalent static seismic loads from the observed response spectrum at the site and an approximation of natural period (T_1) for each overturned building. Therefore, the approximate base shear force (V_B) can be calculated as

$$V_B = 0.816\frac{S_a}{g}W,$$

$$T_1 = (0.02 + 0.01\alpha_h)h,$$

where S_a is the acceleration response at the base of the foundation, g is the gravitational acceleration (9.81 m/s^2), W is building self-weight, and α_h is the ratio of steel-frame height to total building height (h).

The base shear force was compared with total shear capacity of the piles in Section "Tension and Shear Capacities" to investigate the potential shear failure of the piles during ground shaking using a safety factor between total shear capacity ($\sum Q_u$) and base shear force (V_B). Table 4 shows the calculation of total shear capacity and base shear force for Buildings B, C, D, and E. The acceleration response (S_a) was obtained from the response spectrum in Ishinomaki city that was provided by NIED, which is the nearest available data to Onagawa [National Research Institute for Earth Science and Disaster Resilience (accessed 2016)]. Based on the safety factor, all piles of Building D failed during the ground shaking due to large base shear force between the pile heads and pile caps, whereas all piles of Buildings C and E were still in a sufficiently good condition to resist tsunami. In addition, some piles of Building B could have failed during the ground shaking because the safety factor is close to 1.00, as shown in Table 4.

TABLE 4. Total shear capacity ($\sum q_u$) and base shear force (v_b) at pile foundation.

Building	Number of piles	F_u (kN)	Q_u (kN)	$\sum Q_u$ (kN)	T_1 (s)	S_a (gal)	W (kN)	V_B (kN)	Safety factor
B	32	153	65	2,080	0.28	1,050	2,320	2,030	1.02
C	20	306	177	3,540	0.43	1,000	3,340	2,780	1.27
D	24	153	65	1,560	0.21	1,400	7,240	8,430	0.18
E	14	306	177	2,478	0.14	1,100	1,780	1,630	1.52

Overturned Buildings Resisted by Only Building Self-Weight

The possible overturning mechanism of Buildings A and D was investigated by comparing the overturning moment calculated from hydrodynamic force (F_d) and buoyancy force (F_b) to the resisting moment calculated from building self-weight (W). Building A had a mat foundation, and all piles of Building D failed during the ground shaking, as shown in Figure 5.

Figure 6A shows the tsunami inundation depths and flow velocities at Buildings A and D in the 2014 and 2016 simulations. The residual air space ratio (C_b) was estimated based on the condition of larger buoyancy force than building self-weight, as shown in Figure 6B. The residual air space ratios for Buildings A and D were approximately 0.6 and 0.4, respectively. Figure 6C shows the time series of the overturning moment (M_d and M_b) and the resisting moment (M_w).

FIGURE 5. MAT FOUNDATION OF BUILDING A AND PILE FOUNDATION OF BUILDING D. Note: taken by our survey team in March 29, 2011 at the town of Onagawa.

As shown in Table 5, a comparison of Building A based on the 2014 simulation reveals that the peak overturning moment ($M_d + M_b$) that occurred at 15:28 was 1.59 times higher than the resisting moment (M_w), whereas a comparison based on the 2016 simulation suggests that the peak overturning moment that occurred at 15:29 was 1.84 times higher than the resisting moment. The peak overturning moment in the 2014 simulation occurred when the inundation depth was 10.60 m and the flow velocity was 2.83 m/s^2, whereas the maximum depth and velocity of 13.50 m and 4.57 m/s^2, respectively, occurred at different times, as shown in Figure 6A. The peak overturning moment in the 2016 simulation occurred when the inundation depth was 11.43 m and the flow velocity was 3.48 m/s^2, whereas the maximum depth of 16.18 m occurred at different times, as shown in Figure 6A. Building A was only 10.5 m tall; thus, it was overturned when the tsunami flow exceeded the top of the building.

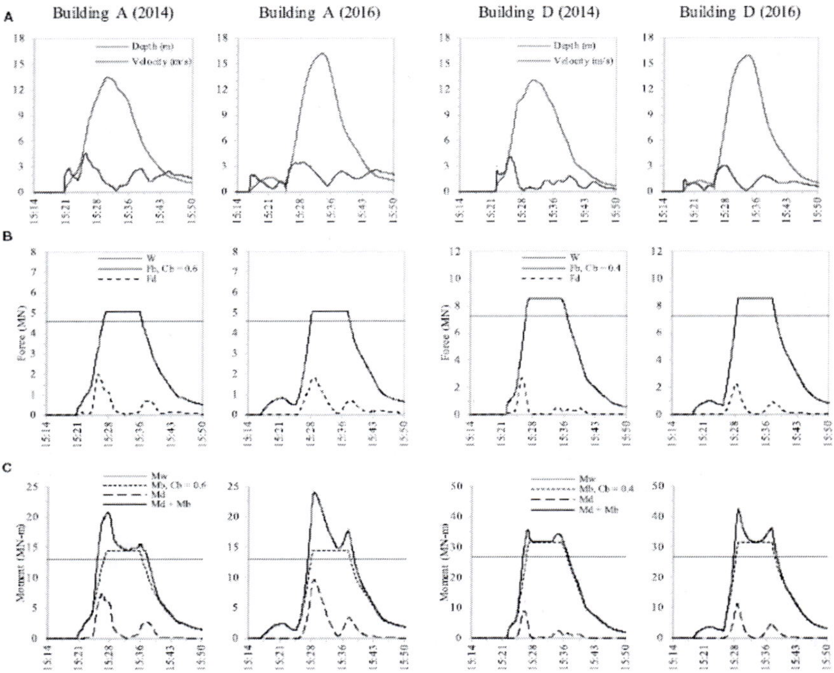

FIGURE 6. ANALYSIS RESULTS OF BUILDINGS A AND D FROM 2014 AND 2016 SIMULATIONS. (A)Tsunami inundation depth and flow velocity. **(B)** Building self-weight, buoyancy force, and hydrodynamic force. **(C)** Resisting moment from building self-weight and overturning moment from hydrodynamic and buoyancy forces.

TABLE 5. Analysis results of all overturned buildings.

Analysis results	Building A		Building B		Building C		Building D		Building E	
	2014	2016	2014	2016	2014	2016	2014	2016	2014	2016
Maximum inundation depth (m)	13.50	16.18	13.19	16.02	13.18	16.27	13.05	15.93	13.63	15.88
Maximum flow velocity (m/s²)	4.57	3.48	4.90	3.03	4.43	2.43	4.07	3.02	3.82	3.15
Residual air space ratio (C_b)	0.6	0.6	0.7	0.7	0.5	0.5	0.4	0.4	0.6	0.6
Max depth (m) at peak overturning moment	10.60	11.43	8.71	12.45	8.61	12.14	9.98	10.43	7.06	10.57
Max vel. (m/s²) at peak overturning moment	2.83	3.48	3.73	2.10	3.77	1.95	2.02	2.71	3.48	3.15
Time at peak overturning moment	15:28	15:29	15:27	15:30	15:27	15:30	15:27	15:29	15:26	15:29
Overturning ratio (OR)	1.59	1.84	1.61	1.59	1.17	1.24	1.33	1.59	1.47	1.34

All piles of Building D failed during the ground shaking by shear failure between the pile heads and pile caps, in which the shear strength of a pile (Q_u) was 65 kN, as shown in Table 4. As shown in Table 5, a comparison of Building D based on the 2014 simulation reveals that the peak overturning moment ($M_d + M_b$) that occurred at 15:27 was 1.33 times higher than the resisting moment (M_w), whereas a comparison based on the 2016 simulation suggests that the peak overturning moment that occurred at 15:29 was 1.59 times higher than the resisting moment. The peak overturning moment in the 2014 simulation occurred when the inundation depth was 9.98 m and the flow velocity was 2.02 m/s^2, whereas the maximum depth and velocity of 13.05 m and 4.07 m/s^2, respectively, occurred at different times, as shown in Figure 6A. The peak overturning moment in the 2016 simulation occurred when the inundation depth was 10.43 m and the flow velocity was 2.71 m/s^2, whereas the maximum depth and velocity of 15.93 m and 3.02 m/s^2, respectively, occurred at different times, as shown in Figure 6A. Building D was 10.5 m tall; thus, it was overturned when the tsunami flow was lower than the top of the building.

Another observation is that buoyancy force could generate an overturning moment larger than hydrodynamic force, particularly for Building D, as shown in Figure 6C. After building overturning occurred at 15:28, these buildings floated and then moved from their original positions such that buoyancy force was immediately greater than building self-weight, as shown in Figure 6B.

Tension Failure of Piles Caused by Overturning Moment

The overturning mechanism of Building C was investigated by comparing the overturning moment calculated from hydrodynamic force (F_d) and buoyancy force (F_b) to the resisting moment calculated from building self-weight (W) and pile resistance force (R_{TC}). Building C had a pile foundation and all of the piles failed by tension failure at the pile heads, except for one pile at the top-right pile cap, as shown in Figure 7. Figure 8A shows the tsunami inundation depth and flow velocity at Building C in the 2014 and 2016 simulations. The residual air space ratio (C_b) was approximately 0.5, as estimated based on the condition of larger buoyancy force than building self-weight, as shown in Figure 8B. Figure 8C shows the time series of the overturning moment (M_d and M_b) and the resisting moment (M_w and M_r).

FIGURE 7. PILE FOUNDATION OF BUILDING C (ONE PILE PULLED OUT OF GROUND). Note: taken by our survey team in March 29, 2011 at the town of Onagawa.

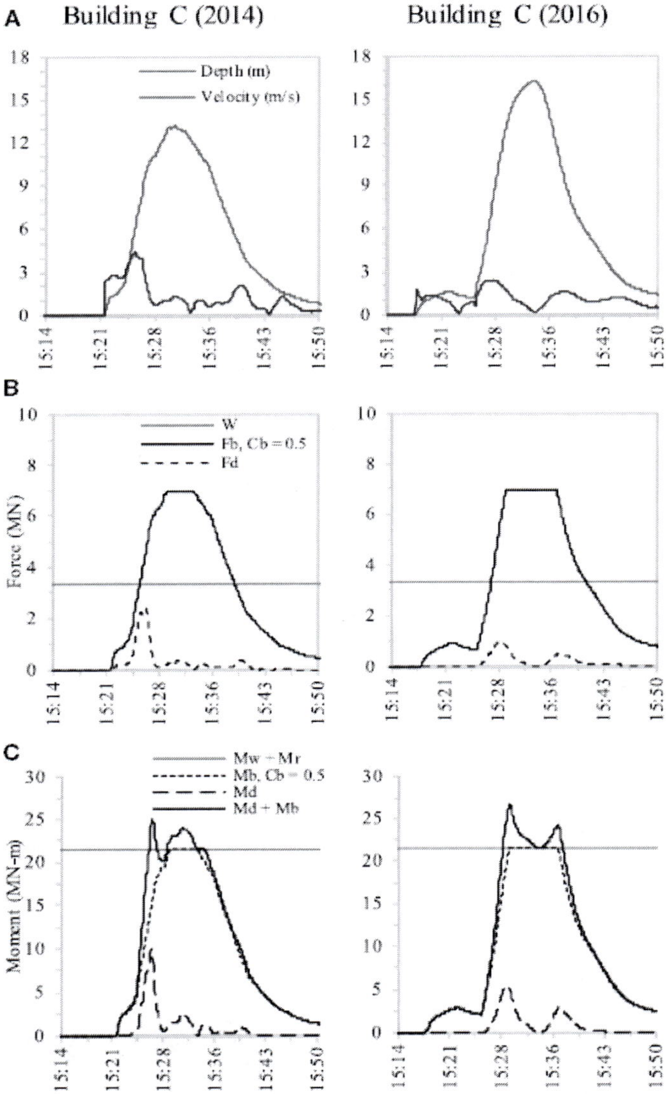

FIGURE 8. ANALYSIS RESULTS OF BUILDING C FROM 2014 AND 2016 SIMULATIONS. (A) Tsunami inundation depth and flow velocity. **(B)** Building self-weight, buoyancy force, and hydrodynamic force. **(C)** Resisting moment from building self-weight and pile resistance force, and overturning moment from hydrodynamic and buoyancy forces.

For Building C, the pile foundation failed during tsunami flow through the observed tension failure between the pile heads and pile caps in which the tensile strength of a pile (F_u) was 307 kN, as shown in Table 4. The effective piles out of the 20 piles were used to calculate the resisting moment (M_r), with 4.0-m moment arm on the pile group equal to the pile length, as shown in Figure 7. As shown in Table 5, a comparison of Building C based on the 2014 simulation reveals that the peak overturning moment ($M_d + M_b$) that occurred at 15:27 was 1.17 times higher than the resisting moment ($M_w + M_r$), whereas a comparison based on the 2016 simulation suggests that the peak overturning moment that occurred at 15:30 was 1.24 times higher than the resisting moment. The peak overturning moment in the 2014 simulation occurred when the inundation depth was 8.61 m and the flow velocity was 3.77 m/s^2, whereas the maximum depth and velocity of 13.18 m and 4.43 m/s^2, respectively, occurred at different times, as shown in Figure 8A. The peak overturning moment in the 2016 simulation occurred when the inundation depth was 12.14 m and the flow velocity was 1.95 m/s^2, whereas the maximum depth and velocity of 16.27 m and 2.43 m/s^2, respectively, occurred at different times, as shown in Figure 8A. Building C was 14.2 m tall; thus, it was overturned when tsunami flow was lower than the top of the building.

Buoyancy force could generate an overturning moment equal to the resisting moment, as shown in Figure 8C. Therefore, although hydrodynamic force could generate a smaller overturning moment than buoyancy force, the additional overturning moment from hydrodynamic force had a significant impact on building overturning.

Pulling out of Piles including Effect of Soil Liquefaction

The overturning mechanisms of Buildings B and E were investigated by comparing the overturning moment calculated from hydrodynamic force (F_d) and buoyancy force (F_b) to the resisting moment calculated from building self-weight (W) and pile resistance force (R_{TC}). Building B had a pile foundation, and 12 piles were pulled out of the ground, as shown in Figure 9, whereas 20 piles might have failed as a result of base shear force during the ground shaking because the safety factor was close to 1.00, as shown in Table 4. Building E had a pile foundation, and all piles were pulled out of the ground except for one pile that failed in tension, as shown in Figure 9. Figure 10A shows the tsunami inundation depths and flow velocities at Buildings B and E in the 2014 and 2016 simulations. The residual air space ratio (C_b) was estimated based on the condition of larger buoyancy force than building self-weight, as shown in Figure 10B. The residual air space ratios for Buildings B and E were approximately 0.7 and

0.6, respectively. Figure 10C shows the time series of the overturning moment (M_d and M_b) and the resisting moment ($M_w$and M_r).

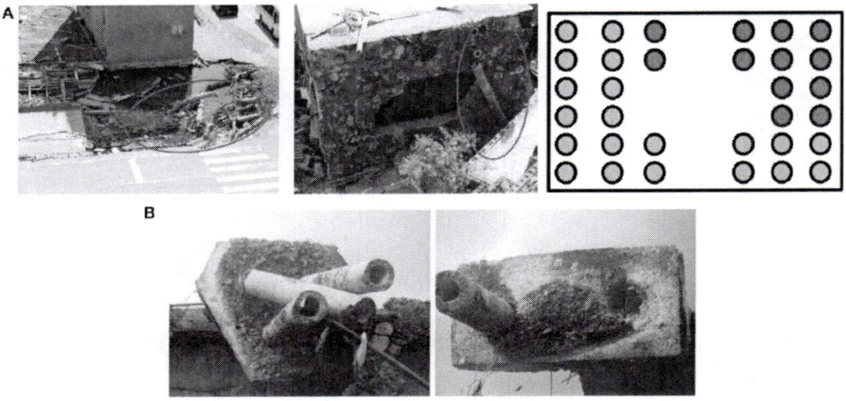

FIGURE 9. PILE FOUNDATION OF BUILDINGS B AND E. **(A)** Red circles are 12 piles pulled out of ground, and gray circles are 20 piles broken during ground shaking. **(B)** Only one pile failed in tension. Note: taken by our survey team in March 29, 2011 at the town of Onagawa.

For Building B, boring No. 11 at the nearest location and the assumed pile length of 6 m were used to calculate skin friction capacity (Q_s) of a pile, which was 38 kN/m, for a pile diameter of 20 cm, as shown in Table 3. As shown in Table 5, the comparison of Building B based on the 2014 simulation reveals that the peak overturning moment ($M_d + M_b$) that occurred at 15:27 was 1.61 times higher than the resisting moment ($M_w + M_r$), whereas the comparison based on the 2016 simulation suggests that the peak overturning moment that occurred at 15:30 was 1.59 times higher than the resisting moment. The peak overturning moment in the 2014 simulation occurred when the inundation depth was 8.71 m and the flow velocity was 3.73 m/s^2, whereas the maximum depth and velocity of 13.19 m and 4.90 m/s^2, respectively, occurred at different times, as shown in Figure 10A. The peak overturning moment in the 2016 simulation occurred when the inundation depth was 12.45 m and the flow velocity was 2.10 m/s^2, as the maximum depth and velocity of 16.02 m and 3.03 m/s^2, respectively, occurred at different times, as shown in Figure 10A. Building B was 14.0 m tall; thus, it was overturned when tsunami flow was lower than the top of the building.

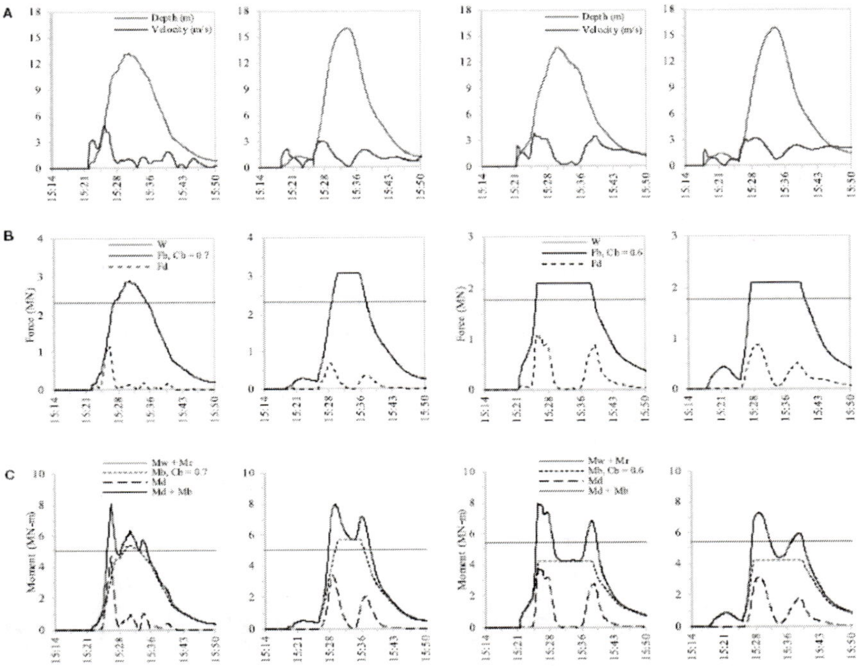

FIGURE 10. ANALYSIS RESULTS OF BUILDINGS B AND E FROM 2014 AND 2016 SIMULATIONS. (A)Tsunami inundation depth and flow velocity. **(B)** Building self-weight, buoyancy force, and hydrodynamic force. **(C)** Resisting moment from building self-weight and pile resistance force, and overturning moment from hydrodynamic and buoyancy forces.

For Building E, boring No. 5 at the nearest location and the assumed pile length of 6 m were used to calculate skin friction capacity (Q_s) of a pile, which was 64 kN/m, for a pile diameter of 25 cm, as shown in Table 3. As shown in Table 5, the comparison of Building B based on the 2014 simulation in reveals that the peak overturning moment ($M_d + M_b$) that occurred at 15:26 was 1.47 times higher than the resisting moment ($M_w + M_r$), whereas the comparison based on the 2016 simulation suggests that the peak overturning moment that occurred at 15:29 was 1.34 times higher than resisting moment. The peak overturning moment in the 2014 simulation occurred when the inundation depth was 7.06 m and the flow velocity was 3.48 m/s^2, whereas the maximum depth and velocity of 13.63 m and 3.82 m/s^2, respectively, occurred at different times, as shown in Figure 10A. The peak overturning moment in 2016 occurred when the inundation depth was 10.57 m and the flow velocity was 3.15 m/s^2, whereas the maximum depth and velocity were 15.88 m and 3.15 m/s^2, respectively, occurred at different times, as shown in Figure 10A. Building E was 7.0 m tall; thus, it was overturned when tsunami flow exceeded the top of the building.

The overturning ratio (OR) in Table 5 can be calculated from the ratio between the peak overturning moment and the peak resisting moment. For each of 18 soil boring data, skin friction capacity (Q_S) is shown in Table 3 for pile lengths of 4.0, 6.0, and 8.0 m and including the effect of soil liquefaction R′TCRTC′. However, the conventional method to evaluate the effect of soil liquefaction tends to overestimate liquefaction hazards (Chen et al., 2016). These 18 boring data were used to evaluate the potential of overturning for Buildings B and E based on the ORs, which can be classified as no possibility, low possibility, medium possibility, or high possibility.

TABLE 6. Overturning ratio (or) of buildings b and e for each soil boring data.

	Boring data	Pile length 4.0 m		Pile length 6.0 m		Pile length 8.0 m	
		OR	OR'	OR	OR'	OR	OR'
Building B 2014	No.1	1.67	1.80	1.21	1.36	1.00	1.33
	No.2	1.23	1.32	1.03	1.32	0.93	1.32
	No.3	1.31	1.62	0.95	1.20	0.89	1.20
	No.4	1.36	1.86	1.21	1.86	1.10	1.86
	No.5	1.61	1.86	1.48	1.86	1.26	1.54
	No.6	1.27	1.27	1.16	1.16	1.09	1.09
	No.7	0.97	0.97	0.91	0.91	0.84	0.84
	No.8	1.49	1.65	1.37	1.65	1.24	1.65
	No.9	1.47	1.58	1.30	1.58	1.17	1.58
	No.10	1.65	1.65	1.54	1.64	1.41	1.42
	No.11	1.69	1.86	1.61	1.86	1.58	1.86
	No.12	1.73	1.86	1.59	1.86	1.42	1.86
	No.13	1.65	1.65	1.55	1.58	1.12	1.17
	No.14	1.75	1.86	1.64	1.81	0.95	1.01
	No.15	1.31	1.59	1.15	1.59	0.77	0.95
	No.16	1.32	1.32	0.69	0.69	0.47	0.47
	No.17	1.12	1.17	0.91	0.94	0.87	0.91
	No.18	1.12	1.33	1.03	1.33	0.83	1.03

	Boring data	Pile length 4.0 m		Pile length 6.0 m		Pile length 8.0 m	
		OR	OR'	OR	OR'	OR	OR'
Building B 2016	No.1	1.65	1.77	1.19	1.34	0.99	1.32
	No.2	1.22	1.30	1.02	1.30	0.92	1.30
	No.3	1.29	1.60	0.94	1.19	0.87	1.19
	No.4	1.34	1.83	1.20	1.83	1.09	1.83
	No.5	1.59	1.83	1.46	1.83	1.25	1.52
	No.6	1.25	1.25	1.14	1.14	1.08	1.08
	No.7	0.95	0.95	0.90	0.90	0.83	0.83
	No.8	1.47	1.63	1.35	1.63	1.23	1.63
	No.9	1.45	1.56	1.29	1.56	1.15	1.56
	No.10	1.63	1.63	1.52	1.52	1.39	1.41
	No.11	1.67	1.83	1.59	1.83	1.54	1.83
	No.12	1.71	1.83	1.57	1.83	1.40	1.83
	No.13	1.63	1.63	1.53	1.54	1.10	1.15
	No.14	1.72	1.83	1.62	1.79	0.94	1.00
	No.15	1.29	1.57	1.13	1.57	0.76	0.93
	No.16	1.30	1.30	0.68	0.68	0.47	0.47
	No.17	1.11	1.16	0.90	0.93	0.86	0.89
	No.18	1.10	1.31	1.02	1.31	0.82	1.02

	Boring data	Pile length 4.0 m		Pile length 6.0 m		Pile length 8.0 m	
		OR	OR'	OR	OR'	OR	OR'
Building E 2014	No.1	1.88	2.10	1.22	1.42	0.96	1.38
	No.2	1.25	1.37	1.00	1.37	0.89	1.37
	No.3	1.35	1.81	0.91	1.21	0.83	1.21
	No.4	1.42	2.20	1.22	2.20	1.08	2.20
	No.5	1.79	2.20	1.59	2.20	1.29	1.67
	No.6	1.29	1.29	1.15	1.15	1.07	1.07
	No.7	0.92	0.92	0.86	0.86	0.78	0.78
	No.8	1.61	1.85	1.43	1.85	1.26	1.85
	No.9	1.57	1.74	1.34	1.74	1.16	1.74
	No.10	1.85	1.85	1.67	1.67	1.49	1.51
	No.11	1.92	2.20	1.79	2.20	1.71	2.20
	No.12	1.89	2.20	1.76	2.20	1.50	2.20
	No.13	1.85	1.85	1.69	1.71	1.10	1.17
	No.14	2.01	2.20	1.84	2.12	0.91	0.97
	No.15	1.35	1.76	1.14	1.76	0.70	0.90
	No.16	1.36	1.36	0.62	0.62	0.41	0.41
	No.17	1.11	1.17	0.86	0.90	0.82	0.85
	No.18	1.10	1.38	1.00	1.38	0.77	1.00

	Boring data	Pile length 4.0 m		Pile length 6.0 m		Pile length 8.0 m	
		OR	OR'	OR	OR'	OR	OR'
Building E 2016	No.1	1.72	1.92	1.12	1.30	0.88	1.26
	No.2	1.14	1.25	0.92	1.25	0.81	1.25
	No.3	1.23	1.66	0.83	1.11	0.76	1.11
	No.4	1.30	2.01	1.12	2.01	0.99	2.01
	No.5	1.64	2.01	1.46	2.01	1.18	1.53
	No.6	1.18	1.18	1.05	1.05	0.98	0.98
	No.7	0.84	0.84	0.78	0.78	0.72	0.72
	No.8	1.47	1.70	1.31	1.70	1.16	1.70
	No.9	1.44	1.60	1.23	1.60	1.07	1.60
	No.10	1.69	1.69	1.53	1.53	1.37	1.38
	No.11	1.76	2.01	1.64	2.01	1.56	2.01
	No.12	1.82	2.01	1.61	2.01	1.37	2.01
	No.13	1.69	1.69	1.55	1.56	1.01	1.07
	No.14	1.84	2.01	1.68	1.94	0.83	0.89
	No.15	1.23	1.61	1.04	1.61	0.64	0.82
	No.16	1.25	1.25	0.57	0.57	0.37	0.37
	No.17	1.02	1.07	0.78	0.82	0.75	0.78
	No.18	1.01	1.26	0.91	1.26	0.70	0.92

NO	OR < 1.00	LOW	1.00 ≤ OR < 1.20
MEDIUM	1.20 ≤ OR < 1.50	HIGH	OR ≥ 1.50

Table 6 shows the ORs of Buildings B and E in the 2014 and 2016 simulations considering the pile length and soil liquefaction. For all pile lengths, only boring No. 7 can provide safety from building overturning because the OR was less than 1.00 and including the effect of soil liquefaction. On the other hand, boring No. 11 provided the high possibility of building overturning for all pile lengths and including the effect of soil liquefaction. For borings No. 16 and No. 17, safety can be obtained using pile lengths of 6.0 and 8.0 m instead of 4.0 m, as was used in boring No. 3. However, building overturning could occur at boring No. 3 with pile lengths of 6.0 and 8.0 m in the case of soil liquefaction. Nevertheless, increasing the pile length can reduce the OR when neglecting the effect of soil liquefaction. In particular, for boring No. 14, increasing the pile length from 4.0 and 6.0 m to 8.0 m can prevent building overturning by changing the classification from high possibility to no possibility.

The effect of soil liquefaction generally increased the ORs for most of 18 soil boring data. However, the ORs for borings No. 6 and No. 16 were the same with and without soil liquefaction because soil liquefaction could not decrease of skin friction capacity. In the case of soil liquefaction, increasing the pile length could not affect the potential for building overturning in several borings, such as No. 2, No. 4, No. 8, No. 9, No. 11, and No. 12, because skin friction capacity is constant for all pile lengths, as shown in Table 3. In some borings, such as No. 2 and No. 3 with a pile length of 8.0 m, the OR exceeded 1.00 when soil liquefaction occurred because skin friction force was small and decreased significantly, as shown in Table 3.

CONCLUSION

Based on the surveyed data, the overturning mechanism of buildings in tsunami can be investigated by comparing the overturning moment induced by hydrodynamic and buoyancy forces and the resisting moment induced by building self-weight and pile resistance force. For a pile foundation, the potential for the shear failure of the piles at the pile head during the ground shaking can be analyzed based on the simplified method in a building design standard. In this study, building overturning was investigated based on three possible mechanisms:

1. Building overturning due to hydrodynamic and buoyancy forces that were resisted by only building self-weight, such as in Buildings A and D.
2. Tension failure of piles at the pile head caused by the overturning moment, such as in Building C.

3. Pulling of the piles from the ground including the effect of soil liquefaction, including in combination with shear failure during the ground shaking in Building B and combination with tension failure during tsunami flow in Building E.

The analysis results of all overturned buildings indicated that buoyancy force could generate a larger overturning moment than hydrodynamic force, particularly for Buildings C and D. Previous studies indicated that the opening ratio had a significant effect on buoyancy force. However, the criterion that uses opening ratio should not be proposed in building design codes because it is difficult to estimate the volume of water inside the building during tsunami flow based on the size of the opening. This study focused on the performance of building foundations during earthquake and subsequent tsunami, including ground shaking, soil liquefaction, and tsunami inundation. The possible failure mechanism of these overturned buildings was investigated based on the residual performance from the earthquake and the sequential damage from the tsunami. The results suggested that a new criterion of building foundation design should be proposed in a building design guideline to prevent building overturning. In this criterion, the building performance should be evaluated from sequential scenarios of an earthquake and tsunami. The building foundation design should consider the states of the art of both earthquake and tsunami engineering. Otherwise, the evaluation of building performance will be misleading.

Soil liquefaction is a consequence of the earthquake that may reduce the performance of building foundation to resist building overturning during the tsunami. Due to soil liquefaction, the loss of skin friction capacity between the pile and soil could occur in most of 18 soil boring data, resulting in a decrease of the resisting moment calculated from pile resistance force. However, these 18 soil boring data are located near the shoreline and not in the precise locations of the overturned buildings, which contain a considerable amount of sand from filling and sediment. This might be a reason why there was a possibility of building overturning at most of 18 soil boring data. In addition, skin friction capacity including the effect of soil liquefaction was calculated by the conventional method in a building design standard. The accuracy of the evaluation of the effect of soil liquefaction could be improved by using a more advanced method, such as soil dynamic analysis.

AUTHOR CONTRIBUTIONS

PL did a contribution on investigating the possible mechanisms of overturned buildings in Onagawa based on the surveyed and simulated data. He wrote all parts in this manuscript. AS did a contribution on collecting the surveyed data of the characteristics of overturned buildings in Onagawa after the 2011 Great East Japan earthquake and tsunami. He wrote some part of the building characteristics. AY did a contribution on studying on building design codes. She suggested a method to evaluate pile resistance force from the Japanese standard for building foundation design and the survey report after the 2011 earthquake and tsunami. BA did a contribution on performing tsunami inundation simulation. He calculated the waveform at each overturned building to evaluate hydrodynamic and buoyancy forces during tsunami flow. SK has supported Bruno for tsunami inundation simulation in Onagawa. He provided the video evidence that was used for verifying the simulation. YK has supported AY because of his expertise in building design engineering. FI has supported PL for overall images of this manuscript and gave very useful comments.

CONFLICT OF INTEREST STATEMENT

The authors declare that the research was conducted in the absence of any commercial or financial relationships that could be construed as a potential conflict of interest.

ACKNOWLEDGMENTS

The authors express their sincere gratitude to the anonymous reviewers for their valuable suggestions, Prof. Nobuo Shuto of Tohoku University and the government office in the town of Onagawa for providing soil boring data, and the cooperation of other members of our survey team, including Dr. Erick Mas and Dr. Hideomi Gokon. This research was funded by the Reconstruction Agency of the Government of Japan, JSPS Grant-in-Aid for Young Scientists (B) "Applying developed fragility functions for the Global Tsunami Model (GTM)" (No. 16K16371), the Willis Research Network, and Tokio Marine & Nichido Fire Insurance Co., Ltd. through IRIDeS, Tohoku University.

REFERENCES

1. Adriano, B., Hayashi, S., Gokon, H., Mas, E., and Koshimura, S. (2016). Understanding the extreme tsunami inundation in Onagawa town by the 2011 Tohoku earthquake, its effects in urban structures and coastal facilities. *Coast. Eng. J.* 58, 1640013.

2. Adriano, B., Mas, E., and Koshimura, S. (2014). "Tsunami inundation features verification through numerical modeling and video analysis: case study of Onagawa town on the 2011 Tohoku earthquake," in *Proc., Asia Oceania Geosciences Society 11th Annual Meeting*, Sapporo.
3. Architectural Institute of Japan. (2001). *Recommendations for Design of Building Foundations*. Tokyo, Japan: Architectural Institute of Japan, 486.
4. Architectural Institute of Japan. (2004). *Chapter 7 Seismic Loads*. Tokyo: AIJ Recommendations for Loads on Buildings, 652.
5. Building Center of Japan (BCJ). (2007). *Technical Standard Manual for Building (In Japanese)*. Tokyo, Japan: Building Center of Japan.
6. Chen, J., Takeyama, T., Otani, H., Fujita, K., and Hori, M. (2016). A framework for assessing liquefaction hazard for urban areas based on soil dynamics. *Int. J. Comput. Methods* 13, 1641011–1641011.
7. Chock, G., Carden, L., Robertson, I., Olsen, M., and Guangren, Y. (2013). Tohoku tsunami-induced building failure analysis with implications for U.S. tsunami and seismic design codes. *Earthquake Spectra* 29, S99–S126.
8. Federal Emergency Management Agency. (2011). *Guidelines for Design of Structures for Vertical Evacuation from Tsunamis*. Washington: FEMA, 158.
9. Fraser, S., Raby, A., Pomonis, A., Goda, K., Chian, S. C., Macabuag, J., et al. (2013). Tsunami damage to coastal defences and buildings in the March 11th 2011 Mw9.0 Great East Japan earthquake and tsunami. *Bull. Earthquake Eng.* 11, 205–239.
10. Ishida, M., Tokimatsu, K., and Inoue, S. (2015). "Overturning of buildings induced by tsunami run-up at Onagawa in the 2011 Great Tohoku Earthquake," in *Proc., 6th International Conference on Earthquake Geotechnical Engineering*, Christchurch.
11. Japanese Geotechnical Society, Tohoku Branch. (1986). *Damage Survey Report of the 1983 Sea of Japan Earthquake (In Japanese)*. Sendai, Japan: Japanese Geotechnical Society, 297.
12. Kabeyasawa, T., Okuda, Y., Fukai, A., Fukuyama, H., Kato, H., Ishihara, T., et al. (2012). "Evaluation of lateral and buoyant forces on reinforced concrete buildings by the tsunami of the 2011 East Japan earthquake," in *Proc., 15th World Conference in Earthquake Engineering*, Lisbon.
13. Latcharote, P., Suppasri, A., Yamashita, A., Adriano, B., Koshimura, S., Kai, Y., et al. (2014). "Mechanism and stability analysis of overturned buildings by the 2011 Great East Japan earthquake and tsunami in Onagawa town," in *Proc., 14th Japan Earthquake Engineering Symposium*. Chiba, 108–117.
14. Mizutani, T. (2008). "Lesson 13 soil liquefaction, principle of natural disaster: disaster case studies (In Japanese)," in *National Research Institute for Earth Science and Disaster Resilience (NIED)*. Available at: http://dil.bosai.go.jp/workshop/
15. National Institute for Land and Infrastructure Management (NILIM) and Building Research Institute (BRI) in Japan. (2012). Report on field surveys and subsequent investigations of building damage following the 2011 off the pacific coast of Tohoku earthquake. *Build. Res. Data* 136, 6.2–32–6.2–40.
16. National Research Institute for Earth Science and Disaster Resilience (NIED). (n.d.). *Strong-Motion Seismograph Networks (K-NET, KiK-net)*. Available at: http://www.kyoshin.bosai.go.jp/ (accessed November 24, 2016).
17. Nikkei BP Company. (2011). Lessons from the Great East Japan earthquake disaster: city and architecture part (In Japanese). *Nikkei Archit. Sect.* 207, 32–38.
18. Onagawa. (1960). *Onagawa Magazine (In Japanese)*. Onagawa, Miyagi Prefecture, 1010.
19. Onagawa. (2013). *Onagawa Guide: Industrial Promotion Division of Onagawa (In Japanese)*. Onagawa Tourism Association.
20. Suppasri, A., Koshimura, S., Imai, K., Mas, E., Gokon, H., Muhari, A., et al. (2012). Damage characteristic and field survey of the 2011 Great East Japan tsunami in Miyagi prefecture. *Coastal Eng. J.* 54, 1250005.
21. Suppasri, A., Shuto, N., Imamura, F., Koshimura, S., Mas, E., and Yalciner, A. C. (2013). Lesson learned from the 2011 Great East Japan tsunami: performance of tsunami

countermeasures, coastal buildings, and tsunami evacuation in Japan. *Pure Appl. Geophys.* 170, 993–1018.
22. Tokimatsu, K., Ishida, M., and Inoue, S. (2016). Tsunami-induced overturning of buildings in Onagawa during 2011 Tohoku earthquake. *Earthquake Spectra* 32, 1989–2007.
23. Tokimatsu, K., Tamura, S., Suzuki, H., and Katsumata, K. (2012). Building damage associated with geotechnical problems in the 2011 Tohoku Pacific Earthquake. *Soils Found.* 52, 956–974.
24. Yeh, H., and Sato, S. (2016). Tsunami effects on buildings and coastal structures. *J. Disaster Res.* 11, 662–669.
25. Yeh, H., Sato, S., and Tajima, Y. (2013). The 11 march 2011 East Japan earthquake and tsunami: tsunami effects on coastal infrastructure and buildings. *Pure Appl. Geophys.* 170, 1019–1031.

CHAPTER 12

Tsunamigenic Earthquakes at Along-dip Double Segmentation and Along-strike Single Segmentation near Japan

Junji Koyama [1,2,†,*], **Motohiro Tsuzuki** [1,†] **and Kiyoshi Yomogida** [1,†]

[1]*Division of Natural History Sciences, Graduate School of Science, Hokkaido University, N10 W8, Kita-ku, Sapporo, Hokkaido 0650810, Japan*

[2]*Hyotanjima Scholorship, Sapporo, Hokkaido 151-1854630, Japan*

ABSTRACT

A distinct difference of the earthquake activity in megathrust subduction zones is pointed out, concerning seismic segmentations in the vicinity of Japan—that is, the apparent distribution of earthquake hypocenters characterized by Along-dip Double Segmentation (ADDS) and Along-strike Single Segmentation (ASSS). ADDS is double aligned seismic-segmentation of trench-ward seismic segments along the Japan Trench and island-ward seismic segments along the Pacific coast of the Japan Islands. The 2011 Tohoku-oki megathrust earthquake of Mw9.0 occurred in ADDS. In the meantime, the subduction zone along the Nankai Trough, the western part of Japan, is the source region of a multiple rupture of seismic segments by the 1707 Houei earthquake, the greatest earthquake in the history of Japan. This subduction zone is narrow under the Japan Islands, which is composed of single aligned seismic-segmentation side by side along the Nankai Trough, which is typical of ASSS. Looking at the world seismicity, the 1960 and 2010 Chile megathrusts, for example, occurred in ASSS, whereas the 1952 Kamchatka and the 1964 Alaska megathrusts occurred in ADDS. These megathrusts in ADDS result from the rupture of strong asperity in the trench-ward seismic segments. Since the asperity of earthquakes in ASSS is concentrated in the shallow part of subduction zones and the asperity of frequent earthquakes in ADDS is in deeper parts of the island-ward seismic segments than those of ASSS, there must be a difference in tsunami excitations due to earthquakes in ADDS and ASSS. An analysis was made in detail of tsunami and seismic excitations of earthquakes in the vicinity of Japan. Tsunami heights of ASSS earthquakes are about two times larger than those of ADDS earthquakes with the same value of seismic moment. The reason for this different tsunami excitation is also considered in relation to the seismic segmentations of ADDS and ASSS.

Keywords: ADDS; ASSS; tsunamigenic earthquake; tsunami magnitude

1. INTRODUCTION

Devastating tsunamis in the last two decades required a new paradigm of the earthquake occurrences and tsunami generations on a geologically extended time-span. The extraordinary tsunami in the Indian Ocean of 2004 was generated by the earthquake of more than 1000km in fault length, which includes historical fault ruptures of smaller scales along the Sumatra-Andaman subduction zone [1]. The 2011 Tohoku tsunami in Japan resulted from a fault rupture developed from the Japan Trench to the coastline of the Tohoku prefecture of Japan along the dip-direction of the Pacific plate as well as along the strike-direction of the Japan Trench. This wide-spread source region of the 2011 megathrust earthquake of Mw9.0 includes trench-ward seismic segments, and the segments have been believed to be an interface of aseismic segments [2,3] without any stress accumulation.

These tsunamis were due to megathrust earthquakes much larger than anticipated sizes in their respective seismological histories, although some geological evidence demonstrated past earthquakes and tsunamis larger than those in historical documents and instrumentally observed records. It has been pointed out recently that observations of their seismic activity in recent years revealed that the megathrust earthquakes occurred in seismic segments of either Along-strike Single Segmentation (ASSS) or Along-dip Double Segmentation (ADDS) [4,5,6]. The different seismic segmentations—ASSS and ADDS—have been identified, referring to their regional seismic-activity, focal mechanisms, rupture patterns, geometry of subduction zones, types of overriding plates and back-arc activity [5].

Previous studies discuss the seismic segmentation along the subduction zones as a more complex structure, being composed of aseismic-zone, unstable zone, conditionally stable zone, and stable sliding zone, e.g. [2]. The present ADDS/ASSS hypothesis would be one where the segmentation relates directly to observable seismic activity in the source regions of great (Mw~8) and megathrust earthquakes (Mw~9). The main purpose of this study is to discuss the tsunami excitation in relation to these segmentations so as to make a comprehensive understanding of source regions of megathrust earthquakes.

Previous studies showed that tsunami excitation in the Japan Sea is larger than that in the Pacific Sea side of Japan [7,8,9]. These studies attributed the difference to the dip angle of earthquake faults or to the rigidity of earthquake source-regions. In this study, a further discussion is made about the tsunami excitation referring to the recent understanding of earthquake sources in different segmentations of ADDS and ASSS in the vicinity of Japan.

2. DIVERSITY OF MEGATHRUST EARTHQUAKES IN THE WORLD

Seismic activity off the Pacific coast of the Tohoku district, Japan has been extensively investigated from historical and instrumental records. It is characterized into regional seismic segments; there is a double aligned seismic-segmentation along the dip-direction of the subduction zone. The segments along the island-arc side of Japan frequently generated earthquakes as large as Mw8, but the segments along the Japan Trench had been considered to be aseismic. This double aligned seismic-segmentation is called Along-dip Double Segmentation (ADDS). The 2011 megathrust earthquake of Mw9.0 ruptured many segments along not only strike- but also dip-directions of the Japan Trench, covering an area of about 200 × 500 km^2, as shown in Figure 1.

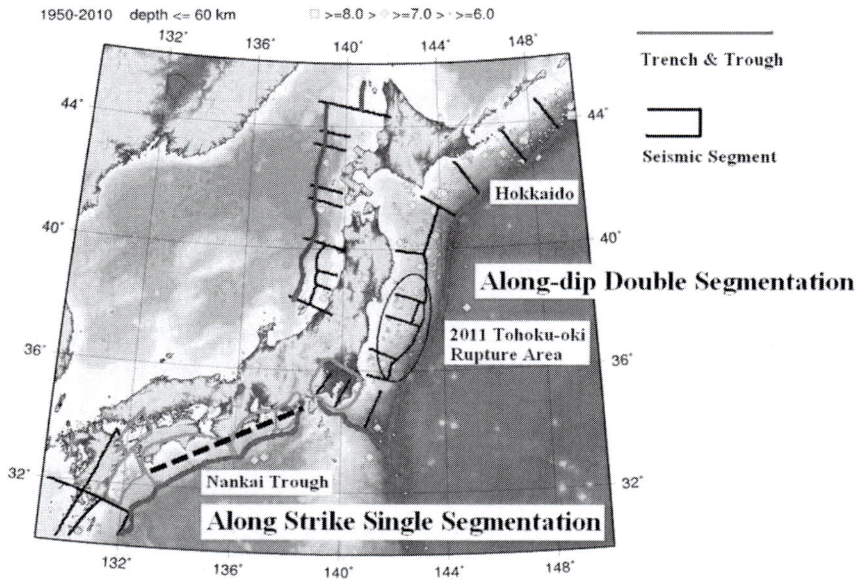

Figure 1. Seismic activity in Japan and in its vicinity. Seismic segments have been used in the official earthquake forecasting [10] of Evaluation of Major Subduction Zone Earthquakes by the Headquarters for Earthquake Research Promotion. Epicenters of earthquakes are plotted by yellow symbols from 1950 to 2010 with magnitudes larger than 5.9 and their focal depths shallower than 61km determined by the Japan Meteorological Agency [11]. Trench and trough near the Japan Islands are illustrated by red curves. The 2011 Tohoku-oki megathrust earthquake ruptured the area circled by a solid ellipse, where is Along-dip Double Segmentation (ADDS). Along-strike Single Segmentation (ASSS) can be found in the Nankai Trough, where little recent seismic activity has been observed. Such regions are often called seismic gaps [12]. The source extent of the 1707 Houei great earthquake in this segmentation is added by a broken line along the Nankai Trough to the original figure on [13].

The best-known megathrust earthquake in Japan is the 1707 Houei great earthquake along the Nankai Trough with three major segments in Figure 1. This historical event exhibits an interaction in the trench-axis direction among adjacent segments. The 1707 Houei great earthquake took place at a very different site from that of the 2011 megathrust event, where little seismic activity is observed in the single aligned seismic-segmentation along the axis of the Nankai Trough (Figure 1). The inactive seismicity in this region not only applies to the period analyzed in Figure

1 but also to the whole period from 1924 to the present, according to the Japan Meteorological Agency [11], except for the enhanced aftershock activity following the 1944 Tonankai earthquake Mw8.2 and the 1946 Nankaido earthquake Mw8.2, which successively occurred along the Nankai Trough. The Houei earthquake of multi-segment rupture is referred to an earthquake in Along-strike Single Segmentation (ASSS), which contrasts to the 2011 Tohoku-oki megathrust earthquake in ADDS.

The reason for a gigantic megathrust earthquake in ADDS to grow up to such the scale is due to the rupture of strongly-coupled asperity in the trench-ward segment with a longer recurrence time (e.g., millennial) after large earthquakes in the land-ward segments repeated with a shorter recurrence time (e.g., centennial). The activity in ASSS is characterized by almost 100% coupled areas of shallow subduction zones, which finally gives rise to a great earthquake. In other words, the difference between these two types of segmentations appears in a seismic gap [12] along the subduction zone in ASSS and in a doughnut pattern [14] of seismic activity in the subduction zone prior to a gigantic megathrust earthquake in ADDS.

These two types of segmentations (*i.e.*, ASSS and ADDS) can be found not only in the vicinity of Japan but also elsewhere in the world. The 1952 Kamchatka earthquake of Mw 9.0 and the 1964 Alaska earthquake of Mw9.2 are pointed out to be of ADDS type, while the 1960 Chile earthquake of Mw9.5 and the 2010 Maule earthquake of Mw8.8 are of ASSS [4,5]. Speaking about the 2004 Sumatra-Andaman earthquake of Mw9.3, the faulting process is peculiar. Judging from the detailed analysis on the seismicity around the source region of the 2004 earthquake [1], the 2004 event started in the ADDS region and extended into the Andaman-Nicobar ASSS region [13]. The latter region is a typical oblique subduction zone, and is quite similar to the source areas of great earthquakes of the 1957 Andreanof Mw8.6 and the 1965 Rat Island Mw8.7 in the Aleutian arc.

General descriptions of proposed ADDS and ASSS are summarized in Table 1, indicating the distinction from the previous seismic segmentations e.g., [2,3]. Seismic segments in ASSS are usually characterized by a narrow subduction zone from the oceanic trench to the island arc. The evidence has been presented, showing the asperity of the 2010 Maule earthquake in accordance with the strongly-coupled plate interface identified by GPS observation beforehand of the 2010 event [15]. Seismic segments in ADDS are, on the other hand, rather wide. The 2011 Tohoku-oki megathrust earthquake revealed that the strongly-coupled segments exist in the trench-ward of ADDS, where the millennial seismic-asperity of the event was observed [16,17]. This is also true in the 1964 Alaska earthquake [18], showing a very strong asperity in the Alaska Bay close to the Aleutian Trench. Large (not as large as gigantic) earthquakes occur repeatedly in the landward segments of ADDS with their foci in the deep part of the subduction zone. As a result, the asperity of repeated large earthquakes in ASSS is distributed in a shallow part of the subduction zone, while that of the repeated large earthquakes in ADDS stay in a rather deeper part. This would result in a different tsunami excitation between ADDS and ASSS earthquakes.

Figure 2 illustrates an image of earthquake cycles in ADDS and ASSS. An earthquake cycle in ASSS is commonly restricted in the shallow part of the subduction zone, where the contact of the plate interface is strong, forming the seismic gap all over the particular subduction zone. Successive rupture of multiple seismic segments along the trench direction results in a megathrust earthquake (metaphoric centennial event) in ASSS. On the other hand, an earthquake cycle in ADDS is characterized by repeated earthquakes as large as Mw8 and many smaller earthquakes are commonly found in the island-ward segments of the subduction zone, though the asperity in the trench-ward stays still, which eventually induces a megathrust earthquake (metaphoric millennial event) in ADDS. At the time of an ADDS megathrust event, the rupture extends in the dip-direction as well as in the strike-direction of the subduction zone.

Table 1. Characteristics of the seismic activity in Along-strike Single Segmentation (ASSS) and Along-dip Double Segmentation (ADDS).

	ASSS	ADDS
Alignment	Single Aligned	Double Aligned
Seismic Zone	Narrow	Wide
Width/Length	1:4	1:2
Interface Contact	Single Whole Contact	Island-ward/Trench-ward
Seismic Activity	Quiet Everywhere Seismic Gap	Active/Quiet Doughnut Pattern
Recurrence Time	A Few Hundred	A Few Hundred/A Thousand
Interface Contact	Whole Zone	Contact/Strong Contact
Previous Events	1960 Chile, 1707 Houei	1964 Alaska, 2011 Tohoku
Possible Region	Cascade, Canada	Hokkaido, Japan [21]

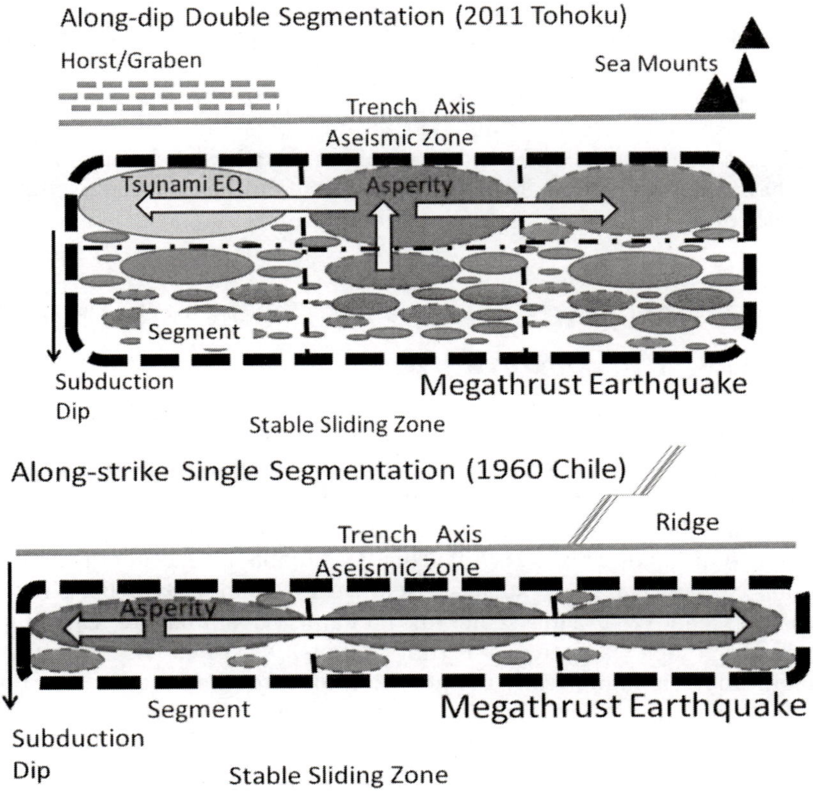

Figure 2. Schematic illustration of Along-dip Double Segmentation (ADDS) and Along- strike Single Segmentation (ASSS). ADDS is characterized by a double aligned seismic-segments along dip direction of subduction, one of which is a strongly-coupled asperity in the trench-ward segments, which eventually ruptures induced by an asperity in the island-ward segment(s) where repeated large earthquakes as large as Mw8 and many smaller-sized earthquakes (solid ellipses) occur. Rupturing all these segments results in a millennial megathrust earthquake (square by thick broken line) like the 2011 Tohoku event. ASSS shows a seismic gap of subduction interface along the trench axis in single aligned seismic-segments side by side. The asperity stays in a narrow part of the subduction interface in ASSS. Horst and Graben and seamounts are considered to be responsible for generating tsunami earthquakes [19,20,21]. As well as Horst and Graben, surface roughness such as seamounts and oceanic ridges plays an important role in blocking seismic segments (dotted broken lines) along the subduction zone [2].

Although the segmentation of ADDS/ASSS is our proposed understanding on the seismic source region of megathrust/great earthquakes, we need to deepen our understanding of these segmentations in order to distinguish megathrust earthquakes in the future and also to reduce the disasters due to such different types of earthquakes and tsunamis. This study is an attempt to enlarge our knowledge on the tsunami excitations in relation to the different seismic segmentations.

3. TSUNAMI MAGNITUDES FOR TSUNAMIGENIC EARTHQUAKES NEAR JAPAN

Tsunamis generated in the vicinity of Japan have been quantified from −1 to 4 using tsunami heights and disaster distribution near respective source regions [22,23]. This is called the Imamura-Iida scale of tsunamis, and it has been determined from historical documents and recorded evidence in Japan since 176 A.D. This is available at present, as listed in major tsunami catalogs, such as Historical Tsunami Data Base for the World Ocean [24]. This scale for tsunamis is similar to earthquake intensity, such as the Mercalli intensity scale. Local tsunami magnitude *m* is numerically defined [25,26], extending the Imamura-Iida scale. The definition of *m* is, using tsunami heights *H* (peak to trough in meter) recorded on tide-gauges and correction for tsunami travel distances Δ (km) as

$$m = 2.7 \log H + 2.7 \log \Delta - 4.3 \quad (1)$$

Since this is the tsunami magnitude scale investigated well in Japan, at first we analyzed this local tsunami magnitude *m*comparing with seismic moment *Mo* of each earthquake that occurred near Japan.

Another magnitude scale, *Mt* is the far-field tsunami magnitude [27], of which definition is

$$Mt = \log H + 9.1 + \Delta C \quad (2)$$

where H is the maximum amplitude of tsunami heights in meter and ΔC is an empirical station correction or regionally-averaged correction. The correction term and the constant 9.1 are introduced so that *Mt* agrees with moment magnitude *Mw* [28] of each earthquake. *Mt* scale is extended to estimate tsunami magnitude *Mt* for local earthquakes near Japan [7] as

$$Mt = \log H + \log\Delta + 5.80 \quad (3)$$

The same analysis is made to study the relation between *Mt* and *Mo* of tsunamigenic earthquakes near Japan, taking into account the seismic segmentations.

Historical tsunami heights until recent years were measured in tide gauge stations at best located in developed harbors, most of which face in the opposite direction to their respective open seas. Others are eyewitness records and traces of inundations and up-streams. All these data and their sources are listed on the worldwide tsunami catalogs [23,24]. Quantification of tsunami excitation and scaling relations of tsunamis to earthquake source parameters have been made based on those observations. After the recent installation of ocean-bottom pressure gauges, the observation makes it possible to study the tsunami generation free from the contaminations of detail oceanic bathymetry and coastal topography. Unfortunately or fortunately, only few records are available now e.g., [17]. Therefore, it is necessary to apply historical tsunami data to describe the general scaling relation of tsunami excitation to earthquakes and to study the generation mechanism of tsunamis.

In the above equations of (1) and (2), the dependence of tsunami magnitudes on tsunami heights are different. This is because local tsunami magnitude *m* is based on the energy of tsunami waves and far-field tsunami magnitude *Mt* on the amplitude of tsunami waves. It is true that these tsunami magnitudes are empirical parameters and needed to be rigorously quantified; however, the present study is interested only in their regional bias due to regional seismic segmentation. The analysis would be granted, since the discussion here is made in a relative manner based on the tsunami magnitudes, which would be free from the absolute uncertainty of tsunami magnitude determination.

4. TSUNAMI MAGNITUDE AND SEISMIC MOMENT OF TSUNAMIGENIC EARTHQUAKES NEAR JAPAN

Local tsunami magnitude *m* and seismic moment *Mo* of 59 tsunamigenic earthquakes in the vicinity of Japan since 1923 have been determined. Table 2 summarizes those data with their references. There is a variety of earthquakes occurring near Japan, which cannot be classified into ADDS nor ASSS, although tsunamis were excited and recorded. Some of them were intra-plate and outer-rise earthquakes of normal-fault type and the others are earthquakes near the shore of the Japan Islands of the strike-slip type [13]. Therefore, we classify them into ADDS, ASSS and NFSS (Normal Fault-Strike Slip). Figure 3 shows the locations of all the earthquakes in Table 2.

Figure 4 shows the least-squares regressions between *m* and *Mo* for ADDS (mD; 22 events in Table 2) and ASSS (mS; 10 events in Table 2).

$$m_D = 1.26 \log Mo - 24.4\ (\pm 0.17)$$
$$m_S = 1.19 \log Mo - 22.3\ (\pm 0.39) \quad (4)$$

The data of which *m* is smaller than 0 in Table 2 is not included in the regression analyses, because the tsunami height is smaller than 15cm at the distance of 100km for m=−0.5 and smaller than −0.5, which we considered large uncertainty in observations. The regression relation for the NFSS category is not calculated, because it is irrelevant in this study. However, it should be noted that the relationship looks similar to that for ADDS in general and deviates from ASSS.

The purpose of above analyses is not to introduce other empirical relationships but to quantify the difference in tsunami excitation due to the seismic segmentation. We find that *m* of ASSS is about 0.7 larger than that of ADDS for the earthquakes with Mo of 1020 Nm and 0.6 for1021 Nm, which are within the 95% confidence interval. Considering the scatter of the data in Figure 4 and the basic data of tsunami heights to determine *m*, the difference of *m* would be about 1.0 at most.

Since the increase of one unit in *m* indicates the increase by the factor of 2.24 in tsunami heights [25,26], the above difference of 1.0 means the tsunami heights of ASSS earthquakes are about two times larger than those of ADDS earthquakes with the same seismic moment.

Figure 5 shows the relations between *Mt* and *Mo* of earthquakes in ADDS (MtD) and ASSS (MtS):

$$Mt_D = 1/1.5 \log Mo - 6.04\ \pm 0.03$$
$$Mt_S = 1/1.5 \log Mo - 5.86 \pm 0.04 \quad (5)$$

where the coefficient of 1/1.5 of the relations is from the definition of *Mw* [28] with the assumption of *Mt* = Mw;

$$Mw = 1/1.5 \log Mo - 6.06 \quad (6)$$

Although there seems to be a little difference in plots between ADDS and ASSS in Figure 5, the difference is within the estimation uncertainty. It is about 0.2 in *Mt* from Equation (5), which would suggest that tsunami heights by ASSS earthquakes are larger by a factor of about 1.8 than those by ADDS earthquakes with the same value of seismic moment. All the discussions here suggest that the tsunami excitations by ASSS earthquakes are larger, by more or less about two times, than those by ADDS earthquakes near Japan.

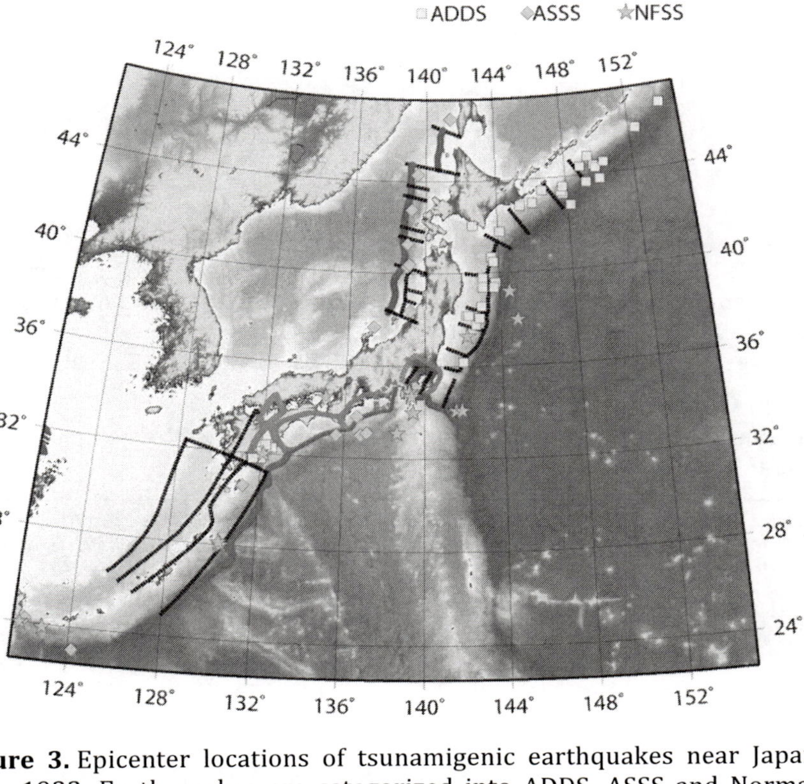

Figure 3. Epicenter locations of tsunamigenic earthquakes near Japan from 1923. Earth-quakes are categorized into ADDS, ASSS and Normal Fault/Strike Slip (NFSS), which are plotted by different symbols. Seismic segments in Figure 1 are also drawn.

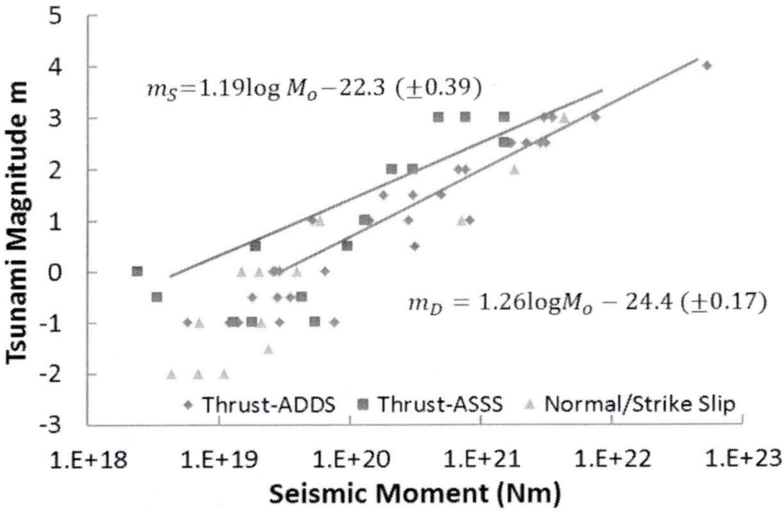

Figure 4. Tsunami magnitude *m* in relation to seismic moment *Mo* of corresponding earthquakes near Japan. Different symbols represent different category of earthquakes. Least squares regressions between *m* (mD and mS) and *Mo* for ADDS and ASSS earthquakes, respectively are derived.

Table 2. Tsunami magnitude and seismic moment of earthquakes near Japan since 1923.

Year	Month	Day	Location	Mw#	Mt$	m%	Mo (N·m)	Dip+	DipSlip*	Segment	Ref
1923	9	1	Kanto	7.9	8.0	2.0	7.60×10^{20}	0.56	0.17	ADDS	[29]
1933	3	3	Sanriku	8.4	8.3	3.0	4.30×10^{21}			NF	[30]
1938	11	5	Fukushima	7.8	7.6	1.0	7.00×10^{20}			NF	[31]
1940	8	2	W Hokkaido	7.5	7.7	2.0	2.10×10^{20}	0.72	0.72	ASSS	[32]
1944	12	7	Tonankai	8.1	8.1	2.5	1.50×10^{21}	0.17	0.17	ASSS	[33]
1946	12	21	Nankaido	8.1	8.1	3.0	1.50×10^{21}	0.17	0.17	ASSS	[33]
1952	3	4	Tokachi-oki	8.1	8.2	2.5	1.70×10^{21}	0.34	0.33	ADDS	[34]
1963	10	13	Kurile	8.5	8.4	3.0	7.50×10^{21}	0.37	0.37	ADDS	[34]
1964	5	7	Oga-oki	7.0	7.1	−0.5	4.30×10^{19}			ASSS	[32]
1964	6	16	Niigata	7.6	7.9	2.0	3.00×10^{20}	0.94	0.94	ASSS	[35]
1968	4	1	Hyuganada	7.4	7.7	1.5	1.80×10^{20}	0.29	0.29	ADDS	[36]
1968	5	16	Tokachi-oki	8.2	8.2	2.5	2.80×10^{21}	0.34	0.21	ADDS	[37]
1968	6	12	Iwate-oki	7.1	7.4	1.0	5.10×10^{19}	0.50	0.24	ADDS	[38]
1969	8	12	Kurile	8.2	8.2	2.5	2.20×10^{21}	0.27	0.27	ADDS	[39]
1970	7	26	Hyuganada	7.0	7.1	−0.5	4.10×10^{19}			ADDS	[36]
1971	9	6	Sakhalin	7.3	7.5	0.5	9.50×10^{19}	0.63	0.62	ASSS	[32]
1973	6	17	Nemuro-oki	7.8	8.1	2.0	6.70×10^{20}	0.45	0.42	ADDS	[40]
1975	6	10	Kurile	7.0	7.9	1.5	3.00×10^{20}	0.22	0.22	ADDS	[41]
1978	1	14	Oshima	6.6	6.7	−2.0	1.10×10^{19}			SS	[42]
1978	6	12	Miyagi-oki	7.6	7.4	0.5	3.10×10^{20}	0.24	0.21	ADDS	[43,44]

1980	6	29	E. Izu	6.4	6.3	−2.0	7.00×10^{18}			SS	[43,44]
1982	3	21	Urakawa	6.9	7.1	0.0	2.60×10^{19}	0.47	0.47	ADDS	[43,44]
1982	7	23	Ibaraki-oki	7.0	7.0	−0.5	2.80×10^{19}			ADDS	[43,44]
1983	5	26	C.Nihonkai	7.9	8.1	3.0	7.60×10^{20}	0.45	0.45	ASSS	[43,44]
1983	6	21	W. Aomori	7.0	7.3	0.5	1.90×10^{19}	0.68	0.68	ASSS	[43,44]
1984	3	24	Etorof-oki	7.1	7.1	0.0	6.40×10^{19}	0.29	0.28	ADDS	[43,44]
1984	8	7	Hyuganada	6.9	6.9	−1.0	2.90×10^{19}			ADDS	[43,44]
1984	9	19	Boso-oki	6.8	7.3	0.0	2.00×10^{19}			NF	[43,44]
1986	11	15	Taiwan-oki	7.3	7.6	1.0	1.30×10^{20}	0.54	0.54	ASSS	[43,44]
1989	10	29	Sanriku-oki	6.9	6.8	−1.0	5.80×10^{18}			ADDS	[43,44]
1989	11	2	Sanriku-oki	7.2	7.5	1.0	1.40×10^{20}	0.23	0.23	ADDS	[43,44]
1990	2	20	N. Oshima	6.2	6.5	−2.0	4.30×10^{18}			SS	[43,44]
1990	9	24	Tokai-oki	6.5	6.8	−1.0	7.10×10^{18}			SS	[43,44]
1991	12	22	Uruppu Isl	7.5	7.5	1.0	2.80×10^{20}	0.28	0.27	ADDS	[43,44]
1992	7	18	Sanriku-oki	6.8	7.2	0.0	2.70×10^{19}	0.19	0.18	ADDS	[43,44]
1993	2	7	Noto Pen.	6.6	6.7	−0.5	3.40×10^{18}			ASSS	[43,44]
1993	7	12	SW Hokkaido	7.7	8.1	3.0	4.70×10^{20}	0.57	0.57	ASSS	[43,44]
1994	10	4	E Hokkaido	8.1	8.2	3.0	3.00×10^{21}	0.66	0.27	ADDS	[43,44]
1994	12	28	Sanriku-oki	7.7	7.7	1.5	4.90×10^{20}	0.21	0.19	ADDS	[43,44]
1995	1	17	S Hyougo	6.8	6.4	−1.5	2.40×10^{19}			SS	[43,44]
1995	10	18	Kikaijima	6.9	7.6	1.0	5.90×10^{19}			NF	[43,44]
1995	10	19	Kikaijima	6.7	7.3	0.0	1.50×10^{19}			NF	[43,44]
1995	12	4	Etorof-oki	7.6	7.6	1.0	8.20×10^{20}	0.21	0.21	ADDS	[43,44]
1996	10	19	Hyuganada	6.6	6.9	−1.0	1.40×10^{19}			ADDS	[43,44]
1996	12	3	Hyuganada	6.7	6.7	−1.0	1.20×10^{19}			ADDS	[43,44]
2001	12	18	Yonagunijima	6.8	6.8	−1.0	2.10×10^{19}			NF	[43,44]
2002	3	31	E Taiwan	7.1	7.2	−1.0	5.40×10^{19}			ASSS	[43,44]
2003	9	26	Tokachi-oki	8.1	8.1	2.5	3.10×10^{21}	0.19	0.14	ADDS	[43,44]
2003	10	31	Fukushima-oki	6.8	7.0	−0.5	3.50×10^{19}			ADDS	[43,44]
2005	8	16	Miyagi-oki	7.1	7.0	−1.0	7.60×10^{19}			ADDS	[43,44]
2005	11	15	Sanriku-oki	6.9	7.3	0.0	3.90×10^{19}			NF	[43,44]
2006	11	15	Kurile	7.9	8.2	3.0	3.50×10^{21}	0.26	0.26	ADDS	[43,44]
2007	1	13	Kurile	8.2		2.0	1.80×10^{21}			NF	[43,44]
2007	3	25	Noto Pen.	6.9		−1.0	1.30×10^{19}			ASSS	[43,44]
2007	8	2	Sakhalin	6.2		0.0	2.40×10^{18}	0.67	0.67	ASSS	[43,44]
2008	7	19	Fukushima-oki	6.9		0.0	2.90×10^{19}	0.28	0.28	ADDS	[43,44]
2008	9	11	Tokachi-oki	7.1		−0.5	1.80×10^{19}			ADDS	[43,44]
2009	10	30	Amamioshima	6.8		−1.0	1.80×10^{19}			ASSS	[43,44]
2011	3	11	Tohoku-oki	9.0	9.1	4.0	5.31×10^{22}	0.17	0.17	ADDS	[43,44]

Mw#: Moment magnitude [45], *Mt*$: Tsunami magnitude [46], *m*%: Tsunami magnitude [8,25,26]; +Effect of fault dip; sin δ, *Effect of fault dip and slip; sinδsinλ.

5. DISCUSSION

The difference in tsunami excitation is obtained by earthquakes in ADDS and ASSS near Japan, as shown in Figure 4 and Figure 5. The difference is by a factor of about two. Similar emphasis has been made for tsunamigenic earthquakes in the Japan Sea through the analyses including those of tsunamis with smaller tsunami magnitude [7,8,9]. Generally speaking, the larger dip angles and dip components of dislocations on the faults of earthquakes in the Japan Sea give rise to the larger tsunami excitation. This is the major reason for the difference stated in the previous studies, in addition to the effect of a soft ocean-bottom structure. However, the previous emphasis should not be restricted to tsunamis in the Japan Sea, since ASSS earthquakes outside of the Japan Sea show larger tsunami excitation similar to those in the Japan Sea (Table 3). Therefore, the difference may not be attributed to the local ocean-bottom structures nor to faulting parameters but to the global characteristics of seismic segmentation or to the geological activity which makes such the surface structures.

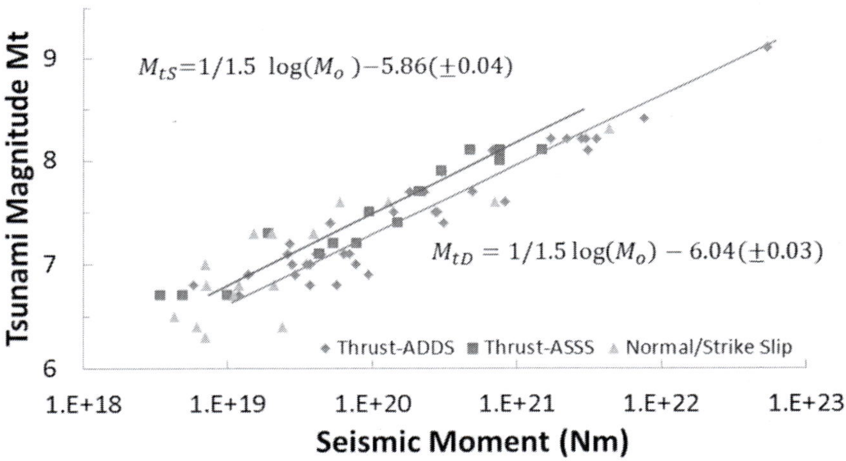

Figure 5. Tsunami magnitude *Mt* [27] in relation to seismic moment *Mo* of each earth-quake near Japan. Others are the same as those in Figure 4.

Table 3. Effect of dip angles and slip angles of dislocations on faults resposible for ocean-bottom vertical deformation. Average values are calculated from Table 2.

	Dip Angle ($\sin\delta$)	Sigma %	Dip and Slip Angle ($\sin\delta\sin\lambda$)	Sigma %
ADDS	0.32	0.13	0.26	0.08
ASSS	0.55	0.24	0.55	0.24
ASSSexJapan *	0.44	0.25	0.44	0.25

% Standard deviation, * Earthquakes classified ASSS but not in the Japan Sea.

Tsunami height is basically responsible for the ocean-bottom vertical deformation due to earthquake faults [47], so that the difference obtained indicates larger ocean-bottom vertical deformation for ASSS earthquakes than that for ADDS earthquakes of the same seismic moment. There would be many reasons to lead to such the difference as follows;

(A) Since the focal depth of ASSS earthquakes is in general shallower than those of ADDS earthquakes, the dislocation by ASSS earthquakes on particular underground faults produces larger effect on ocean-bottom deformation.

(B) Since the dip angle of faults by ASSS earthquakes is larger than those by ADDS earthquakes, the focal mechanism affects larger ocean-bottom deformation

(C) Slip angle on the faults specifies the partition of along dip and along strike components of the dislocation on a fault. Focal mechanism of ASSS earthquakes provides more along dip components than that of ADDS earthquakes.

(D) Due to three dimensional structures of ocean-bottom topography, such as trench and continental shelf, tsunami excitations become large in ASSS.

(E) Focusing effect of tsunamis due to ocean-bottom topography near ASSS earthquake sources results in larger tsunamis.

(F) Soft skin layer in the ocean bottom enhances the vertical deformation, therefore, the soft sediment in the Japan Sea side would enhance the tsunami excitation there.

The reason (A) is the result of the character of ASSS asperities and/or coupled zones existing in the shallower depth of plate interfaces than that of ADDS asperities, as described in Section 2 and Figure 2. The reason (B) is related to (F), that is, dip angles of earthquake faults in the Japan Sea are generally larger than those of the subduction zone earthquakes in the Pacific Ocean side (Table 3). However, ASSS earthquakes in this study include earthquakes in the Nankai Trough and the Okinawa Trough along the Philippine Sea Plate subduction as well as earthquakes in the west coast of Sakhalin. It is also apparent from the result in Table 3 that the effect of dip and slip angles of faults classified into ASSS earthquakes excluding the events in the Japan Sea is consistent with that of ASSS earthquakes. Smaller tsunamis ($m \leq -0.5$) which were not included in the present analysis do show the similar tendency of the larger tsunami excitations in ASSS compared to those in ADDS (Figure 4). Furthermore, a theoretical study on the deformation due to earthquake faults gives that the difference in the maximum deformation on the surface between dip angles with 30 and 60 degrees is within about 30% [48,49,50]. That is too small to comprehend the observed difference in Equation (4) between the Japan Sea and the Pacific Sea side of Japan. For (B) and (C), dip angles and slip angles of faults in ADDS and ASSS are evaluated and listed in Table 2. It is found that there are larger values for the earthquakes in the Japan Sea side than those in the Pacific Ocean side, but these earthquakes are not restricted in the Japan Sea as is mentioned in (B).

Regarding reason (D), ASSS earthquakes include those occurring in the Japan Sea side, in the back-arc basin without apparent trenches, as well as along the trench with developed continental shelves in the Nankai Trough. They generate larger tsunamis than those by ADDS earthquakes elsewhere. This eliminates the possibility of (D) for the different tsunami excitation. Speaking about (E), the results from both the local and far-field tsunami magnitude give consistent results in Equations (4) and (5). Therefore, the effect of local ocean-bottom structures alone is not the reason. Considering the reason (F) in the above, we should remind readers that tsunami earthquakes, earthquakes with slow-slip characteristics along the Sanriku Coast of the Tohoku, have occurred, where the source region of the 2011 megathrust earthquake located. The subduction zone shows a developed Horst and Graben structure [19,20,21], which is formed by sedimentation and related to the generation of tsunami earthquakes. The source region is not similar to the sedimentary layer in the Japan Sea.

6. CONCLUSIONS

The relationship has been derived between moment magnitude *Mw* of earthquake sources and local tsunami magnitude *m* [25,26] for the tsunamigenic earthquakes near Japan, and it turns out to be similar to the relation between *Mt* and *Mw*[27]. These two tsunami magnitude scales suggest that the tsunami excitations in ASSS are about two times larger than those of ADDS near Japan. Observations of tsunamis are of special importance in measuring the strength of historical large earthquakes before the dawn of instrumental seismometry in the 19th century. Previous studies pointed out that there are larger tsunami excitations in the Japan Sea compared to those in the Pacific Ocean side of Japan; however, it was found in this study that the larger excitation of tsunamis is not only restricted for earthquakes in the Japan Sea but also for those in the extended seismic region of Along-strike Single Segmentation (ASSS). The reason inducing this difference has been discussed in detail and the most probable reason is suggested that the asperity of earthquakes in ASSS spreads over the whole shallow area of the subduction interface. The rupture of the shallow asperity causes large ocean-bottom deformation. On the other hand, an asperity in the deeper seismic segments of the subduction interface of ADDS exists as well as an everlasting asperity in the shallower seismic segments, which form the double aligned seismic segments along the dip direction of the subduction. In the deeper part, repeated large earthquakes occur. A rupture of such asperity induces a smaller amount of the ocean-bottom deformation compared to that by ASSS events. The asperity in the shallower part of ADDS ruptures only in the case of a megathrust event as a millennial event. This rupture process has been observed in the 2011 Tohoku-oki megathrust and the 1964 great Alaska earthquake, and the other candidate for this type of a gigantic tsunami is the 17th century tsunami in Hokkaido, Japan [21]. The global importance of different tsunami excitation due to the seismic segmentation of ADDS or ASSS is stressed and is now open to future studies.

It is necessary to take into account the different excitation of tsunamis due to the seismic segmentation pointed out in this study and to consider the effect of dip and slip angles to retrieve the information on the source size of historical earthquakes. This is also important in estimating the strength of tsunamis for future earthquakes judging from so-called seismic gaps and/or doughnut patterns as well as from geodetic deformations by GPS observations.

ACKNOWLEDGMENTS

A.R. Gusman kindly discussed with us on the tsunami excitation, showing many of his synthetic calculations to us. We thank K.T. Ginboi for his thorough review of the galley proof of the manuscript.

AUTHOR CONTRIBUTIONS

All co-authors contributed to the present study. J. Koyama performed the data analysis and prepared the manuscript. M. Tsuzuki performed the data collection and analysis. K. Yomogida took part in discussions regarding the analyses, reviewed and edited the manuscript.

CONFLICTS OF INTEREST

The authors declare no conflict of interest.

REFERENCES

1. Lay, T.; Kanamori, H.; Ammon, C.J.; Nettles, M.; Ward, S.N.; Aster, R.; Beck, S.L.; Bilek, S.L.; Brudzinski, M.R.; Butler, R.; et al. The great Sumatra-Andaman earthquake of 26 December, 2004. Science 2005, 308, 1127–1138.
2. Bilek, S.L. Invited review paper: Seismicity along the South American subduction zone: Review of large earthquakes, tsunamis, and subduction zone complexity. Tectonophysics 2010, 495, 2–14.
3. Peng, Z.; Gomber, J. An integrated perspective of the continuum between earthquakes and slow-slip phenomena. Nature Geosci. 2010, 3, 599–607.
4. Yomogida, K.; Yoshizawa, K.; Koyama, J.; Tsuzuki, M. Along-dip segmentation of the 2011 off the Pacific coast of Tohoku earthquake and comparison with other megathrust earthquakes. Earth Planets Space 2011, 63, 697–701.
5. Koyama, J.; Yoshizawa, K.; Yomogida, K.; Tsuzuki, M. Variability of megathrust earthquakes in the world revealed by the 2011 Tohoku-oki earthquake. Earth Planets Space 2012, 64, 1189–1198.
6. Kopp, H. The control of subduction zone structural complexity and geometry on margin segmentation and seismicity. Tectonophysics 2013, 589, 1–16.
7. Abe, K. Quantification of major earthquake tsunamis of the Japan Sea. Phys. Earth Planet. Inter. 1985, 38, 214–223.
8. Hatori, T. Characteristics of tsunami magnitude near Japan. Rep. Tsunami Eng. 1996, 13, 17–26.
9. Watanabe, H. Regional variation of the formulae for the determination of tsunami magnitude in and around Japan. J. Seism. Soc. Japan 1995, 48, 271–280.
10. Available online: http://www.j-shis.bosai.go.jp/map/?lang=en (accessed on 21 September 2015).
11. Available online: http://www.jma.go.jp/jma/en/Activities/earthquake.html (accessed on 21 September 2015).
12. Kelleher, J.; Savino, J. Distribution of seismicity before large strike slip and thrust-type earthquakes. J. Geophys. Res.1975, 80, 260–271.
13. Koyama, J.; Tsuzuki, M. Activity of significant earthquakes before and after megathrust earthquakes in the world. J. Seism. Soc. Jpn. 2014, 66, 83–95.
14. Mogi, K. Seismicity in western Japan and long-term earthquake forecasting. In Earthquake PredictionMaurice Ewing Series; 4th ed.; Simpson, D.W., Richards, P.G., Eds.; AGU: Washington, DC, USA, 1981; pp. 43–51.
15. Moreno, M.; Rosenau, M.; Oncken, O. Maule earthquake slip correlates with pre-seismic locking of Andean subduction zone. Nature 2010, 467, 198–202.
16. Kohketsu, K.; Yokota, Y.; Nishimura, N.; Yagi, Y.; Miyazaki, S.I.; Satake, K.; Fujii, Y.; Miyake, H.; Sakai, S.; Yamanakae, Y.; et al. A unified source model for the 2011 Tohoku earthquake. Earth Planet. Sci. Let. 2011, 310, 480–487.
17. Maeda, T.; Furumura, T.; Sakai, S.; Shinohara, M. Significant tsunami observed at ocean-bottom pressure gauges during the 2011 off the Pacific coast of Tohoku earthquake. Earth Planet Space 2011, 63, 803–808.
18. Ruff, L.; Kanamori, H. The rupture process and asperity distribution of three great earthquakes from long-period diffracted P-waves. Phys. Earth Planet. Inter. 1983, 31, 202–230.

19. Tanioka, Y.; Ruff, L.; Satake, K. What controls the lateral variation of large earthquake occurrence along the Japan trench. The Island Arc 1998, 6, 261–266.
20. Bell, R.; Holden, C.; Power, W.; Wang, X.; Downes, G. Hikurangi margin tsunami earthquake generated by slow seismic rupture over a subducted seamount. Earth Planet. Sci. Let. 2014, 397, 1–9.
21. Ioki, K.; Tanioka, Y. Re-evaluated fault model of the 17th century great earthquake off Hokkaido using tsunami deposit data. Earth Planet. Sci. Lett. 2015. submitted.
22. Imamura, M. Chronological table of tsunamis in Japan. J. Seism. Soc. Japan 1949, 2, 23–28.
23. Iida, K. Magnitude and energy of earthquakes associated by tsunami, and tsunami energy. J. Earth Sci. Nagoya Univ.1958, 6, 101–112.
24. Available online: http://www.ngdc.noaa.gov/nndc/struts/form?t=101650&s=70&d=7 (accessed on 21 September 2015).
25. Hatori, T. Relation between tsunami magnitude and wave energy. Bull. Earthq. Res. Inst. Tokyo Univ. 1979, 54, 531–541.
26. Hatori, T. Classification of tsunami magnitude scale. Bull. Earthq. Res. Inst. Tokyo Univ. 1986, 61, 473–515.
27. Abe, K. Size of great earthquakes of 1837–1974 inferred from tsunami data. J. Geophys. Res. 1979, 84, 1561–1568.
28. Hanks, T.; Kanamori, H. A moment magnitude scale. J. Geophys. Res. 1979, 84, 2348–2350.
29. Kanamori, H. Faulting of the great Kanto earthquake of 1923 as revealed by seismological data. Bull. Earthq. Res. Inst. Tokyo Univ. 1971, 49, 13–18.
30. Kanamori, H. Seismological evidence for a lithospheric normal faulting—The Sanriku earthquake of 1933. Phys. Earth Planet. Int. 1971, 4, 289–300.
31. Abe, K. Tectonic implications of the large Shioya-oki earthquakes of 1938. Tectonophysics 1977, 41, 269–289.
32. Fukao, Y.; Furumoto, M. Mechanism of large earthquakes along the eastern margin of the Japan Sea. Tectonophysics1975, 25, 247–266.
33. Kanamori, H. Tectonic implications of the 1944 Tonankai and the 1946 Nankaido earthquakes. Phys. Earth Planet. Int.1972, 5, 129–139.
34. Kanamori, H. Synthesis of long-period surface waves and its application to earthquake source studies—Kurile Islands earthquake of October 13, 1963. J. Geophys. Res. 1970, 75, 5011–5027.
35. Aki, K. Generation and propagation of G waves from the Niigata earthquake of June 16, 1964, Part 1. A statistical analysis. Bull. Earthq. Res. Inst. Tokyo Univ. 1966, 44, 33–72.
36. Shiono, K.; Mikumo, T.; Ishikawa, Y. Tectonics of the Kyushu-Ryukyu arc as evidenced from seismicity and focal mechanism of shallow to intermediate-depth earthquakes. J. Phys. Earth 1980, 28, 17–43.
37. Kanamori, H. Focal mechanism of the Tokachi-Oki earthquake of May 16, 1968: Contortion of the lithosphere at a junction of two trenches. Tectonophysics 1971, 12, 1–13.
38. Yoshioka, N.; Abe, K. Focal mechanism of the Iwate-oki earthquake of June 12, 1968. J. Phys. Earth 1976, 24, 251–262.
39. Abe, K. Tsunami and mechanism of great earthquakes. Phys. Earth Planet. Int. 1973, 7, 143–153.
40. Shimazaki, K. Nemuro-oki earthquake of June 17, 1973: A lithospheric rebound at the upper half of the interface. Phys. Earth Planet. Int. 1974, 9, 315–327.
41. Takemura, M.; Koyama, J.; Suzuki, Z. Source process of the 1974 and 1975 earthquakes in Kurile Islands in special relation to the difference in excitation of tsunami. Tohoku Geophys. J. 1977, 24, 113–132.
42. Shimazaki, K.; Sommerville, P. Static and dynamic parameters of the Izu-Oshima, Japan, earthquake of June 14, 1978. Bull. Seism. Soc. Am. 1979, 69, 1343–1378.
43. Available online: http://earthquake.usgs.gov/earthquakes/eqarchives/sopar/ (accessed on 21 September 2015).
44. Available online: http://www.globalcmt.org/CMTsearch.html (accessed on 21 September 2015).

45. Available online: http://earthquake.usgs.gov/earthquakes/world/historical_country.php (accessed on 21 September 2015).
46. Available online: http://wwweic.eri.u-tokyo.ac.jp/tsunamiMt.html (accessed on 21 September 2015).
47. Gusman, A.R.; Tanioka, Y. W phase inversion and tsunami inundation modeling for tsunami early warning: Case study for the 2011 Tohoku event. Pure Appl. Geophys. 2013, 170.
48. Matsu'ura, M.; Sato, R. Displacement fields due to the faults. J. Seism. Soc. Jpn. 1975, 28, 429–434.
49. Himematsu, Y. Hokkaido Univ. prepared numerical calculations on the displacements fields based on Okada, Y. Bull. Seism. Soc. Am. 1992, 82, 1018–1040.
50. Available online: http:/www.bosai.go.jp/study/application/dc3d/DC3Dhtml_J.html (accessed on 21 September 2015).

CHAPTER 13

The Tsunami Vulnerability Assessment of Urban Environments through Freely Available Datasets: The Case Study of Napoli City (Southern Italy)

Ines Alberico [1,*,†], Vincenzo Di Fiore [1,†], Roberta Iavarone [2,†], Paola Petrosino [3,†], Luigi Piemontese [2,†], Daniela Tarallo [1,†], Michele Punzo [1,†] and Ennio Marsella [1,†]

[1]*IAMC—Istituto per l'Ambiente Marino Costiero, CNR, Napoli 80133, Italy*

[2]*Dipartimento di Architettura, Università degli Studi di Napoli Federico II, Napoli 80134, Italy*

[3]*Dipartimento di Scienze della Terra, dell'Ambiente e delle Risorse Università degli Studi di Napoli Federico II, Napoli 80138, Italy*

ABSTRACT

The analysis of tsunami catalogues and of data published on the NOAA web site pointed out that in the Mediterranean basin, from 2000 B.C. to present, about 480 tsunamis occurred, of which at least a third involved the Italian peninsula. Within this framework, a GIS-aided procedure that takes advantage of spatial analysis to apply the Papathoma Tsunami Vulnerability Assessment model of urban environments is presented, with the main purpose of assessing the vulnerability of wide areas at spatial resolution of the census district. The method was applied to the sector of Napoli city enclosed between Posillipo Hill and the Somma-Vesuvio volcano because of the high population rates (apex value of 5000 inh/km^2) and potential occurrence of hazardous events such as earthquakes, volcanic eruptions and mass failures that can trigger tsunamis. The vulnerability status of the urban environment was depicted on a map. About 21% of the possibly inundated area, corresponding with the lowlands along the shoreline, shows a very high tsunami vulnerability. High vulnerability characterizes 26% of inundable zones while medium-low vulnerability typifies a wide area of the Sebeto-Volla plain, ca 800 m away from the shoreline. This map represents a good tool to plan the actions aimed at reducing risk and promoting resilience of the territory.

Keywords: tsunami; vulnerability; hazard; urban environment; GIS procedure; Napoli city

1. INTRODUCTION

The tsunami was considered as a secondary hazard included with earthquake [1,2] or volcanic eruption archives [3,4]. The first tsunami catalogue of the Eastern Mediterranean Sea was published by Antonopoulos *et al.* [5] and only at the end of the 1990s, thanks to the results of the European projects GITEC (1992–1995) and GITEC TWO (1998–2001), a database of tsunamis that occurred in the Mediterranean basin was published.

The analysis of available data [6,7,8] revealed that from 2000 B.C. to present, about 476 tsunamis occurred in the Mediterranean Sea; as far as their causes are concerned, 380 were triggered by earthquakes and only 20 by volcanic eruptions, the remaining were ascribed to mass failure, earthquakes, landslides, volcanic eruptions or unknown causes.

These natural events become dangerous when the tsunamis hit a densely inhabited area, where they can pose hazards and turn into disasters. Risk is defined by UNISDR [9] as the combination of the probability of occurrence of an event and its negative consequences. Mostly for geophysical risks [10,11,12,13,14,15,16,17], the notation proposed by UNESCO [18] and Fournier d'Albe [19], Risk = Hazard × Vulnerability × Exposure, is adopted. Hazard is the likely frequency of occurrence of a dangerous event in a fixed future time, Exposure measures people, property, systems, or other elements present in hazard zones that are thereby subject to potential losses, Vulnerability is the proportion of lives or goods likely to be lost, and accounts for the features of a system or asset that make it susceptible to the damaging effects of a hazard. This definition identifies vulnerability as a characteristic of the element of interest (community, system or asset) which does not depend on its exposure [9]. Ultimately, the notation points out that the risk can be reduced both by lowering exposure and acting on vulnerability.

Traditionally, the vulnerability of a territorial system affected by tsunami events was assumed to be invariable within the flood zone. e.g., [20,21]. Recent papers, thanks to the analysis of damages caused by tsunamis in many areas around the world, evidenced that tsunami vulnerability changes within the flood zone in response to several parameters mainly reflecting the features of urban environment [22,23,24,25,26,27].

Starting from this idea, to evaluate the territorial vulnerability, Papathoama *et al.* [22] developed the Papathoma Tsunami Vulnerability Assessment Model (PTVA). They built a database recording information on building features, sociological, economic, environmental and physical data and assigned to each a weight to quantify its contribution to the vulnerability assessment. Two indices, the building vulnerability and human vulnerability, defined as weighted average of data recorded into the database were assessed to show into maps the spatial and temporal variability of tsunami vulnerability over the inundated area [22,24,28].

Dall'Osso *et al.* [29] improved the PTVA Model, renamed PTVA-3, introducing a new set of attributes, related to the water intrusion effects, and the use of the Analytic Hierarchy Process (AHP) to evaluate the building vulnerability. This multi-criteria approach limits the subjective ranking and displays the contribution of the single attribute to the overall buildings vulnerability.

In this paper, the PTVA-3 model was partly modified to implement it in a Geographic Information System (GIS), as a procedure to assess the Relative Vulnerability Index (RVI) of an urban environment *(i.e.*, of wide zones with high building density) without losing the information associated with individual buildings, to which the procedure had originally been applied. The implemented workflow takes advantage of the experiences gained in the areas recently affected by tsunamis and of freely available datasets as the ISTAT (Italian National Institute for Statistics) [30] data and the Google Earth satellite images.

The main outcome of the present research is the urban vulnerability map; it represents a good tool to limit the physical damage of tsunami propagation and to improve social preparedness, essential to favor the development of a resilient community. The method was tested on Napoli megacity because it is exposed to many natural events (seismic, volcanic, landslide) that could trigger tsunamis.

2. METHOD FOR TSUNAMI VULNERABILITY ASSESSMENT AT URBAN SCALE

In the present paper, the PTVA-3 model previously applied to the single buildings of several areas [29,31,32,33,34] was partially modified to work at the regional scale, assessing the urban vulnerability of wide areas by using remote sensing and National Institute for Statistics data. The procedure was implemented in a GIS environment to take full advantage of the spatial analysis algorithms for analyzing and integrating data with different spatial resolutions.

Figure 1 shows the input data and the sequence of steps required for calculation of Relative Vulnerability Index (RVI); the ISTAT [30] geodataset and the satellite images represent the main source of data needed to implement the proposed procedure.

The censual district (ISTAT 2011, [30]) was assumed as the smallest geographical feature (spatial resolution) while the associated dataset (number of buildings, typologies of buildings, number of floors, age of building) provided part of the parameters required for the vulnerability assessment. The distance from the shoreline, shape and orientation of buildings, building rows and the presence of seawalls were identified thanks to the analysis of Google Earth satellite images of 2015 (Figure 1). The method adopted to calculate the inundation height is described in the hazard assessment section.

The RVI was defined as the sum of two factors:

$$B_h = \frac{\sum_{i=1}^{n} B\, f_n\, h_i}{N} \tag{1}$$

The *WV* (Water Vulnerability) is the vulnerability of buildings linked to the impact of water and *SV* (Structural Vulnerability) is the carrying capacity of buildings associated with horizontal hydrodynamic force of water flow. The *WV*factor was calculated by comparing the mean building elevation with the mean inundation height.

The mean building elevation at census district scale (B_h) was calculated for the four building categories defined by the ISTAT [30] (buildings with one floor, two floors, three floors, four and more than four floors) and assuming a mean floor height (h_i) of 3 m, as follows:

$$Bv = \frac{(w_1 * s + w_2 * m + w_3 * g + w_4 * so + w_5 * y)}{\sum_{i=1}^{n} N_i} \quad (2)$$

where *B* represents the number of buildings pertaining to the four ISTAT [30] categories, f_n is the number of floors and *N* is the total number of buildings located in the census district.

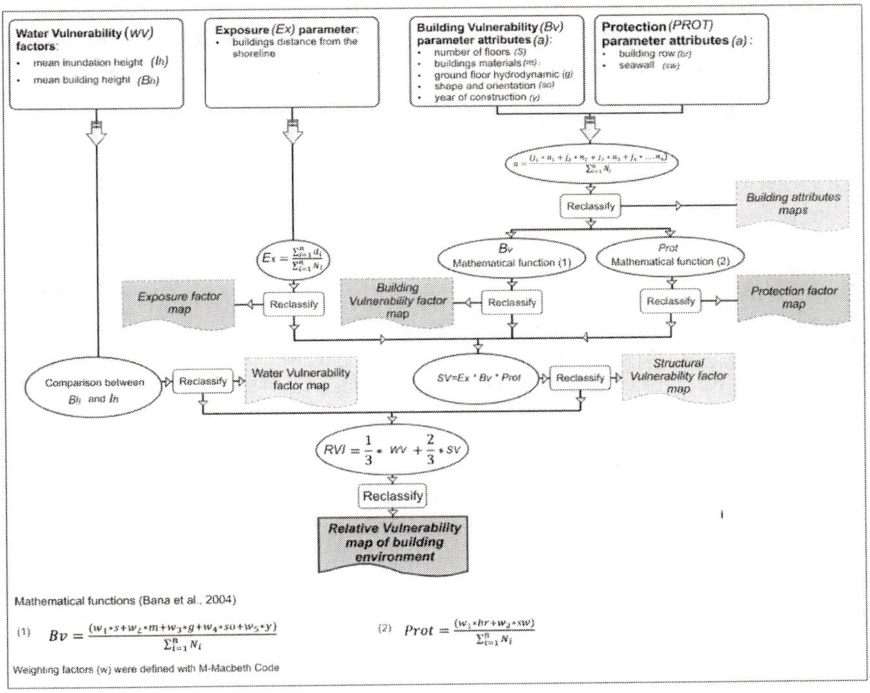

Figure 1. Flow chart. Diagram reporting the procedure for census districts vulnerability assessment: white rounded rectangles enclose the factors and the attributes; ellipses indicate the algorithm adopted; the colored boxes indicate the maps drawn in the different steps of the procedure.

Similarly to the last parameter, the inundation height was calculated, at census district scale, as mean of inundation values assigned to each building from the inundation dataset.

The water vulnerability was evaluated through the comparison of the mean inundation height and the mean building elevation; the census districts with inundation height higher than building elevation were identified as the most vulnerable. Following this premise we identified four categories of vulnerability (1 floor flooded /4 floors, 2 floors flooded /4 floors, 3 floors flooded/4 floors, buildings totally flooded) with a weight varying linearly from 0.25 to 1.

The *SV* factor of (1), was defined as the product of *Ex*, *Bv* and *Prot* parameters.

The *Ex* (Exposure) parameter depends on the water depth and on the location of the building.

The *Bv* (Building Vulnerability) parameter depends on the building features and reflects the resistance to flooding.

The *Prot* (Protection) parameter expresses the capability of manmade structures to protect the coastal zone from the tsunami during its inland propagation.

For the *Ex* definition, in accordance with Fritz *et al.* [35], the assumption that exposure is directly related to the depth of the flow and decreases away from the coast was adopted. This parameter was calculated as average of building distance from the shoreline assigned to each census district.

For the *Bv* and *Prot* assessment were used the following relations:

$$Bv = \frac{(w_1 * s + w_2 * m + w_3 * g + w_4 * so + w_5 * y)}{\sum_{i=1}^{n} N_i} \quad (3)$$

$$Prot = \frac{(w_1 * br + w_2 * sw)}{\sum_{i=1}^{n} N_i} \quad (4)$$

where *s* is the number of floors, *m* is the building materials, *g* is the ground floor hydrodynamics, *so* is the shape and orientation, *y* is the age of building, *br* is the building row, *sw* is the seawall, are the weighting coefficients of each attribute and *N* is the total number of buildings pertaining to each census district.

The weighting coefficients (*w*) reported in Table 1, the only available for the building vulnerability assessment, are those defined by Dall'Osso *et al.* [29] through the Analytic Hierarchy Process (AHP) accessible in the M-Macbeth environment [36,37]. The attributes *s*, *m*, *y*, were derived from the ISTAT geodataset (2011, [30]) and are linked to geographical element of census districts, whereas *g*, *so*, *br sw* attributes were acquired through the visual analysis of Google Earth satellite images and are related to the single buildings (Table 2). Since the census district was assumed as the smallest geographical element available for analysis at urban scale, a procedure to summarize the different features of single attributes *(i.e.,* for *so* there are four classes: regular floor plan inscribed in a square, regular floor plan with the short side oriented perpendicularly to the flow direction, *etc.*) and transfer the information from the buildings to the censual district was defined.

Table 1. Weights of attributes. List of attributes weighting factors used for the *Bv* and *Prot* assessment (from Dall'Osso *et al.* [29]).

Building Vulnerability (*Bv*)		Protection (*Prot*)	
Attributes	Weighting Factors (%)	Attributes	Weighting Factors (%)
s (number of floors)	$w_1 = 100$	*br* (building row)	$w_1 = 100$
m (building materials)	$w_2 = 80$	*sw* (seawall)	$w_2 = 73$
g (ground floor hydrodynamics)	$w_3 = 60$		
so (shape and orientation)	$w_4 = 46$		
y (year of construction)	$w_5 = 23$		

Three main steps were followed: (a) the *so, br, g* and *sw* attributes were identified for the single building; (b) these values were successively ranked and to each class an index j_i ranging between 0.25 and 1 was assigned (Table 2), where the higher index accounts for the higher vulnerability; (c) these indices were introduced in the following formula to calculate the value of each attribute as arithmetic mean:

$$a = \frac{(j_1 * n_1 + j_2 * n_2 + j_3 * n_3 + j_4 * \ldots . n_4)}{\sum_{i=1}^{n} N_i} \qquad (5)$$

where (a) is one of the seven attributes listed above (for example: shape and orientation), (n) symbolizes the number of buildings with specific features (for example: regular floor plan inscribed in a square, regular floor plan with the short side oriented perpendicularly to the flow direction, *etc.*) and N is the number of buildings pertaining to each census district. The procedure is used to define the average of ISTAT [30] attributes as well, because the single census district has different classes for each attribute type *(i.e.,* for the number of floors).

Table 2. Indices of building vulnerability attributes. List of indices assigned to the single attribute influencing the structural vulnerability of a building (*Bv*). The values ranging from 0.25 to 1 indicate an increase of the average building vulnerability ([29]; modified).

Attributes	Indices			
	0.25	0.50	0.75	1.00
number of floors (*s*)	≥4	3	2	1
building materials (*m*)	reinforced concrete	double brice	single brice (tuff)	timber
ground floor hydrodynamics (*g*)	open plan	50% open plan	not open plan but many windows	not open plan
shape and orientation (*so*)	regular floor plan inscribed in a square or with the short side oriented perpendicularly to the direction of flow	regular floor plan, which does not fit into a square, with the long side oriented perpendicularly or diagonally to the direction of flow	irregular complex floor plan, with the short side oriented perpendicularly to the direction of flow	irregular complex floor plan, with the long side oriented perpendicularly or diagonally to the direction of flow
year of construction (*y*)	before 1982	between 1962 and 1981	between 1919 and 1961	after 1919

The criteria used to define the j_i indices are:

- For the "*s*" attribute - starting from the idea that multi-floors buildings have a good possibility of not being completely submerged by water and could promote the vertical evacuation, the lowest vulnerability was assigned to buildings with four and more than four floors and the highest to buildings with one floor (Table 2).
- For the "*m*" attribute - the reinforced concrete buildings resist the tsunami effects better than the other building typologies (Table 2).
- For the "*g*" attribute - the survey of damages to buildings as a consequence of several tsunamis, points out that buildings with

openings on the ground floor suffered a lower damage than the other buildings typologies [38] (Table 2).

- For the "*so*" attribute - the buildings with long rectangular shape or "L" shaped with the main wall oriented perpendicular to the flow direction showed high vulnerability [24,39]. The buildings with the lowest damages are those having a shape fitting into a square, because they react equally to the flows notwithstanding the direction of impact [40]. Furthermore, the degree of vulnerability decreases progressively from buildings diagonal to perpendicular to parallel to the coastline [24].
- For the "*y*" attribute - since the older a building the higher is its vulnerability, the elapsed time from its construction was used as indicator of good/worst condition (Table 2).
- For the "*br*" attribute - the number of structures located between each building and the coastline is one of the most important protection factors against the tsunami impact. The lower the value of this attribute, the higher is vulnerability (Table 3).
- For the "*sw*" attribute - Darlymple and Kriebel [38] stated that buildings protected by enclosure walls were significantly less damaged (Table 3).

Moreover, for the formulae (3) and (4) some simplifications were needed at the scale adopted in the present work. In detail, the sub-parameter age of building, displaying the state of preservation of buildings, substitutes the foundation type and the preservation condition considered by Dall'Osso *et al.* [29] because this data is not available at census district scale. The attribute "seawall" was not considered because this information was too detailed for an analysis at urban scale.

3. THE CASE STUDY OF THE CITY OF NAPOLI

Napoli megacity, with about one million people, has one of the highest population rates of all of Europe (population rate reaching apex values of 5000 inh/km^2). The long history of this city can be read by both punctual architectures and large zones displaying the historical transition from ancient to modern time. We can read the signs of the first settlements at Echia Mount (Pizzofalcone), the site of the first Greek inhabitants of Parthenope and Neapolis; the Roman town; the modern city, which grew in size and population density; Napoli as capital in Byzantine age and the urban zones built after the Second World War [41]. Today, these territorial transformations result in a complex urban texture and space configuration that schematically divide the city into five macro-zones. The old town includes all districts sourced until the end of last century. The consolidated urban zones coincide with the hills bordering the core of the city where the population has moved from the overcrowded central urban area after the First World War. The western area includes the districts falling inside the Campi Flegrei volcanic area, whose original agricultural vocation was gradually transformed into industrial and then tertiary activities in the last decade. The suburbs include the northern and eastern zones of the city, the former as a symbol of the urban distress characterizing the megacity while the latter as a symbol of degradation mainly linked to the progressive abandon of the factories once well developed in these zones [41].

This urban landscape, representing about 78% of Napoli city, was conditioned by the morphology (Figure 2). It is quite various, since several monogenetic volcanic edifices like Nisida Island, La Pietra, Monte Spina and Fossa Lupara characterize the area. The highest peak of the city is the Camaldoli hill (454 m), bordered to the west by Pianura (175 m a.s.l.) and Soccavo (100 m a.s.l.) inland plains, and gradually dipping towards the Agnano-Bagnoli coastal plain. The Camaldoli hill passes to the East to the Vomero area, where the narrow tuffaceous belt of the Posillipo hill (150 m a.s.l) starts, and gently dips towards E-NE in the Sebeto-Volla plain [42]. This morphology is linked to the growth of Napoli city in a complex multi-source active volcanic area sandwiched between the Campi Flegrei volcanic field and the Somma-Vesuvio district. It is situated in a large graben, broken into smaller horst and graben structures, connected to the Pliocene-Quaternary Tyrrhenian margin evolution [42,43,44,45,46,47]. In particular, the western sector of the city is set on the horst-type structure of Campi Flegrei caldera, the eastern one is located in a smaller graben, bounded to the east by another horst which hosts the Somma-Vesuvio volcano. The substratum of the city is made up of volcanic rocks, marine and alluvial deposits in response to regional, volcano tectonics and sea level changes occurred during the Holocene [48].

Neapolitan Yellow Tuff (NYT, 15 ky, [49]) forms the framework of most of the hills in Napoli and is overlain by the fall products of the last 10 ka explosive activity of Campi Flegrei. The same products are found in sparse boreholes of the Agnano, Bagnoli-Fuorigrotta, Pianura and Soccavo plains, with a mean thickness of 30 m [50,51]). The Agnano plain, generated by a volcano-tectonic collapse following the Agnano-Monte Spina eruption (4.1 ka [52]), has a poly-crateric morphology linked to several eruptive events [53]. In the subsurface also marine and swampy deposits are found, related to the volcano-tectonic evolution which occurred in the middle of the Holocene, when the northern half of the caldera was uplifted, while the southern half remained under the sea level to form the present bay of Pozzuoli [54]. The Bagnoli-Fuorigrotta volcano-tectonic plain is filled with the products of the post-NYT explosive activity. In the southern part of the Bagnoli-Furigrotta plain, as in the Agnano plain, the volcanic products are interbedded to marine deposits [48,50,51].

Several drill holes performed in the eastern part of the city, corresponding to the Sebeto-Volla Holocene coastal plain, encountered the effusive and explosive products of Somma-Vesuvio and the explosive products of Campi Flegrei piled up during the last 15 ka. The stratigraphic successions of over one hundred drill holes were investigated [51,55] and made it possible to deduce the presence of marine and alluvial deposits embedded to volcanic products. In the western sector of Sebeto-Volla plain prevails the NYT that thins toward the eastern sector, where lava flows and pyroclastic flow deposits of the Somma-Vesuvio eruptions are more common.

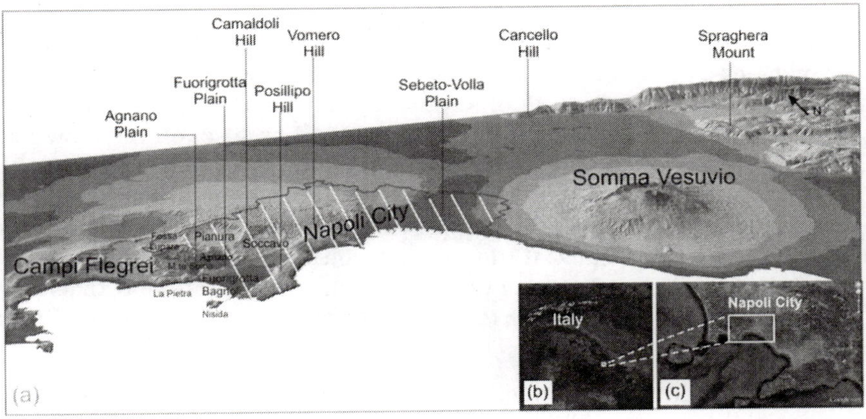

Figure 2. Location of study area. In the 3D view the morphology of area close to the Napoli municipality is reported (red line) (**a**); The Napoli city is drawn at national (**b**) and regional (**c**) scale.

This complex geological and morphological on-set coupled with several sources of hazard (earthquakes, landslides, pyroclastic flows) that could trigger the tsunamis as cascade events and the intense urbanization make the Napoli city a good test area for the RVI assessment. For the case study area, firstly the tsunami hazard zone was defined by applying a deterministic approach based on a worst case scenario [56] and then the urban vulnerability was assessed.

3.1. Tsunami Hazard in Napoli Gulf

The occurrence of past tsunami events in the Napoli Bay is testified to in historical sources. In his *Geographica* (source in Buchner, [57]), the Greek writer Strabo (64 B.C.–21 A.D.) gives the description of a sudden collapse of a sector of the Ischia island with an associated tsunami wave aged before the Greek period and recently dated between ~3 ka B.P. and 2.4 ka B.P. [58]. The possible consequences of this event were assessed by Tinti *et al.* [59], suggesting that this could have produced a tsunami that propagated in the Gulf of Napoli in 8 to 15 minutes time and reached the present city site with a water elevation of 20 m.

The mass failures may be responsible for the generation of tsunami waves [60]; the largest events may have an effect at the scale of the whole Mediterranean basin and bear a high tsunamigenic potential regardless of the water depth at which they occur. Landslides exceeding 10 km^3 in volume can still bear significant tsunamigenic potential at regional scale, and therefore their probability of occurrence should be carefully considered.

Tsunamis linked to the volcanic activity of Somma-Vesuvio are rare, nevertheless, the historical documents report the occurrence of anomalous waves in the Gulf of Napoli not only in correspondence of the largest explosive eruptions (79 A.D., 1631) but also of the low-size eruptive events of 14 May 1698, 17 May 1813 and 4 April 1906 [61,62]. Tinti *et al.* [62] analyzed through numerical simulations the sea perturbation provoked by the lightest component of the pyroclastic flows travelling down the flank of the Vesuvio volcano. They pointed out that the size of tsunamis that could reach Napoli is moderate and that they would produce effects similar to those triggered by the pyroclastic flows of the 1631 eruption [63,64,65,66].

Tsunami hydrodynamic modeling includes three main steps: generation, propagation and inundation [67]. The thorniest issue in tsunami calculation is not deep water propagation, but rather the point where the waves are generated and the point where the waves run into shallow water and then onto land. The first part is critical to assess the intrinsic level of hazard, the second is important to map site-specific hazard [68].

The present work mainly deals with the inland propagation of water. The results of numerical model applied by Tinti *et al.* [59] that explores the consequences of the Ischia Debris Avalanche (IDA) which occurred in historical times, was used as input data to assess the maximum inland inundation. A giant sea wave triggered by the flank collapse of Ischia Island may have affected all the coasts of Ischia and of the Gulf of Napoli. It can be considered an extreme event for the Napoli Bay, in fact those occurred in historical times were characterized by a lower wave height [59,61,62,63,64,65,69]. The mechanism that has led to overload-oversteepening, and thus to failure, is the volcano-tectonic uplift of the Mt Epomeo and the recurrent seismicity of the area may ultimately have triggered the mass failure [58].

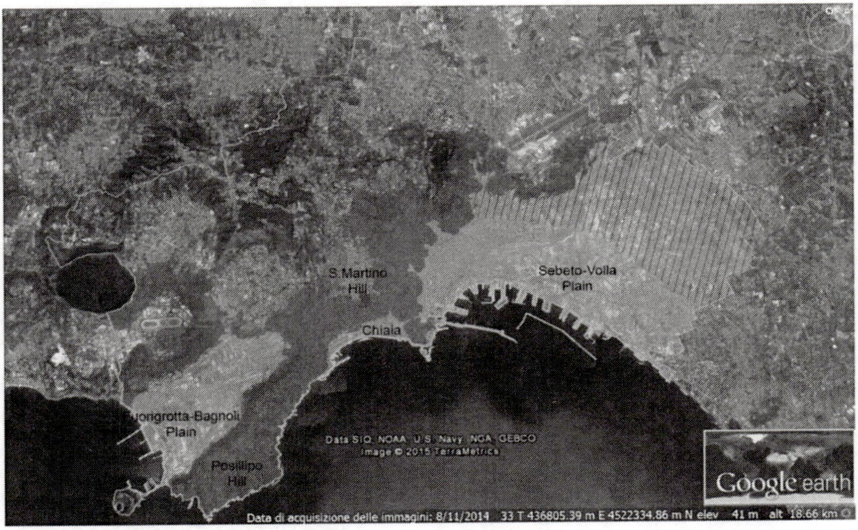

Figure 3. Topographic barriers and areas prone to inundation. The zones with slope exceeding 5° (brown color) and those prone to inundation (white color, slope lower than 5°) are shown. The dashed lines indicate the areas more than 1200 m away from the shoreline.

This event, which occurred about 36 km away from Napoli city, may have determined a maximum wave height of about 20 m along the coast of the study area. The identification of areas likely to be flooded was carried out assuming that the first wave reaches the maximum height and keeps the free horizontal surface condition. This hypothesis, as a first approach to the problem, acts for safety advantage and completely neglects the loss of energy to which the mass of water undergoes during propagation on the mainland. Accepting this hypothesis, the inland tsunami propagation was defined through a run up formula *[Math Processing Error]* [70], where "*I*" is the maximum inland inundation, R_u is the maximum water height calculated by Tinti *et al.* [59] and α is the slope characterizing the coastal zones of the study area. In order to consider the influence of inland morphology on the maximum inland propagation [71,72,73,74], a slope zoning of coastal zone was performed from a detailed Digital Elevation

Model (DEM) [75]. Firstly a Boolean map was drawn to point out the area with high spatial probability to be inundated (white zones in Figure 3), and those with slope exceeding 5° that represent topographic barriers against the tsunami inland propagation (dark brown in Figure 3); in fact, since a value of α = 5° and a water elevation of 20 m, deduced for IDA [29], were taken into account the maximum depth is about 200 m. For the Sebeto-Volla and Chiaia zones a mean slope of 1° and 2° was respectively calculated and used in the run-up formula to delineate the maximum inundation depth (Figure 4). Moreover, the inundation height (I_h) was assessed by applying the relation $I_h = W_h - T_h$, where W_h is the water height and T_h is the topographic height [70,76], the inundability becomes zero when topographic height is larger than the water height.

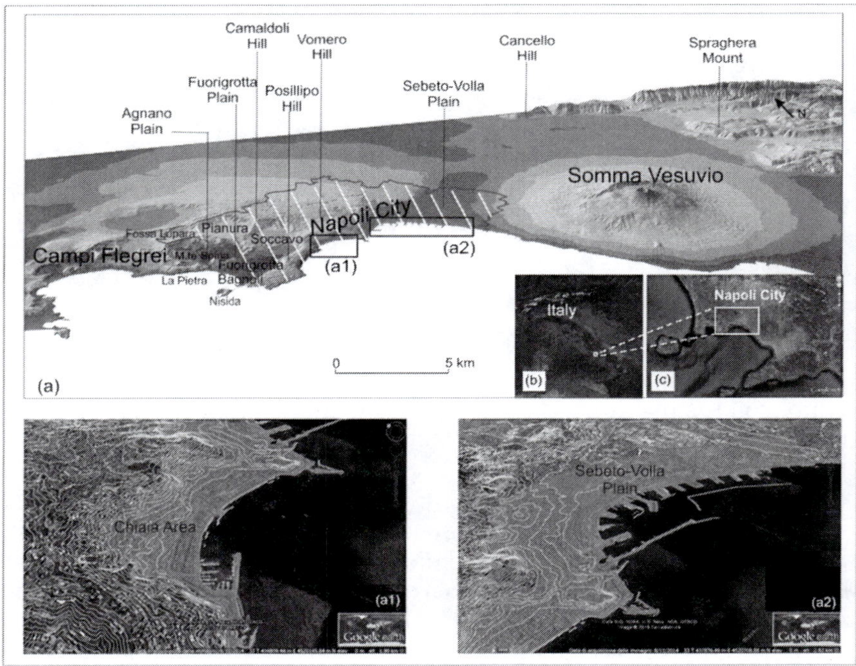

Figure 4. Location of zones that could be reached by a tsunami. The black boxes in (**a**) indicate the two inland zones prone to tsunami inundation. In (**a₁**) and (**a₂**) the light blue zones point out the maximum tsunami inland propagation and the green lines represent the topographic contours with 10 m interval. The Napoli city is drawn at national (**b**) and regional (**c**) scale.

3.2. Urban Vulnerability of Napoli City

Starting from the method described in section 2, in the following the criteria adopted to define the role of attribute, parameters and factors necessary to calculate the RVI for urban areas of Napoli city are explained.

For the number of floors (*s*), the lowest vulnerability was assigned to the buildings with four and more than four floors and the highest to buildings with one floor considering that several floors of a multi-floor building could be safe from inundation and consequently host the people climbing upstairs from the lower floors (Table 2, Figure 5A1).

As far as the building materials (*m*) are concerned, Napoli is characterized by reinforced concrete and masonry buildings; a level of vulnerability higher than reinforced concrete was assigned to masonry because of its excellent resistance to vertical stresses, but poor ability to react to horizontal forces (Table 2, Figure 5A2).

As to the ground floor hydrodynamics (*g*), buildings with openings on the ground floor showed lower damage than the other building typologies. In the study area, there are few buildings displaying these features, therefore a unique index was assigned to all buildings (Table 2, Figure 5A3).

For the shape and orientation of buildings (*so*), an index value that increases from buildings inscribed into a square to buildings with irregular floor plan and with the long side normal to the direction of flow was defined (Table 2, Figure 5A4).

For the year of construction (*y*), an index taking into account the age of the building (Table 2, Figure 5A5) was calculated.

For building row (*br*), the number of structures located between each building and the coastline is one of the most important protection factors against the impact of tsunami. For the "Porto di Napoli" and the area close to "Via Marina", the buildings are spread over the territory and are directly exposed to the tsunami propagation; the maximum vulnerability was assigned to them. To each building located at the back of Via Marina a decreasing vulnerability was assigned in response to the idea that moving away from the coast the buildings are protected by an increasing numbers of building rows (Table 3, Figure 5B1).

Table 3. Indices of protection attributes. List of indices assigned to the single attribute influencing the level of protection of a building (*Prot*). The values from 0.25 to 1 indicate an increase of the average building vulnerability ([29], modified).

Attributes	Indices			
	0.25	0.50	0.75	1.00
building row (*br*)	>VII	IV–VI	II–III	I

seawall (*sw*)	height exceeding 3 m	height between 1.5 and 3 m	height between 0 and 1.5 m	Absent

For seawall (*sw*), in the Gulf of Napoli there are just two breakwaters with rectangular shape with the main wall perpendicular to the flood. The height of these protections was used to calculate the related index value, the higher the protection the lower is the vulnerability. The *sw* was the only attribute defined at a scale larger than the census district. The area exposed to tsunamis was divided into three sectors: (a) the zone without seawall protection; (b) the zone protected by a seawall with a mean height of 1.20 m a.s.l.; (c) the zone protected by a seawall with mean height of 0.9 m a.s.l. (Table 3, Figure 5B2).

The attributes were ranked into several classes and labeled with specific indices to figure out their contribution to the vulnerability assessment at census scale (Table 4). The ranking of values was performed according to the natural break method; it identifies groups and patterns of data in a set by defining the separation points among the frequency values and minimizing the variance in each class [77]. These values and the weighting factors (w_i) were entered in equations (3) and (4) to define the *Bv* and *Prot* parameters and consequently the *Sv* factor of equation (1).

Moreover, to easily and entirely visualize the spatial distribution of the attributes, the sub-parameters and factors used in the *RVI* assessment were depicted into easily readable maps.

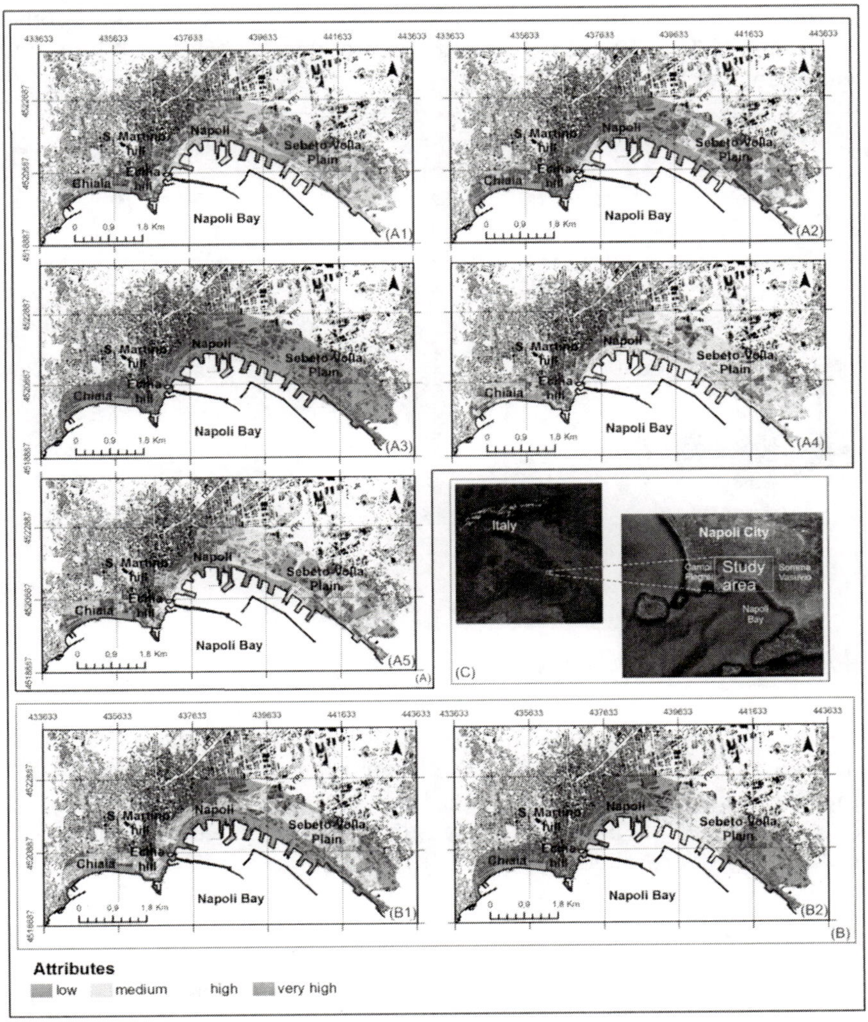

Figure 5. Attributes maps. Frameworks displaying the spatial distribution of building vulnerability (A1–A5) and protection factors (B1-B2): (**A1**) number of floors; (**A2**) buildings material; (**A3**) capability of the ground floor to resist to water flow pressure; (**A4**) shape and the orientation of building; (**A5**) year of building; (**B1**) buildings row; (**B2**) seawall protection; In the picture (**C**) the framework of the study area is shown.

Table 4. List of attributes and parameters structural vulnerability assessment. The attributes and parameters were classified into four vulnerability classes.

Parameters and Attributes	Range			
	Low (1)	**Medium (2)**	**High (3)**	**Very high (4)**
Building vulnerability (Bv)	1.00–2.25	2.25–2.83	2.83–3.45	3.45–4.00
Number of stories (*s*)	0.00–0.35	0.35–0.45	0.45–0.79	0.79–1.00
Building materials (*m*)	0.00–0.32	0.32–0.54	0.54–0.68	0.68–0.75
Ground floor hydrodynamics (*g*)				1.00
Shape and orientation (*so*)	0.00–0.54	0.54–0.72	0.72–0.89	0.89–1.00
Year of construction (*y*)	0.00–0.58	0.58–0.81	0.81–0.93	0.93–1.00
Protection (Prot)	0.00–1.43	1.43–2.30	2.30–3.14	3.14–4.00
Building row (*br*)	0.00–0.35	0.35–0.59	0.59–0.82	0.82–1.00
Seawall (*sw*)		0.5	0.75	1.00
Exposure (Ex)	1200–885	885–590	590–295	≤ 295

4. RESULTS AND DISCUSSION

The main goal of this study was to draw a map showing the vulnerability of a wide urban environment to the occurrence of tsunamis thought as a tool for the territorial planning. In the following the spatial distribution of all attributes and parameters used to assess the tsunamis vulnerability of Napoli city is recalled.

4.1. Attributes Maps

In Figure 5A,B the spatial distribution of attributes required to define the building vulnerability (frames A1 to A5) and protection parameters (frames B1 and B2) are respectively shown.

The frames from A1 to A5 point out some differences between the western part of the city, which represents the oldest suburb, and the eastern one, mainly characterized by industrial activities.

The frame A1 refers to the number of floors attribute. It shows that the western area has a medium-low vulnerability (yellow to green color) due to the presence of many buildings with four or more floors; the eastern area, on the counterpart, is characterized by buildings with few floors (orange to red color = medium-high vulnerability).

The frame A2 shows the distribution of type of material attribute, and a high vulnerability results for the western part of the city as a consequence of the wide presence of masonry buildings. The eastern zone is characterized by the presence of both reinforced concrete and masonry buildings; this condition causes a level of vulnerability variable and proportional to the type of buildings characterizing the single census district.

The frame A3 depicts the distribution of hydrodynamics of the ground floor attributes; a unique value of high vulnerability was assigned to the whole study area because most buildings have the ground floors closed by thick walls.

The frame A4 provides information on the distribution of shape and orientation of buildings attribute. The western part of the study area is more vulnerable than the eastern one, because in the former the buildings with the long side normal to water flow prevail.

The frame A5 shows the spatial distribution of the year of construction attribute; the census districts with low vulnerability (shades of green) are mostly made up of edifices built after 1982 and are concentrated in the eastern area.

The frame B1 outlines the areas with a different number of building rows attributes. Four zones with a level of vulnerability decreasing proportionally to the increasing distance from the shoreline are identified. The highest vulnerability pertains to the buildings located in the first row facing the shoreline, conversely the buildings far away from the shoreline (green color) are protected from the tsunami effect by a high number of building rows.

The frame B2 points out the protection role of seawalls. The western and the eastern zones have the highest vulnerability because of the absence of a seawall, whereas the central area shows a good level of protection due to the presence of two man-made protection structures.

4.2. The Bv, Prot and Ex Parameters Maps

The maps of the single attributes were spatially combined, following the relation (3) and (4), to outline the distribution of *Bv* and *Prot* (Figure 6 A,B) parameters into two new maps. Figure 6A shows the vulnerability of urban environment in response to buildings features. Most of the investigated area shows a medium to high level of vulnerability, mainly due to the diffused presence of large masonry buildings. In detail, the low vulnerability of the western zone is linked to the number of floors and the medium-high vulnerability to shape and orientation and year of construction, whereas the eastern sector shows opposite conditions.

Figure 6B summarizes the effects of number of building rows and seawall attributes on the protection parameter assessment. The map points out that the more a vulnerable area is close to the shoreline and that at the same distance from it the central area of the city is less vulnerable (orange color) than the two side zones (red color) due to the presence of protection barriers. Yellow and green colors characterize the zones far from the shoreline.

Figure 6C depicts the contribution of water flow (*Ex* parameter) at the site where the building is located; since the pressure applied to the building decreases with distance from the coastline, the census district vulnerability decreases with increasing distance from it.

4.3. The Sv and Wv Factors Maps

The structural vulnerability map (Figure 7A) zones the territory through the combined effect of buildings vulnerability, protection and exposure parameters. The highest *Sv* values characterize the Chiaia area close to the shoreline and the eastern sector of Sebeto-Volla plain. A sea protection favored the occurrence of a medium vulnerability class in the central sector of the city. Moving away from the shoreline medium and low structural vulnerability characterizes all the territory. In detail, the percentage of area pertaining to the four *Sv* classes evidenced that 36% of the inundated area is characterized by low vulnerability, 48% by high and very high vulnerability and only 16% is characterized by medium values.

The territorial zoning of water vulnerability factor (*Wv*) evidences that the buildings close to the shoreline, where the inundation height is the highest, are the most vulnerable (red color in Figure 7B). This factor is linked to the magnitude of the tsunami event simulated to define a vulnerability scenario. Considering a water elevation of 20 m [59], all the buildings along the shoreline are submerged by the tsunami wave propagation. The Figure 7B points out the different extent of water vulnerability classes. In the Sebeto-Volla Plain the area displaying the highest vulnerability (red zone) extends inland for about 1 km while in the western sector this class has an extent of about 400 m. The distributions of medium and low water vulnerability classes are similar, occupying 14 and 8% of inundated zone, respectively.

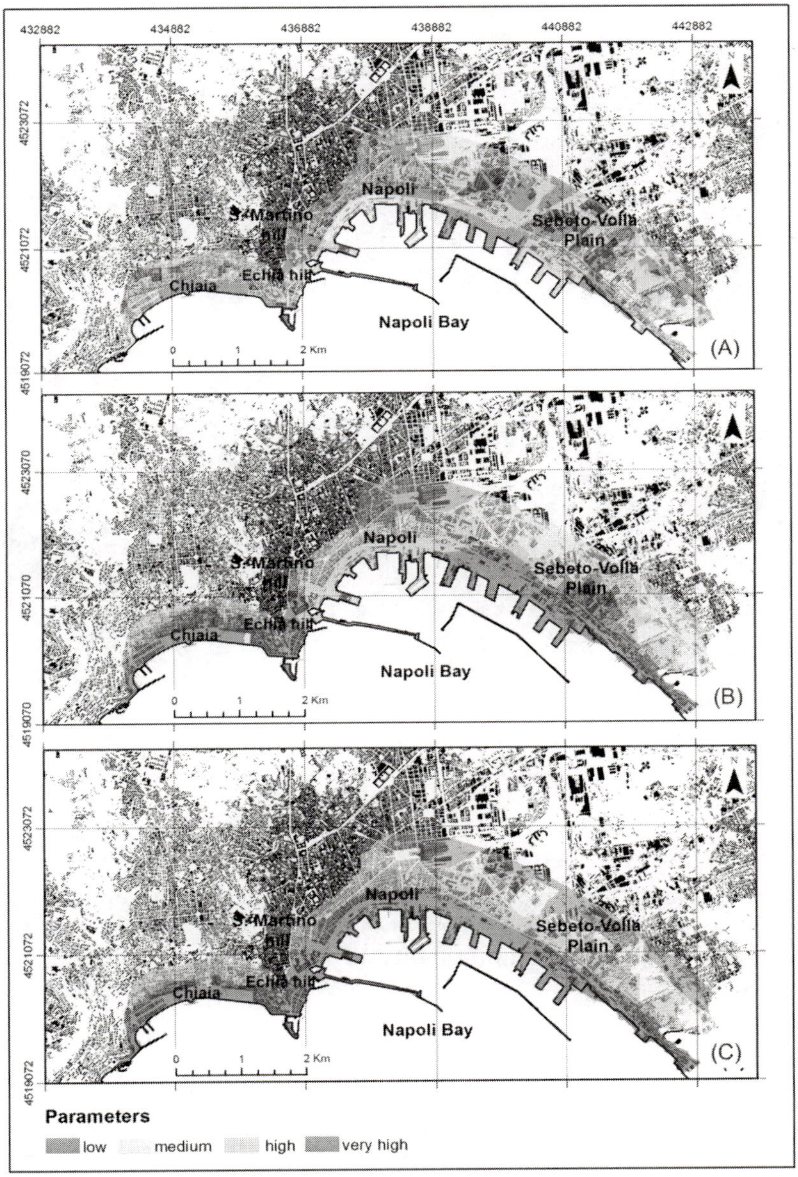

Figure 6. Parameters maps. The maps show the distribution of classes for building vulnerability (**A**), protection (**B**) and exposure (**C**) parameters.

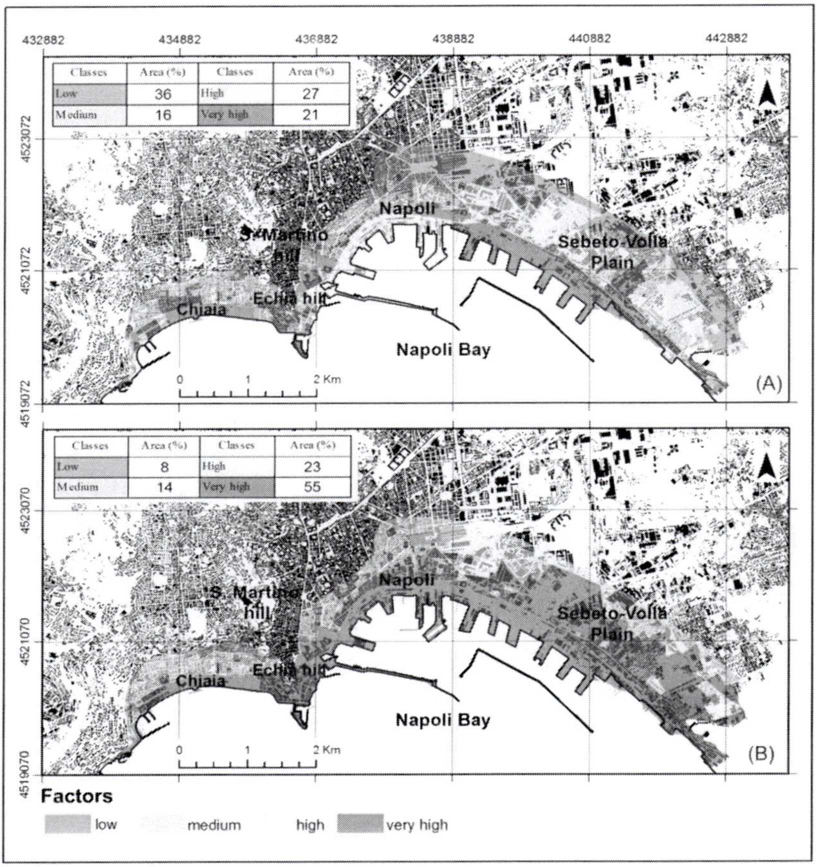

Figure 7. Maps of vulnerability factors. In **A** the vulnerability of the carrying capacity of the buildings structures (Structural Vulnerability) is depicted while in **B** the vulnerability of buildings as a consequence of their contact with tsunami waves (Water Vulnerability) is shown.

4.4. The RVI Map

The Relative Vulnerability Index map summarizes the contribution of structural and water vulnerability factors. In fact, in the territorial zoning shown in Figure 8 several sectors of coastal zones display medium structural vulnerability (Figure 8) are classified as very highly vulnerable in the *RVI* map thanks to the contribution of the *Wv* factor. The coastal zones show the highest vulnerability of the whole study area, with a depth that increases moving from Chiaia to Sebeto-Volla plain. A similar spatial distribution characterizes the other vulnerability classes evidencing the key role of morphology. The gradient of area close to the narrow plain zone of Chiaia and the slopes of San Martino hill prevent the propagation of the tsunami that remains confined within the flat zone. In contrast, the gentle slope of Sebeto-Volla plain favors the propagation of tsunamis for a depth exceeding 1 km.

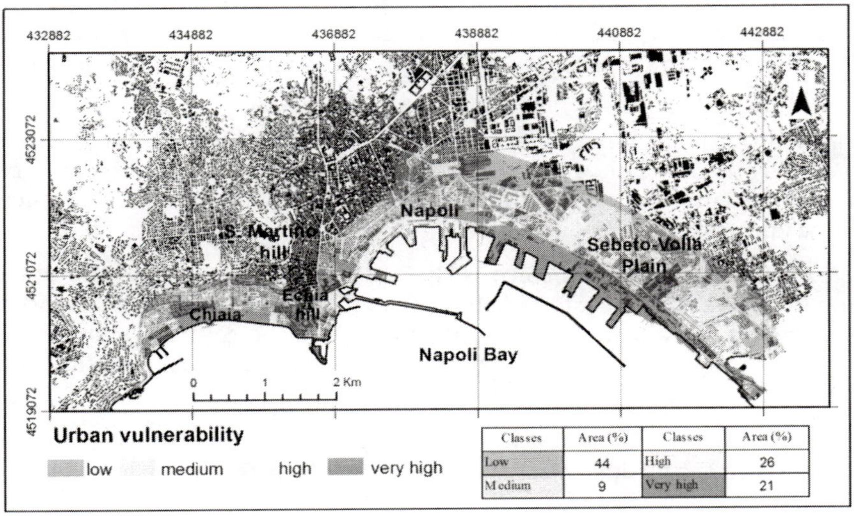

Figure 8. The Relative Vulnerability Index Map of the study area. The colors from red to green indicate a decreasing level of vulnerability.

5. CONCLUSIONS

This paper proposes a workflow able to define the tsunami vulnerability of wide urban zones taking advantage of freely available datasets as the ISTAT [30] urban dataset and the regularly up-dated Google Earth satellite images. The census district was assumed as the smallest geographical element for RVI (Relative Vulnerability index) assessment, which summarizes the contribution of two factors, the Structural Vulnerability (*Sv*) and the Water Vulnerability (*Wv*). The application of the procedure to the censual districts ensures a speedy data processing and a fast map visualization and can hence be successfully used both on the long time for territorial planning and on the short time for possible crisis management.

The method was applied to the case study of Napoli megacity because it is exposed to both volcanotectonic earthquakes, sourcing at Campania active volcanoes, and to tectonic high magnitude earthquakes, sourcing in the Apennine chain. Mostly the landslide cascade events of those earthquakes, implying the remobilization of huge quantities of rocks and their interaction with sea-water, could cause tsunami events. Apart from the high hazard related to tsunamis, the peculiar topography of the city, forcing urban zones to grow in the flat areas along the coastline, makes it even more vulnerable. Furthermore, the intense urbanization and the high population rates contribute to increase the risk of the city and make it worth a specific investigation of urban vulnerability in case of a future tsunami event.

The analysis make it possible to infer that all the buildings located within the narrow land belt along the coastline display a high vulnerability mainly because of their proximity to the sea and of the main occurrence of masonry building. In particular:

- A high value of *Sv* characterizes the Chiaia area close to the shoreline and the eastern sector of the Sebeto-Volla plain; sea protections favored the presence of a medium vulnerability class in the central sector of the city and moving away from the shoreline the medium and the low structural vulnerability prevails.

- A high value of *Wv* typifies a wide area of the Sebeto-Volla Plain, at depth more than 1 km, whereas in the western part *Wv* is controlled by the presence of a 70 m high hill, located only at 400 m from the coastline, that prevents the inland propagation of the tsunami wave.

- The same spatial distribution of high vulnerability class is shown by the medium and low vulnerability classes that are progressively located farther from the coastline.

- *RVI* map, as the sum of *Sv* and *Wv*, shows that very high and high vulnerability (47%) typifies the areas along the shoreline for a depth of about 300 m in the western sector and of about 700 m in the eastern one. This final *RVI* map reflects the magnitude of the type-event on which the procedure is based, that for the Napoli city case study is a high magnitude event (IDA, [59]), but one of the advantages of the workflow implemented in a GIS environment is to make it possible to routinely change the tsunami scenario, by changing the *Wv* factor, and rapidly redraw a new urban vulnerability map.

The RVI map here proposed is conceived as a basic tool for territorial planning in areas exposed to tsunami hazard, where laws and regulations are needed to promote the use of a building code with the aim of increasing durability of residential buildings and reducing the economic losses in case of a destructive event. In the present situation, for Napoli city, there is no tsunami risk program, either in the short term, finalized to a possible crisis management, or in the long term, for increasing resilience in population.

The possibility to identify the tsunami building vulnerability at the censual district scale, supplied by the noticeable flexibility of the GIS-aided procedure here coded, offers vital tools for spatial planning purposing to:

- identify the buildings with features that make them suitable for the vertical evacuation.

- hypothesize for the single areas a building code to reduce the future damage linked to a tsunami phenomenon (e.g., habitable space in building structures must be elevated above the regulatory flood elevation, walls and partitions are required to break away so as not to induce excessive loads on the structural frame, a stable refuge for the vertical evacuation should be provided, *etc.*).

- address resource allocation towards specific interventions aimed at reducing the urban vulnerability.

ACKNOWLEDGEMENTS

The authors wish to thank three anonymous reviewers whose suggestions greatly improved the manuscript. A grateful acknowledgement goes to the editor of the journal for his valuable suggestions and kind availability.

AUTHOR CONTRIBUTIONS

Ines Alberico and Paola Petrosino outlined the working methodology and assessed the areas exposed to tsunami hazard. Roberta Iavarone and Luigi Piemontese carried out the detailed characterization of urban environment in the light of exposure to tsunamis hazard. Vincenzo Di Fiore, Daniela Tarallo and Michele Punzo dealt with historical accounts and reviewed scientific literature to assess tsunami hazard in the Napoli area. Ines Alberico and Ennio Marsella coordinated the research group.

CONFLICTS OF INTEREST

The authors declare no conflict of interest.

REFERENCES

1. Mallet, R. Catalogue of recorded earthquakes from 1606 B.C. to 1850 A.D. Available online: http://www.biodiversitylibrary.org/item/46640#page/5/mode/1up (accessed on 26 August 2015).
2. Baratta, M. I terremoti d'Italia; Fratelli Bocca: Torino, Italy, 1901.
3. Mercalli, G. Vulcani e fenomeni vulcanici in Italia; Francesco Vallardi: Milano, Italy, 1883.
4. Baratta, M. Il Vesuvio e le sue eruzioni. Dall'anno 79 d.C. al 1896; Società Editrice Dante Alighieri: Roma, Italy, 1897.
5. Antonopoulus, J. Data for investigating tsunami activity in the Mediterranean Sea. Int. J. Tsu. Soc. 1990, 8, 38–53.
6. Tinti, S.; Maramai, A.; Graziani, L. The New Catalogue of Italian Tsunamis. Nat. Hazards 2004, 33, 439–465.
7. Pasaric, M.; Brizuela, B.; Graziani, L.; Maramai, A.; Orlic, M. Historical tsunamis in the Adriatic Sea. Nat. Hazards2012, 61, 281–316.
8. NOAA, (National Geographic Data Center). Available online: http://www. ngdc.noaa.gov/hazard/tsu_db.shtml(accessed on 16 December 2012).
9. Terminology on Disaster Risk Reduction; UNISDR United Nations International Strategy for Disaster Reduction: Genevra, Swetzerland, 2009.
10. Sleiko, D. Considerations on seismic risk. Ann. Geofis. 1993, 36, 169–175.
11. Glade, T. Landslide occurrence as a response to land use change: A review of evidence from New Zealand. Catena2003, 51, 297–314.
12. Petrosino, P.; Alberico, I.; Caiazzo, S.; Dal Piaz, A.; Lirer, L.; Scandone, R. Volcanic risk and evolution of the territorial system in the volcanic areas of Campania. Acta Vulcanol. 2004, 16, 163–178.
13. Marzocchi, W.; Selva, J.; Sandri, L. Probabilistic Volcanic Hazard Assessment and Eruption Forecasting: The Bayesian Event Tree approach. In Conception, Verification and Application of Innovative Techniques to Study Active Volcanoes; INGV: Bologna, Italy, 2008; pp. 77–91.
14. Pesaresi, M.; Gerhardinger, A.; Haag, F. Rapid damage assessment of build-up structures using VHR Satellite Data in Tsunami Affected Area. Int. J. Remote Sensing 2007, 28, 3013–3036.
15. Lirer, L.; Petrosino, P.; Alberico, I. Volcanic hazard assessment at volcanic fields: The Campi Flegrei case history. J. Volcanol. Geotherm. Res. 2001, 101, 55–75.
16. Lirer, L.; Petrosino, P.; Alberico, I. Hazard and risk assessment in a complex multi-source volcanic area: the example of the Campania region, Italy. Bull. Volcanol. 2010, 72, 411–429.
17. Grezio, A.; Gasparini, P.; Marzocchi, W.; Patera, A.; Tinti, S. Tsunami risk assessments in Messina, Sicily, Italy. Nat. Hazards Earth Syst. Sc. 2012, 12, 151–163.
18. Consultative Meeting of Experts on the Statistical Study of Natural Hazards and Their Consequences; UNESCO: Paris, France, 1972; pp. 1–11.
19. Fournier d'Albe, E.M. Objective of volcanic monitoring and prediction. J. Geol. Soc. 1979, 136, 321–326.
20. Papadopoulos, G.A.; Dermentzopoulos, T. A tsunami risk management pilot study in Heraklion, Crete. Nat. Hazards1998, 18, 91–118.
21. Ganas, A.; Nikolaou, E.; Dermentzopoulos, Th.; Papadopoulos, G.A. A GIS for tsunami risk mapping and management. In Presented at NATO Advanced Research Workshop "Underwater Ground Failures on Tsunami Generation Modeling, Risk and Mitigation", Istanbul, Turkey, 23–26 May 2001; pp. 249–252.
22. Papathoma, M.; Dominey-Howes, D. Tsunami vulnerability assessment and its implications for coastal hazard analysis and disaster management planning, Gulf of Corinth, Greece. Nat. Hazards Earth Syst. Sci. 2003, 3, 733–747.
23. Leone, F.; Denain, J.C.; Vinet, F.; Bachri, S. Analyse spatiale des dommages au bâti de Banda Aceh (Sumatra, Indonésie), Contribution à la connaissance du phénomène et à

l'élaboration de scénarios de risque tsunami, in LAVIGNE F. & PARIS R. (dir.); Rapport scientifique du programme Tsunarisque (2005–2006) sur le tsunami du 26 décembre 2004 en Indonésie—Délégation Interministérielle pour l'aide Post-Tsunami (DIPT) Ambassade de France en Indonésie, Paris; CNRS: Paris, France, 2006; pp. 69–94.
24. Dominey-Howes, D.; Papathoma, M. Validating a Tsunami Vulnerability Assessment Model (the PTVA Model) Using Field Data from the 2004 Indian Ocean Tsunami. Nat. Hazards 2007, 40, 113–136.
25. Reese, S.; Cousins, W.J.; Power, W.L.; Palmer, N.G.; Tejakusuma, I.G.; Nugrahadi, S. Tsunami vulnerability of buildings and people in South Java—Field observations after the July 2006 Java tsunami. Nat. Hazards Earth Syst. Sci.2007, 7, 573–589.
26. Garcin, M.; Desprats, J.F.; Fontaine, M.; Pedreros, R.; Attanayake, N.; Fernando, S.; Siriwardana, C.H.E.R.; de Silva, U.; Poisson, B. Integrated approach for coastal hazards and risks in Sri Lanka. Nat. Hazards Earth Syst. Sci. 2008, 8, 577–586.
27. Omira, R.; Baptista, M.A.; Miranda, J.M.; Toto, E.; Catita, C.; Catalao, J. Tsunami vulnerability assessment of Casablanca—Morocco using numerical modeling and GIS tools. Nat. Hazards 2010, 54, 75–95.
28. Dominey-Howes, D.; Dunbar, P.; Verner, J.; Papathoma, M. Estimating probable maximum loss from a Cascadia tsunami. Nat. Hazards 2010, 53, 43–61.
29. Dall'Osso, F.; Gonella, M.; Gabbianelli, G.; Withycombe, G.; Dominey-Howes, D. A revised (PTVA) model for assessing the vulnerability of buildings to tsunami damage. Nat. Hazards Earth Syst. Sci. 2009, 9, 1557–1565.
30. ISTAT. Available online: http://www.istat.it/it/ (accessed on 26 August 2015).
31. Dall'Osso, F.; Gonella, M.; Gabbianelli, G.; Withycombe, G.; Dominey-Howes, D. Assessing the vulnerability of buildings to tsunami in Sydney. Nat. Hazards Earth Syst. Sci. 2009, 9, 2015–2026.
32. Dall'Osso, F.; Dominey-Howes, D. Public assessment of the usefulness of "draft" tsunami evacuation maps from Sydney, Australia—Implications for the establishment of formal evacuation plans. Nat. Hazards Earth Syst. Sci. 2010, 10, 1739–1750.
33. Dall'Osso, F.; Maramai, A.; Graziani, L.; Brizuela, B.; Cavalletti, A.; Gonella, M.; Tinti, S. Applying and validating the PTVA-3 Model at the Aeolian Islands, Italy: Assessment of the vulnerability of buildings to tsunamis. Nat. Hazards Earth Syst. Sci. 2010, 10, 1547–1562.
34. Tarbotton, C.; Dominey-Howes, D.; Goff, J.R.; Papathoma-Kohle, M.; Dall'Osso, F.; Turner, I.L. GIS-Based Techniques for Assessing the Vulnerability of Buildings to Tsunami: Current Approaches and Future Steps; The Geological Society: London, UK, 2012.
35. Fritz, M.; Borrero, J.C.; Synolakis, C.E.; Yoo, J. 2004 Indian Ocean tsunami flow velocity measurements from survivors videos. Geophys. Res. Lett. 2006, 33.
36. Bana e Costa, C.A.; Chargas, M.P. An example of how to use MACBETH to build a quantitative value model based on qualitative value judgements. Eur. J. Oper. Res. 2004, 153, 323–331.
37. Bana e Costa, C.A.; Da Silva, P.A.; Correia, F.N. Multicriteria Evaluation of Flood Control Measures: The case of Ribeira do Livramento. Water Resour. Manag. 2004, 18, 263–283.
38. Dalrymple, R.A.; Kriebe, D.L. Lessons in Engineering from the Tsunami in Thailand. Bridge 2005, 35, 4–13.
39. Warnitchai, P. Lessons Learned from the 26 December 2004 Tsunami Disaster in Thailand. In Proceedings of the 4th International Symposium on New Technologies for Urban Safety of Mega Cities in Asia, Singapore, Asia, 18–19 October 2005.
40. Van de Lindt, J.W.; Gupta, R.; Garcia, R.A.; Wilson, J. Tsunami bore forces on a compliant residential building model. Eng. Struct. 2009, 31, 2534–2539.
41. PUC of Napoli city. 2004. Available online: http://www.comune.napoli.it/ flex/cm /pages/ServeBLOB.php/L/IT/IDPagina/1022 (accessed on 26 August 2015).
42. Carrara, E.; Iacobucci, F.; Pinna, E.; Rapolla, A. Gravity and magnetic survey of the Campanian Volcanic Area, Southern Italy. Boll. Geofis. Teor. Appl. 1973, 15, 39–51.
43. Carrara, E.; Iacobucci, F.; Pinna, E.; Rapolla, A. Interpretation of gravity and magnetic anomalies near Naples, Italy using computer techniques. Bull. Volcanol. 1974, 38, 458–467.

44. Ippolito, F.; Ortolani, F.; Russo, M. Struttura marginale tirrenica dell'Appennino Campano: Reinterpretazione di dati di antiche ricerche di idrocarburi. Mem. Soc. Geol. Ital. 1973, 7, 227–250.
45. D'Argenio, B.; Pescatore, T.S.; Scandone, P. Schema geologico dell'Appennino Meridionale. Accad. Naz. Lincei Quad.1973, 183, 49–72.
46. Finetti, I.; Morelli, C. Esplorazione sismica a riflessione dei golfi di Napoli e Pozzuoli. Boll. Geofis. Teor. Appl. 1974, 16, 175–222.
47. Geologia e geofisica del Sistema geometrico dei Campi Flegre; AGIP: Rome, Italy, 1987.
48. Cinque, A.; Rolandi, G.; and Zamparelli, V. L'estensione dei depositi marini olocenici nei Campi Flegrei in relazione alla vulcano-tettonica. Boll. Soc. Geol. Ital. 1985, 104, 327–348.
49. Deino, A.L.; Orsi, G.; de Vita, S.; Piochi, M. The age of the Neapolitan Yellow Tuff caldera-forming eruption (Campi Flegrei caldera—Italy) assessed by 40Ar/39Ar dating method. J. Volcanol. Geotherm. Res. 2004, 133, 157–170.
50. Di Vito, M.A.; Isaia, R.; Orsi, G.; Southon, J.; de Vita, S.; D'Antonio, M.; Pappalardo, L.; Piochi, M. Volcanism and deformation since 12,000 years at the Campi Flegrei caldera (Italy). J. Volcanol. Geotherm. Res. 1999, 91, 221–246.
51. Alberico, I.; Petrosino, P.G.; Zeni, F.; Lirer, L. Geocity: A drill hole geodatabase as a tool to investigate geological hazard in Napoli urban area. In Proceedings of the EUG Joint Assembly, Nice, France, 6–11 April 2003.
52. de Vita, S.; Orsi, G.; Civetta, L.; Carandente, A.; D'Antonio, M.; Deino, A.; di Cesare, T.; Di Vito, M.A.; Fisher, R.V.; Isaia, R.; et al. The Agnano-Monte Spina eruption (4100 years BP) in the restless Campi Flegrei caldera (Italy). J. Volcanol. Geotherm. Res. 1999, 91, 269–301.
53. Di Girolamo, P.; Ghiara, M.R.; Lirer, L.; Munno, R.; Rolandi, G.; Stanzione, D. Vulcanologia e Petrologia dei Campi Flegrei. Boll. Soc. Geol. Ital. 1984, 103, 349–413.
54. Cinque, A.; Aucelli, P.P.C.; Brancaccio, L.; Mele, R.; Milia, A.; Robustelli, G.; Romano, P.; Russo, F.; Russo, M.; Santangelo, N.; et al. Volcanism, tectonics and recent geomorphological change in the bay of Napoli. Suppl. Geogr. Fis. Dinam. Quat. 1997, 3, 123–141.
55. Bellucci, F. Nuove conoscenze stratigrafiche sui depositi vulcanici del sottosuolo del settore meridionale della Piana Campana. Boll. Soc. Geol. Ital. 1994, 113, 395–420.
56. Eckert, S.; Jelinek, R.; Zeurg, G.; Krausmann, E. Remote sensing-based assessment of tsunami vulnerability and risk in Alexandria, Egypt. Appl. Geog. 2012, 32, 714–723.
57. Buchner, G. Eruzioni vulcaniche e fenomeni vulcano-tettonici di età preistorica e storica nell' isola di Ischia. In Tremblements de terre, éruptions volcaniques et vie des hommes dans la Campanie antique; Albore-Livadie C.: Naples, Italy, 1986.
58. De Alteriis, G.; Insinga, D.; Morabito, S.; Morra, V.; Chiocci, F.L.; Terrasi, F.; Lubritto, C.; Di Benedetto, C.; Pazzanese, M. Age of submarine debris avalanches and tephrostratigraphy offshore Ischia island, Tyrrhenian sea, Italy. Mar. Geol. 2010, 278, 1–18.
59. Tinti, S.; Latino Chiocci, F.; Zaniboni, F.; Pagnoni, G.; de Alteriis, G. Numerical simulation of the tsunami generated by a past catastrophic landslide on the volcanic island of Ischia, Italy. Mar. Geophys. Res. 2011, 32, 287–297.
60. Roger Urgeles, R.; Camerlenghi, A. Submarine landslide softhe Mediterranean Sea: Trigger mechanisms, dynamics, and frequency-magnitude distribution. J. Geophys. Res. 2013, 118, 2600–2618.
61. Tinti, S.; Saraceno, A. Tsunamis related to volcanic activity in Italy. In Tsunamis in the World; Springer: Houten, The Netherlands, 1993; Volume 1, pp. 43–63.
62. Tinti, S.; Maramai, A. Catalog of tsunami generated in Italy and in Cote d'Aruz, France a step towards a unified calalogue of tsunamis in Europe. Ann. Geophys. 1996, 39, 6.
63. Relazione dell'incendio del Vesuvio del 1631. In Documenti inediti; Riccio, L., Giannini, e Figli, Eds.; ETH-Bibliothek Zürich: Napoli, Italy, 1631; pp. 513–521.
64. Braccini, G.C. Dell'incendio fattosi sul Vesuvio a XVI Dicembre MDCXXXI e delle sue cause ed effetti, con la narrazione di quanto `e seguito in esso per tutto marzo 1632 e con la storia di tutti gli altri incendi, nel medesimo monte avvenuti, Secondino Roncagliolo; ETH-Bibliothek Zürich: Napoli, Italy, 1631.

65. Giuliani, G. Trattato del Monte Vesuvio e de' suoi incendi; ETH-Bibliothek Zürich: Napoli, Italy, 1632.
66. Mormile, J. L'incendio del Monte Vesuvio, e delle stragi, e rovine, che ha fatto ne' tempi antichi e moderni, infine a 3 di marzo 1632, con nota di tutte le relazioni stampate fino ad oggi del Vesuvio, raccolte da Vincenzo Bove; ETH-Bibliothek Zürich: Napoli, Italy, 1632.
67. Liu, P.L.F.; Cho, Y.S.; Woo, S.B.; Seo, S.N. Numerical simulations of the 1960 Chilean tsunami propagation and inundation at Hilo, Hawaii. In Tsunami: Progress in Prediction, Disaster Prevention and Warning; Springer: Houten, The Netherlands, 1994; pp. 99–115.
68. Ward, S.N.; Day, S. Tsunami Balls: A Granular Approach to Tsunami Runup and Inundation. Commun. Comput. Phys.2008, 3, 222–249.
69. Tinti, S.; Pagnoni, G.; Piatanesi, A. Simulation of tsunamis induced by volcanic activity in the Gulf of Naples (Italy). Nat. Hazards Earth Syst. Sci. 2003, 3, 311–320.
70. Khomarudin, R.M.; Strunz, G.; Ludwig, R.; Zobeder, K.; Post, J.; Kongko, W.; Pranowo, W.S. Hazard analysis and estimation of people exposure as contribution tsunami risk assessment in the West Coast of Eumatra, the South Coast of Java and Bali. Z. fur Geomorphol. 2010, 5, 337–356.
71. Szczuciński, W.; Chaimanee, N.; Niedzielski, P.; Rachlewicz, G.; Saisuttichai, D.; Tepsuwan, T.; Lorenc, S.; Siepak, J. Environmental and Geological Impacts of the 26 December 2004 Tsunami in Coastal Zone of Thailand—Overview of Short and Long-Term Effects. Pol. J. Environ. Studies 2006, 15, 793–810.
72. Nakamura, Y.; Nishimura, Y.; Putra, P.S. Local variation of inundation, sedimentary characteristics, and mineral assemblages of the 2011 Tohoku-oki tsunami on the Misawa cost, Aomori, Japan. Sediment. Geol. 2012, 282, 216–277.
73. Kazuhisa, G.; Koji, F.; Daisuke, S.; Shigehiro, F.; Kentaro, I.; Ryouta, T.; Tomoya, A.; Tsuyoshi, H. Field measurements and numerical modeling for the run-up heights and inundation distances of the 2011 Tohoku-oki tsunami at Sendai Plain, Japan. Earth Planets Space 2012, 64, 1247–1257.
74. Gopinath, G.; Løvholt, F.; Kaiser, G.; Harbitz, C.B.; Srinivasa Raju, K.; Ramalingam, M.; Singh, B. Impact of the 2004 Indian Ocean tsunami along the Tamil Nadu coastline: Field survey review and numerical simulations. Nat. Hazards2014, 72, 743–769.
75. Alberico, I.; Lirer, L.; Petrosino, P.; Zeni, G. Geologia e geomorfologia dell'area urbana di Napoli. In Proceedings of the Geoitalia 2001—Terzo Forum Italiano di Scienze Della Terra, Chieti, Italy, 5–8 settembre 2001; pp. 359–360.
76. Federici, B.; Bacino, F.; Cosso, T.; Poggi, P.; Rebaudengo Landó, L.; Sguerso, D. Analisi del rischio tsunami applicata ad un tratto della costa Ligure. Geomat. Workb. 2006, 6, 53–57.
77. Jenks, G.F.; Caspall, F.C. Error on choroplethic maps: Definition, measurement, reduction. Ann. Assoc. Am. Geogr.1971, 61, 217–244.

CHAPTER 14

Lessons Derived from Two High-Frequency Sea Level Events in the Atlantic: Implications for Coastal Risk Analysis and Tsunami Detection

Begoña Pérez-Gómez[1*], Fernando Manzano[1], Enrique Alvarez-Fanjul[1], Carlos González[2], Juan V. Cantavella[2] and François Schindelé[3]

[1]*Puertos del Estado, Madrid, Spain*

[2]*Instituto Geográfico Nacional, Madrid, Spain*

[3]*Commissariat á l'énergie Atomique et Aux Énergies Alternatives (CEA), Direction des Applications Militaires Ile-de-France (DAM, DIF), Arpajon, France*

ABSTRACT

The upgrade and enhancement of sea level networks worldwide for integration in sea level hazard warning systems have significantly increased the possibilities for measuring and analyzing high frequency sea level oscillations, with typical periods ranging from a few minutes to a few hours. Many tide gauges now afford 1 min or more frequent sampling and have shown such events to be a common occurrence. Their origins and spatial distribution are diverse and must be well understood in order to correctly design and interpret, for example, the automatic detection algorithms used by tsunami warning centers. Two events recorded recently in European

Atlantic waters are analyzed here: possible wave-induced "seiches" that occurred along the North coast of Spain during the storms of January and February of 2014, and small sea level oscillations detected after an earthquake in the mid-Atlantic the 13th of February of 2015. The former caused significant flooding in towns and villages and a huge increase in wave-induced coastal damage that was reported in the media for weeks. The latter was a smaller signal present in several tide gauges along the Atlantic coast that coincided with the occurrence of this earthquake, leading to a debate on the potential detection of a very small tsunami and how it might yield significant information for tsunami wave modelers and for the development of tsunami detection software. These kind of events inform us about the limitations of automatic algorithms for tsunami warning and help to improve the information provided to tsunami warning centers, whilst also emphasizing the importance of other forcings in generating extreme sea levels and their associated potential for causing damage to coastal infrastructure and flooding.

Keywords: sea level, warning, high frequency, detection, automatic algorithms, tsunami

INTRODUCTION

Sea level-related hazards have become an important concern in recent decades due partly to the impact of climate change on mean sea level rise and its potential for increasing the number of storm surge extreme events (Church et al., 2001; Woodworth and Blackman, 2002, 2004). Additionally, several catastrophic tsunamis have occurred since the beginning of this century, raising awareness of society to this risk. Since 2000 more than 380.000 people died during natural disasters related to coastal inundation or sudden sea level rise: Indian Ocean tsunami (2004), Chile tsunami (2010), Japan tsunami (2011), Hurricane Katrina (US, 2005), Hurricane Sandy (US, 2012), Cyclon Nargis (Myanmar, 2012), etc., This terrible number of casualties demonstrates the great vulnerability of the coastal zone, which has seen a population increase of 35% since 1995 and is today inhabited by 23% of the world population.

For this reason, real, or near-real time sea level data are critical to the design of early sea level and tsunami warning systems. The latter are now being implemented in practically all the main basins, following the Indian Ocean tsunami in 2004 and according to the recommendations of the Intergovernmental Coordination Groups (ICG's) established by the UNESCO Intergovernmental Oceanographic Commission (IOC), such as the

NEAMTWS (North East Atlantic, Mediterranean and Adjacent Seas Tsunami Warning System) ICG in Europe (IOC UNESCO, 2007). One of these recommendations is the need of sea level data with 1 min or less sampling and latency, which has motivated the upgrade of existing tide gauge networks around the world. To make best use of these data new software must be developed to automatically quality control and process data, to issue sea level alert messages, and to implement algorithms for automatic tsunami detection. Examples of the algorithms applied to offshore pressure data from DART buoys are described by Rabinovich et al. (2011) and Rabinovich and Eblé (2015), whilst other authors focus on the application of this kind of algorithm to coastal tide gauge data such as Beltrami et al. (2011), Bressan et al. (2013) and Pérez-Gómez et al. (2013).

Several years ago, Puertos del Estado (hereafter PdE) implemented such software for the Spanish REDMAR tide gauge network, in response to the small tsunami generated by the Algerian earthquake, in May 2003, in the Western Mediterranean (Alasset et al., 2006; Sahal et al., 2009; Vich and Monserrat, 2009; Vela et al., 2014). The objective was real time tsunami detection and the transmission of alert messages to the network and harbor operators during future events. Nowadays, due to the recent establishment of the National Tsunami Warning System in Spain, run by the National Geographic Institute (IGN: Instituto Geográfico Nacional) in collaboration with Spanish Civil Protection, real time sea level data and the mentioned alert messages from the REDMAR network are also received by the IGN. However, this algorithm for tsunami detection has also proven useful for identifying other more frequent high-frequency harbor oscillations such as "meteotsunamis" (related to atmospheric pressure) or infragravity waves (generated by wind waves), which present strong similarities to seismically generated tsunamis (same periods and physical properties), in such a way that it can be difficult to recognize one from another. Testing the software with data from a seismic tsunami is of course difficult because these are fortunately rare events in our region. For this reason, any high frequency sea level oscillations are helpful in assessing the skills of these algorithms.

Monserrat et al. (2006a) compared the characteristics of tsunamis and "meteotsunamis," a term suggested by different authors (Nomitsu, 1935; Defant, 1961; Rabinovich and Monserrat, 1996; Vilibić et al., 2005; Rabinovich, 2009) for those atmospherically generated sea level oscillations with periods of a few minutes to a few hours that may affect the coast at particular bays or harbors in the same way than a tsunami generated by an earthquake. Although less catastrophic than major seismic tsunamis (the spatial scale is smaller: local or regional), they are related to atmospheric forcing and mainly to moving pressure disturbances (atmospheric gravity waves, pressure jumps, frontal passages, or squalls).

These events and their worldwide occurrence have been widely studied, described and acknowledged as potentially-hazardous sea level phenomena by many authors, aside from those already mentioned: (Orlić, 1980; Hibiya and Kajiura, 1982; Pattiaratchi and Wijeratne, 2015; Šepić et al., 2015a,b). In Spain, they are particularly common in the Balearic Islands, where they are named "rissagas" (Gomis et al., 1993; Jansà et al., 2007; Marcos et al., 2009), and on the Western Mediterranean Spanish coast. In fact, two special issues on "meteotsunamis" have been published: Physics and Chemistry of the Earth (Rabinovich et al., 2009) and Natural Hazards (Vilibić et al., 2014). Recent studies for the UK and North Sea coast have also been published by Tappin et al. (2013) and Sibley et al. (2016).

"Infragravity waves," a phenomenon first described in 1950 by Walter Munk (the name was coined by Kinsman in 1965, according to Pugh and Woodworth, 2014), are also high frequency sea level oscillations generated by non-linear interactions of swell waves arriving at the coastline during a storm (Munk, 1949, 1962). These waves, with shorter periods than meteotsunamis (between 30 s and 5 min), may sometimes also be amplified through resonance in bays and harbors (Longuet-Higgins and Stewart, 1962; Wu and Liu, 1990; Herbers et al., 1995). Such oscillations are common along the North coast of Spain. Infragravity waves have been reported to be responsible for flooding in other regions (Sheremet et al., 2014) due to their "tsunami-like" behavior and their contribution to increasing the coastal impact of wind waves.

The aim of this paper is to analyze and discuss the characteristics and possible physical sources of two events that occurred in the Atlantic recently: the sea level oscillations recorded along the North Spanish coast in January and February 2014 and the feasibility of a small tsunami being recorded after an earthquake in the Atlantic ridge the 13th of February 2015. Although in principle different in origin, their effects on sea level records from tide gauges are not that different and can be detected with the same algorithms developed for tsunami warning. Understanding the problems and limitations of existing algorithms in identifying and distinguishing different types of high-frequency phenomena is important for issuing appropriate alerts and is in fact the main objective of this paper. At the same time, as new tide gauge technology allows us to measure other sea level processes apart from the tide and storm surge contribution, the study extends our existing knowledge on the physical sources of these higher frequency oscillations and their associated coastal risks. Finally, the paper provides also an assessment of the response of the upgraded REDMAR network and PdE multi-parameter alert system during those dates.

The paper is structured as follows: after description of the data and methods, the results of the analysis of the two events are presented, followed by the discussion and conclusions on the main findings about these particular events and the consequences for future detection and interpretation of sea level oscillations data.

DATA AND METHODS

The 1 min sampled sea level time series available today in the region are the main source of information used in the analysis of these two events, most of them from PdE REDMAR sea level network, composed of 37 tide gauge stations at the main harbors of Spain. Wind wave, atmospheric pressure and wind data from wave buoys deployed near the Spanish coast are also used to determine the influence of these parameters on the observed sea level oscillations. These buoys belong to PdE deep water buoy network (Álvarez-Fanjul et al., 2002), and are located at around 400 m depth and 10–20 miles away from the coast. Outputs from the wave and sea level forecasting systems of PdE (Álvarez-Fanjul et al., 2001; Gómez-Lahoz and Carretero-Albiach, 2005) were used to understand the physical conditions relating to wind waves, the tide and storm surges during the dates of the events. Estimated arrival times of tsunami wave propagation were computed by the Spanish Geographic Institute (IGN) for the earthquake of 13th February 2015 and used for the analysis of the sea level records in this study.

This work also takes advantage of some specific characteristics of the REDMAR tide gauge network: with a raw sampling interval of 2 Hz, this array of Miros radar sensors additionally provides wind wave parameters at the tide gauge site (significant wave height, maximum wave height, mean period, and peak period) with a sampling interval of 20 min. These local wind wave parameters are transmitted to PdE alongside the records of 1-min averaged sea level data. Furthermore, a decision was taken several years ago to upgrade the stations with atmospheric pressure and wind sensors with 1-min sampling resolution, to facilitate their future use in "meteotsunamis" alert systems—an important improvement that was recommended by the scientific community working on "meteotsunamis" (Vilibić et al., 2016). From the 37 stations in REDMAR, 17 already have these new meteorological sensors transmitting their data in real time to PdE system (13 of these were installed during the last year within the national project SAMOA: *Sistema de Apoyo Meteorológico y Oceanográfico a las Autoridades Portuarias:* System of Oceanographic and Meteorological Support to the Harbour Authorities). Unfortunately for the events analyzed here, 1-min atmospheric pressure was only available for the Vigo tide gauge, on the Galician coast.

Total 1-min sea level data from the REDMAR are automatically filtered before being passed through the tsunami detection algorithm and the multi-parameter alert system of waves, sea level, and high frequency oscillations (including tsunamis) implemented by PdE. A detailed description of this system can be found in Pérez-Gómez et al. (2013). The recent upgrade to 1-min sampling and transmission latency has led to a new strategy for sea level data quality control and processing in real-time, including its contribution to the mentioned alert system that sends an email to the network operators when one of the main parameters is over a predetermined threshold, with three levels of alert: 2: "warning" (yellow), 3: "risk" (orange) and 4: "danger" (red).

The algorithm for tsunami detection, an important element of the software, is based on: (1) the elimination of low frequency sea level oscillations (periods larger than 3 h such as the tide) by means of a FIR filter with a Kaiser window of 15 points, (2) the computation of the variance of the filtered signal in a moving window, and (3) its evaluation with respect to predefined thresholds for each harbor. The algorithm has been in operation since 2008, having been tested and validated using the tsunami of May 2003 in the Balearic Islands and other tsunamis in the Indian Ocean. The filtered signal and the alert "level" are stored in the PdE data bank. To date, the performance of the algorithm has been very good, which is in part due to the near-real time automatic quality control of sea level that is undertaken every 15 min. This quality control in real time is a complex task: a tide gauge malfunction will almost always generate an initial false alert when the first 1-min observation is received, and this is automatically canceled when detailed analysis and quality control running a few minutes later flags the suspect data. For this reason, the alerts are received solely by experienced operators at PdE and the IGN who make informed decisions based upon comparisons with other parameters and stations in the system (the IGN will only pay attention to these messages if an earthquake has been detected by their seismic network).

Data from non-REDMAR tide gauges were downloaded from the data portals of the IOC Sea Level Station Monitoring Facility (SLSMF: http://www.ioc-sealevelmonitoring.org/) or the IBIROOS *In-Situ* Tac (http://www.ibi- roos.eu/Access-to-data/The-IBI-Portal). As high-passed filtered data were not available from these sites, we applied the same filter used in REDMAR to the 1-min sea level data from other institutions.

In order to characterize the 1-min high-pass filtered oscillations in this study, we have implemented a scheme of data processing for a specific period, which consists of the following steps:

- Interpolation of short gaps (<4 min)

- Estimation of *seiche* wave amplitude and period by means of the zero-up crossing procedure
- Temporal evolution of the spectra: spectrogram

The zero-up crossing procedure is usually adopted for wind waves, where a wave is defined as the portion of a record between two successive zero-up crossings. This generates a reasonable estimate of the variation in amplitude of these sea level oscillations by computation of the significant wave height [H_s: mean of wave height (trough to crest) of the highest third of the waves] and maximum wave height (H_{max}) in the recorded burst. The algorithm is applied here to a moving window of 6 h (360 min) with an overlap of the last 20 min. From a mathematical point of view, this approach should be applied on stationary sea states. Since this may not be the case for these 1 min sea level data the significant wave height computed here is not perfectly equivalent to the one obtained for wind waves and should be considered just an estimation of the oscillation amplitude ($H_s/2$ or $H_{max}/2$).

The spectrograms of the 1-min filtered signal were obtained by means of a fast Fourier transform applied to windows of around 8 h (512 points) or 1 day (1536 points) with an overlap of around 2 h (128 points).

In order to identify clearly the periods of the infragravity waves and the presence of wind waves or swell, we also performed spectral analysis of the 2 Hz data during the events at some of the stations. After applying a high-pass filter to eliminate periods larger than 45 min, the power spectral density is obtained by calculating the auto-covariance function smoothed with a Parzen lag window (Jenkins and Watts, 1968). In the future, additional data processing scripts will be needed to improve the characterization in near-real time of these high-frequency sea level oscillations from 2 Hz data, especially for those periods of the infragravity waves (30 s to 5 min), which are not well-resolved with 1 min sampling. To address this, we plan to analyze original 2 Hz raw data from the tide gauges in the REDMAR network, that are currently being transmitted hourly to PdE.

Sea Level Oscillations along the Spanish Coast During the Storms of January–February 2014

From early January to end February 2014 the Northeast Atlantic suffered the impact of a sequence of unprecedented extreme storms characterized by huge wind waves and severe coastal flooding. These storms were caused by a powerful jet stream driving low pressure systems and associated winds and waves across the Atlantic. Apart from the great damage to the coastline recorded in other countries such as the UK (BBC, 2014), where storm surge magnitude is usually larger than in the Spanish coast, several coastal villages, beaches, and harbors of the North Spanish coastline were this time also inundated by the extreme waves, resulting in the destruction of maritime promenades and reaching houses at waterfront areas (Figure 1). These events had therefore a tremendous impact on the media (El País, 2014) that put the focus on the huge wind waves seen by the population and recorded in the open waters by PdE deep water buoy network (Álvarez-Fanjul et al., 2002).

FIGURE 1. LEFT: people running away from the waves at Esteiro beach (Xove, Lugo, Northern Galicia) during the storm of February 2nd 2014 (Nadja storm). Source: La Voz de Galicia, photographer: Pepa Losada: http://www.lavozdegalicia.es/album/galicia/2014/02/02/danos-temporal/0110139132886851159 0785.htm. **RIGHT:** impact on Coruña maritime promenade after the same Nadja storm. Source: El Ideal Gallego: http://www.elidealgallego.com/album/coruna/efectos-temporal-coruna/20140202182014171505.html.

The multi-parameter alert system worked well during these storms, providing red alerts (level 4) of wind waves, sea level and oscillations from all the existing buoys and tide gauges along the North Spanish coast, from Galicia to the Basque Country (see stations location in Figure 2). Significant wave heights and mean and peak periods at the buoys are displayed in Figure 3. Sea level oscillation warnings were also correctly issued, indicating that the tsunami detection algorithm was performing well. In the particular case of the storm of January 6th, we received additional red alerts of sea level oscillations from stations as far away as the Canary Islands.

Observational Data

As can be seen in Figure 4 (left) the high-pass filtered 1-min time series from the tide gauges along the Spanish coast exhibit significant sea level oscillations simultaneously with this sequence of storms along almost all the coastline. A detailed study of the sea level data and a description of the atmospheric and oceanographic conditions during these days are presented below. The REDMAR tide gauges also provide information on wind waves recorded locally in the harbors and the significant wave heights (H_{m0}) from these are also displayed in Figure 4 (right). We can see that during these storms there were also important wind waves (0.4–1 m height is considered important inside a harbor) even at the quays where the tide gauge station is located. These wind waves (mostly swell) are usually a problem for harbor operations and confirm the critical situation at the coast and harbors these days: the local wind waves along with sea level oscillations and their associated currents may increase the damage to infrastructures and the possibility of flooding.

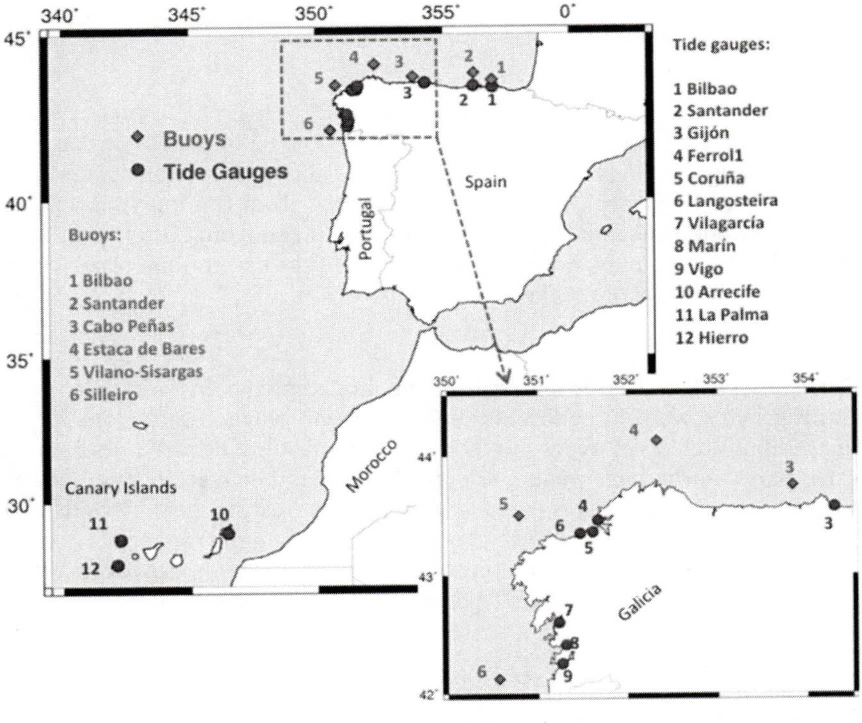

FIGURE 2. POSITION OF THE REDMAR TIDE GAUGES AND DEEP WATER BUOYS FROM PUERTOS DEL ESTADO. Santander buoy belongs to the Spanish Oceanographic Institute.

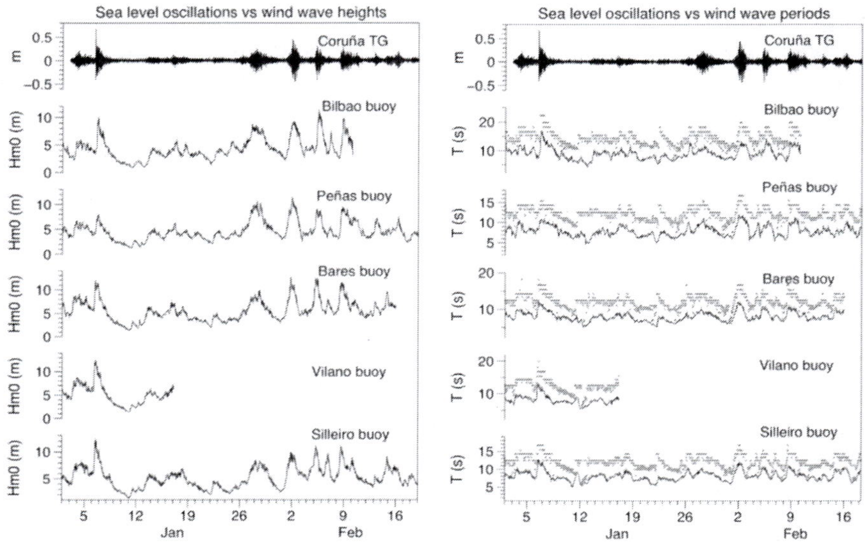

FIGURE 3. SIGNIFICANT WAVE HEIGHT H_{M0} (LEFT) and periods **(RIGHT:** mean period T_m, black line, peak period T_p, blue dots) recorded in open waters by the PdE deep water buoys during the stormy period of January and February 2014. Sampling interval from the buoys is 1 h. Sea level oscillations (1 min high-pass filtered sea level) from Coruña tide gauge at the top of both plots for comparison: the larger oscillations occur during the extreme events of wind waves.

These observed sea level oscillations are thought in principle to be "infragravity waves" generated by non-linear wave interactions of the incredibly high swell waves arriving at the coastline during these storms. The particularly long peak periods of the wind waves in open waters (Figure 3, right) make this event special and interesting: peak periods of up to 22.3 s were recorded at Bilbao buoy the 6th of January. Spectral analysis of the sea level time series should provide information, however, on the possibility of a "meteotsunami" generation as well. It is interesting to note the spatial scale of these events, along hundreds of kilometers of coast.

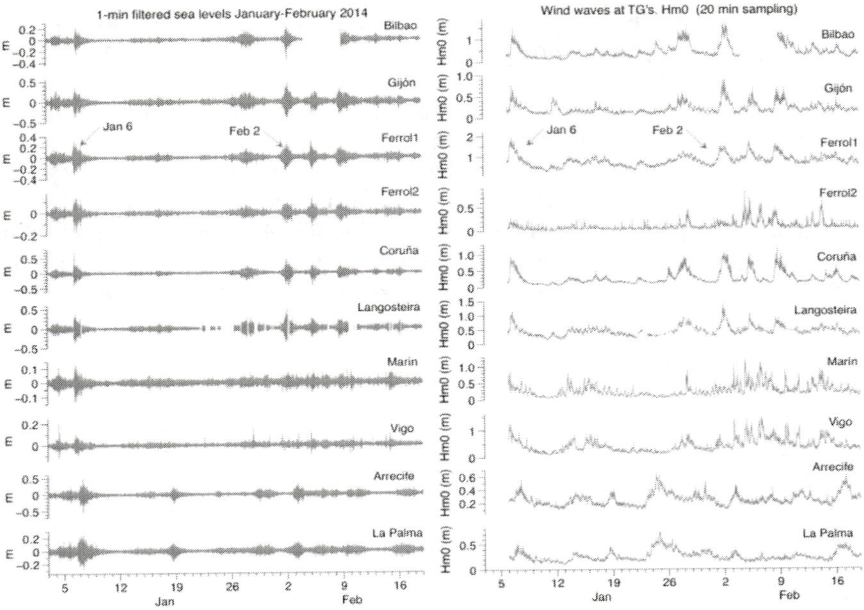

FIGURE 4. LEFT: 1 min high-pass filtered sea level data from the REDMAR network during the stormy period January- February 2014, showing the occurrence of significant high-frequency sea level oscillations along the Spanish coast; **RIGHT:** significant wave height (H_{m0}) computed by these Miros tide gauges for the same period.

The largest sea level oscillations are observed on January 6th and February 2nd, as well as the most extreme wind waves recorded by the deep water buoys. During the first one, maximum significant wave heights of 12.4 m and 12.0 m were recorded by Vilano-Sisargas and Estaca de Bares buoys respectively (PdE, 2014). On February 2nd, however, the latter recorded an even larger significant wave height: 12.8 m, very close to its historical record (12.9 m). Maximum wave heights may be estimated roughly by multiplying these values by 1.6, which would result in individual wind waves of 20 m or more in open waters. This was confirmed by the delayed data processing of raw data from Estaca de Bares buoy, where an individual freak wave of 29 m height was found, becoming the highest wave height ever measured along the Spanish coast (M.I. Ruiz, personal communication). An alert level "4" for wind waves was issued by PdE alert system. As already mentioned these waves presented as well extremely long periods with peak periods reaching 22.3 s (Figure 3, right); this was a very distinctive characteristic of these storms, possibly related to the excited frequencies on the sea level oscillations, as will be shown later.

On both dates we also received level "4" alerts for sea level oscillations from most of the tide gauges shown in Figure 4; however, total sea level alerts (highs and lows) did not exceed level "2" on January 6th, and level "3" around February 2nd. This difference is due to the more precise coincidence of the second storm with the spring tide (Figure 5). In fact, this spring tide played a key role in the flooding, overtopping of waves and extreme sea levels recorded during this storm. The astronomical tide was particularly large that day due to the coincidence of the new moon and the lunar perigee (January 30th), its proximity to the perihelion (January 4th) and to zero lunar declination (February 2nd). Coincidence of lunar perigee and zero lunar declination occurs only every 6 years (Pugh, 2004).

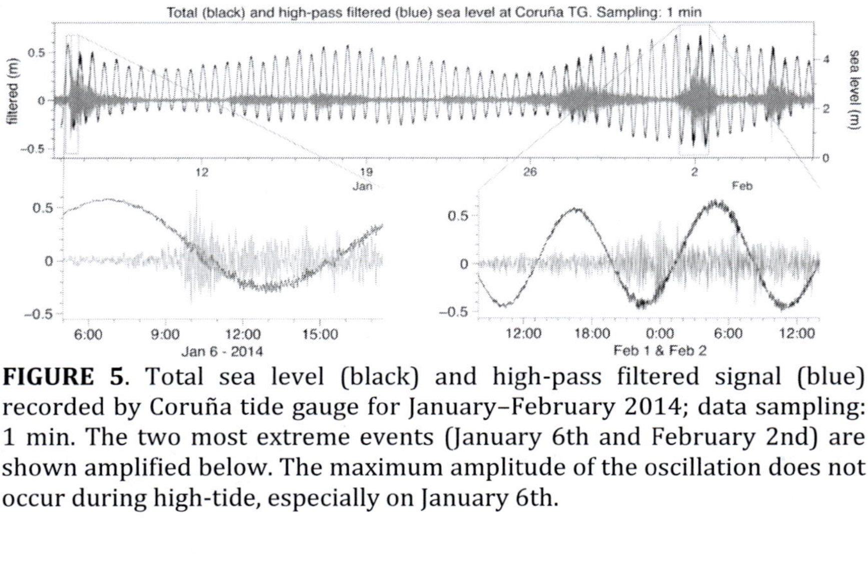

FIGURE 5. Total sea level (black) and high-pass filtered signal (blue) recorded by Coruña tide gauge for January–February 2014; data sampling: 1 min. The two most extreme events (January 6th and February 2nd) are shown amplified below. The maximum amplitude of the oscillation does not occur during high-tide, especially on January 6th.

Origin of the Oscillations

High frequency sea level oscillations were the second important contributor to the total extreme sea levels on these days. It is necessary to determine, first, the origin of these oscillations: in the absence of an earthquake and the coincidence with such big waves the first impression is that they are infragravity waves, very common in the Spanish Atlantic harbors, as already mentioned; nevertheless, they might also be meteotsunamis like those described in Frère et al. (2014). That paper described a meteotsunami event which occurred on June 26–28, 2011 in this region and which was recorded by all the tide gauges presented here, as well as at other stations from Portugal to England. It is interesting to notice the differences with our case: first, weather conditions in June 2011 were rather calm (summer time): wind waves at Peñas buoy (near Gijón tide gauge) were just around 1 m height between the 26th and 28th of June 2011. This is a typical situation during meteotsunamis on the Mediterranean Spanish coast, which usually occur in summer (Tintoré et al., 1988; Gomis et al., 1993; Monserrat et al., 2006b; Marcos et al., 2009). Second, analysis of the atmospheric pressure data from buoys and tide gauges revealed for this event that the oscillations occur when the atmospheric pressure is at a relative low level. This does not seem to be exactly the situation in 2014: Figure 6 shows that the larger sea level oscillations in our case are not exactly coincident with lower levels of pressure at the nearby deep water buoy. In fact, they seem to occur approximately 1 day later and coincide instead on February 2nd with a relatively higher value of pressure.

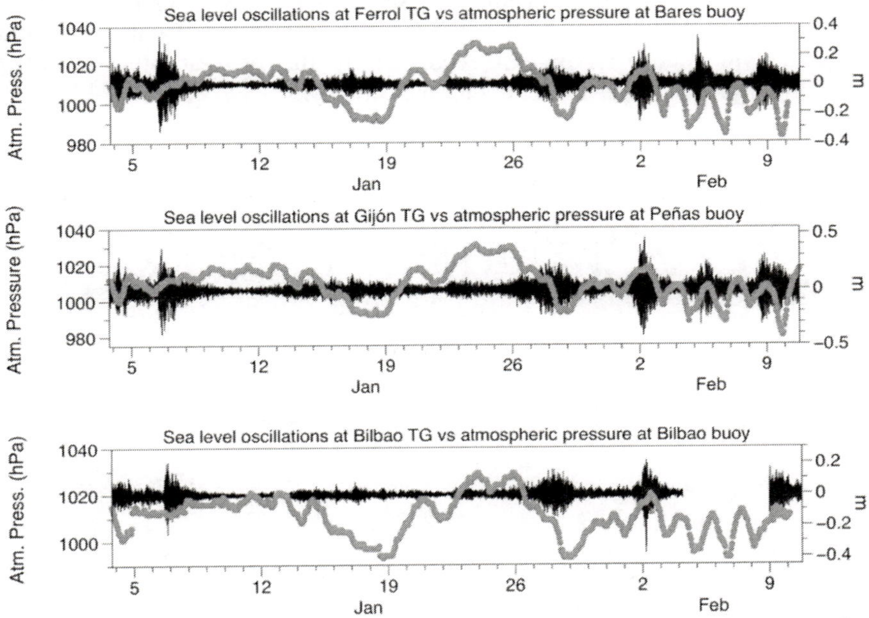

FIGURE 6. Sea level oscillations at Ferrol, Gijón, and Bilbao tide gauges for January–February 2014 (black lines), and atmospheric pressure recorded by the nearest deep water buoy (blue lines: Bares, Peñas, and Bilbao, respectively). Data sampling for sea level oscillations is 1 min, data sampling for atmospheric pressure recorded by buoys is 1 h.

Nevertheless, a meteotsunami could still have been generated for example by a sudden change in atmospheric pressure. Detection of fast atmospheric pressure jumps is usually difficult from existing meteorological networks as the sampling of most of the stations is too infrequent (10 min to 1 h). This is the case for the buoys of PdE network but also for others as described in Frère et al. (2014). The sampling interval of these buoys is 1 h, far from the 1 min temporal resolution required for implementation of future meteotsunami warning systems. As mentioned in previous section the REDMAR tide gauge network already takes into account this requirement and several microbarographs are now in operation with 1-min sampling and a resolution of 0.1 HPa (the resolution at the buoys is 1 HPa). In 2014 one of these sensors was already functioning at Vigo harbor tide gauge. Figure 7 (top) shows in detail the changes in atmospheric pressure recorded at this tide gauge on January 6th: although the resolution is clearly better than the one from the buoy (displayed also in the figure), it is difficult to conclude that there is a change capable of generating a meteotsunami. The relation of the oscillations with the sudden increase in wind waves is however more evident from middle and lower panels of Figure 7, which show the oscillations of sea level at Coruña plotted along with the wind waves significant wave height recorded by the nearest offshore buoy (Vilano) and by the tide gauge itself, respectively.

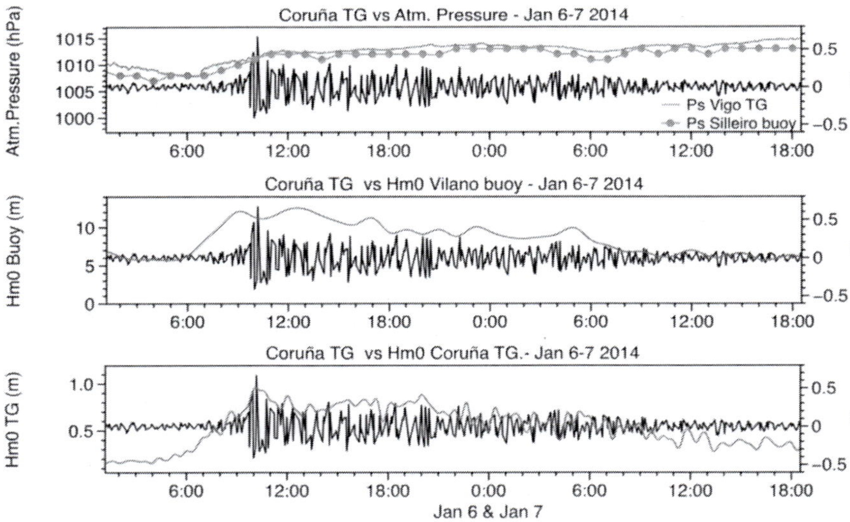

FIGURE 7. Sea level oscillations on January 6th 2014 at Coruña harbor vs.: top:atmospheric pressure data from Silleiro buoy (hourly values, blue dotted line) and from Vigo tide gauge (1 min values, blue line); middle: significant wave height recorded by Vilano buoy (red line); bottom: significant wave height recorded by Coruña tide gauge itself (red line). There is not atmospheric pressure data available from the nearest buoy (Vilano).

The amplitudes of these sea level oscillations (significant and maximum wave heights as explained in Section Data and Methods) are shown in the bottom panels of Figure 8 for La Coruña and La Palma tide gauges. Notice the resemblance of the peaks with the ones in significant wave height of wind waves in Figures 3, 4. This supports the idea of the swell wind waves being the forcing mechanism of these oscillations, which reach H_s and H_{max} of 0.5 and 0.9 m respectively in La Coruña, and of 0.3 and 0.5 m in La Palma, well far away from the center of the storms. Figure 8 (top panels) displays as well the spectrogram or temporal evolution of the spectra of 1 min data for these two stations (1536 points window with 128 points overlap): an increase in the energy content at higher frequencies (2 min or 0.5 cpm to 7 min or 0.15 cpm) is observed when each event starts. These periods are very close to the range of frequencies of the infragravity waves (30 s–5 min).

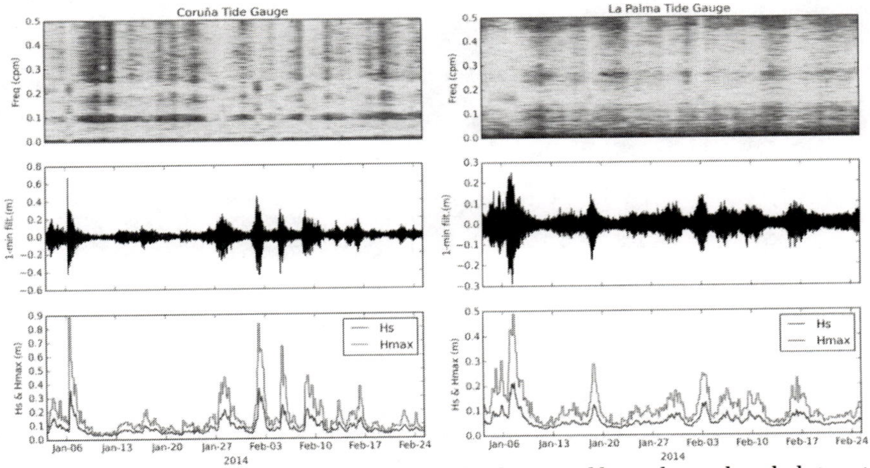

FIGURE 8. Top: spectrogram of 1 min high-pass filtered sea level data at Coruña (left) and La Palma (right) tide gauges, showing the daily variation of the spectra along the stormy period January–February 2014. Applied to 1536 points windows with 128 points overlap (colors in variance units). The storms coincide with the increase of energy on higher frequencies of sea level (periods 2′–6′); middle and bottom: 1 min high-pass filtered time series and estimated “seiche” amplitudes (half the significant wave height Hs and maximum wave height Hmax) at these stations during this period (zero-crossing procedure).

We cannot discard the possibility of one of these frequencies being related to the excitation of the natural eigen periods at some harbors. Left panels on Figure 9 display for Coruña and La Palma tide gauges, the spectra for one specific day of 1 min data when maximum oscillations were recorded (red line) and the spectra for another day with no significant oscillations (blue line). The latter represent the spectral content of the background noise and therefore the natural frequencies or eigen periods of each location. A vertical logarithmic scale is used for an easier comparison of the two spectra. The frequency peaks on red and blue lines seem to be practically the same on both stations, what would confirm that we are observing an excitation of the natural eigen periods during the storm of January 6th, most evident at Coruña tide gauge.

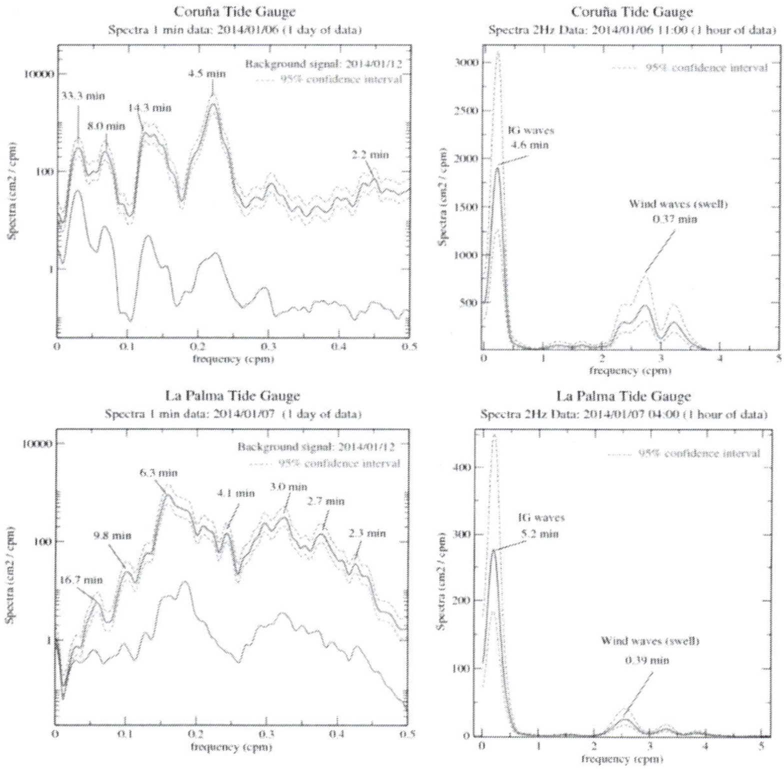

FIGURE 9. Spectral analysis for Coruña and La Palma tide gauges, during the storm of January 6–7, 2014. Left: spectra applied to 1 day of 1-min data, right: spectra applied to 1 hour of 2 Hz data. The 1- min and 2 Hz data have been high-pass filtered (periods <3 h and periods <45 min respectively). Red lines: spectra for 1 day/hour of maximum sea level oscillations (and 95% confidence intervals). Spectra from the background signal was added to the 1-min spectra (blue line) and a logarithmic vertical scale used in this case for easier comparison of the peaks. Spectra method: auto-covariance function smoothed with a Parzen lag window (40 degrees of freedom). Swell waves and infragravity waves are observed in the 2 Hz spectra.

Interestingly, Figure 8 (upper panel) shows a permanent signal in the range of 19–25 min (0.05–0.04 cpm) in the Coruña time series. It seems the signal is present before and after the increase of the oscillations amplitude and is also clear in the background noise spectra of 1 min data in Figure 9. This signal is not present or is very small in La Palma tide gauge (Canary Islands), but it is close to one found for the event of June 2011, on almost all the stations (Frère et al., 2014); this would discard a local topographic feature being responsible for this frequency peak although Frère et al. did not find a clear explanation about their origin. A spectrogram of the 1-min atmospheric pressure from Vigo tide gauge (not shown) revealed that this was the only band of frequencies with significant variance in the high temporal resolution atmospheric pressure data for all these weeks.

As the periods excited during 2014 events are very close to the Nyquist frequency of 1 min data (2 min), spectral analysis of the original 2 Hz data was performed also for the Coruña and La Palma tide gauge records during the storm of January 6th (Figure 9, right panels). In this case we have used a window of 1 h of data with maximum amplitude of the oscillations. The background noise at very high frequencies was so small that its spectra is not shown in this case. These 2 Hz spectra allow a clear distinction between the frequencies of the infragravity waves we are discussing here and the wind-waves at the tide gauge locations: we can see for the latter that most of the energy is concentrated on the swell band, with peaks at 0.37 min (22.2 s) and 0.39 min (23.4 s) for Coruña and La Palma, respectively. These very long peak periods are also present for this event at the open waters buoys, as already mentioned. This would explain therefore that associated infragravity waves were generated at a low frequency range, 4–6 min, and that their wave peaks were much narrower than usual for this particular event. According to the 1-min spectra (Figure 9, left panels), these waves were then further amplified at the resonance eigen periods at Coruña and La Palma, 4.5 min and about 6 min, respectively, becoming a peculiar and interesting case of resonance amplification of infragravity waves.

Contribution to Extreme Sea Levels

A better understanding of the contribution of these oscillations to total sea level is possible from the data presented in Table 1. Column 2 shows the maximum sea levels recorded by the tide gauges of Bilbao, Santander, Gijón, and Coruña during the storm of February 2nd, when the coastal inundation affected a largest area of the coast. These extreme sea levels are derived from the total 1-min sea level data. Column 3 contains the height of the tide at that instant, computed with the harmonic constants derived from harmonic analysis of more than one year of hourly sea level data in each harbor. The magnitude of the storm surge forecasted by the Nivmar system in operation at PdE is presented in column 4 for the point closer to each tide gauge, and column 5 contains the contribution from the "seiche" oscillation, estimated as half the significant wave height H_s (of the "seiche") obtained in the way described in Section Data and Methods. The most important conclusion is that the extreme sea levels recorded during this event were significantly affected by these infragravity waves that had the same or even larger magnitude than the storm surge itself. In fact, the storm-surge component was rather moderate or small as was confirmed from analysis of the storm surge forecasted by the Nivmar system (the value forecasted for Coruña harbor was even negative). The addition of the values in columns 3, 4, and 5 is exactly the observed extreme value in column 2 for Santander and Gijón, and very close to it in Coruña and Bilbao (only 1 and 2 cm larger respectively).

TABLE 1. Column 2: extreme total sea levels (m) / time (hh:mm gmt) recorded the 2nd of february by the tide gauges of bilbao, santander, gijón and coruña (sampling interval: 1 min); columns 3–5: contribution of the tide, surge and high frequency oscillation (considered here as half the significant wave height), the two latter at 05:00 gmt; columns 6 and 7: 99.95 percentile, and maximum sea level of the historical record (1992 to present).

Station	Max/Time (1 min) (m)	Tide (m)	Surge (m) (Nivmar)	Seiche (m) (Hs/2)	Percentile 99.95 (m)	Historical record (m)
Bilbao	4.86/05:01	4.65	0.10	0.13	4.85	4.99
Santander	5.36/04:55	5.15	0.11	0.10	5.34	5.38
Gijón	5.30/04:35	4.96	0.09	0.25	5.23	5.40
Coruña	5.01/04:49	4.85	−0.08	0.25	5.10	5.30

Reference of heights: REDMAR datum for each station (more information on http://portus.puertos.es/Portus_RT/?locale = en).

The individual maximum wave heights (H_{max} of the "seiche") recorded during this event reached 0.56 m in Bilbao, 0.45 m in Santander, 0.77 m in Gijón and 0.83 m in Coruña. The extreme sea level could have been larger in fact in Bilbao and Santander, as this maximum amplitude of the oscillations took place a couple of hours after the high tide in these harbors.

According to Table 1 we can see also that although the extreme high waters were below the historical record or maximum sea level recorded in each tide gauge since 1992, the 99.5 percentile of high waters was exceeded at all these harbors during this particular storm except at Coruña (precisely the one with a smallest storm surge component).

Therefore, the extreme sea level values recorded were in great part caused by the large spring tide added to a moderate or negligible storm surge, in combination with "seiches" of H_s reaching up to 0.5 m (0.25 m amplitude) in some cases. Since these oscillations are stochastic in nature, (and similar to wind waves in this sense) we need to take account of the probability of an individual maximum wave coinciding with the high tide.

Sea Level Data Recorded after the Atlantic Ridge Earthquake, February 13th, 2015

An earthquake with magnitude Mw = 7.1 took place in the middle of the Atlantic ridge on February 13th, 2015 at 18:59:12 (UTC). The earthquake epicenter was located at 52.649°N, 31.902°W, well within the Earthquake Source Zone monitored by the NEAMTWS Candidate Tsunami Service Providers (CTSP's). The exact position of the epicenter in this region determines the kind of message provided by the NEAMTWS CTSPs in the Atlantic; in this case it was in the region where messages must be issued for earthquakes with magnitude larger than 5.5.

In spite of the relatively high magnitude of the earthquake the focal mechanism was a strike-slip one in a transform fault (with centroid depth: 25.2 km, strike: 277, dip: 88, slip: −170, according to the Global Centroid Moment Tensor Catalog: Dziewonski et al., 1981; Ekström et al., 2012); a tsunami wave is not expected for this kind of earthquakes due to the lack of significant vertical movement of the ocean bottom. For this reason, NEAMTWS warning center provided just an information message for the region 12 min after the earthquake.

Interestingly, about 3 h after, a small increase of variability was observed in the 1-min filtered data from Langosteira tide gauge, a station located on the Northwest corner of the Spanish coast, at the external harbor of La Coruña (Figure 1). This signal could have been interpreted by mistake as the record of a small tsunami by a technician on duty on a tsunami warning center. In fact, this information was leaked out to press by an expert external to NEAMTWS and the National Tsunami Warning Center in Spain. The automatic algorithm for tsunami detection in PdE triggered an alert level "2" message for the station of Langosteira; the first alert was issued at 22:54:00 UTC, and several times thereafter until the morning of the 14th of February. It is important to stress that the thresholds of the algorithm are fixed according to the local variability of sea level at each harbor. As described in the previous section, oscillations similar to these ones are common along the North Spanish coast and are usually caused by wind waves in this region.

Looking for more tsunami signals after the earthquake of February 13th, 2015, a review was made through the IOC Sea Level Station Monitoring Facility (IOC/SLSMF) and other data portals such as the IBIROOS *In-Situ* Tac (that includes all *in-situ* marine observational platforms, including tide gauges, in the Atlantic European coast). A significant improvement on the availability of 1 min sea level time series during the last 10 years has been possible thanks to the new requirements established for tsunami warning: the percentage of tide gauges with 1 min sampling in this region has increased from 0% in 2006 to a 61% of the existing stations in 2015. As a result, records were available from the 50 locations shown in Figure 10(left), with stations from Iceland (Reykjavik) to Cape Verde (Palmeira), and a couple of stations in the Canadian coast. Other stations on the US coastline and Canada that did not appear to display any significant signal or that had less frequent data sampling were not included in this analysis. Similarly, French stations are not included here as CEA (France) advised that a significant signal was not detected along the French coast.

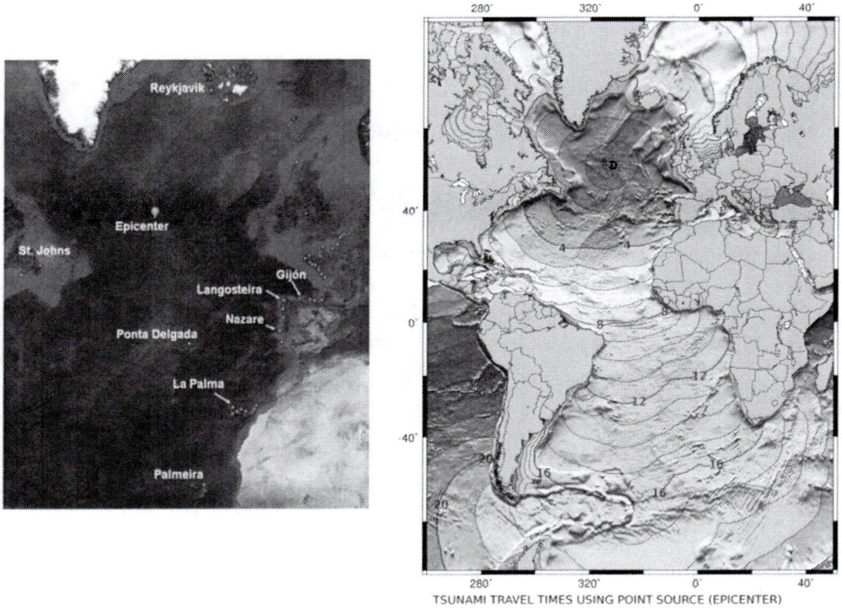

FIGURE 10. Left: up to 50 tide gauges (red dots) with 1-min data were available during the 13th of February 2015 earthquake event. Only a subset of them presented a possible small signal; the most evident was observed in Langosteira tide gauge; right: tsunami propagation (1 h contour line) generated at the epicenter of the earthquake of 13th February 2015, 18:59:12 UTC, in the Atlantic ridge, as provided by the model TTT SDK v3.3.

The filtered signal obtained from the rest of stations revealed different magnitudes of potentially normal variability at each harbor whilst those stations along the Iberian Peninsula North and North-West coasts displayed a possible earthquake-related signal. These stations were: Bilbao, Santander, Gijón, Coruña, Ferrol, and Langosteira, from the REDMAR network, and Peniche, Nazare, and Leixoes in the Portuguese coast, north of Lisbon. A possible signal could also be perceived at Reykjavik (Iceland) and Palmeira (Cape Vert). However, there were not significant oscillations in the filtered time series at the Canary Islands stations.

The arrival time of the potential tsunami wave was obtained by IGN for each tide gauge station from the numerical model TTT (Travel Time Software) SDK v3.3 (Table 2). The program calculates propagation velocities based on an input bathymetry grid and uses Huygens constructions to propagate the wave front from the epicenter to all nodes on the grid (Wessel, 2009). A map of this theoretical tsunami propagation time is shown in Figure 10 (right), with contour lines of 1 h.

TABLE 2. Estimated arrival times of the first tsunami wave from numerical simulations of the earthquake of february 13th, 2015 in the atlantic ridge.

Station	Tsunami arrival time	UT Time
Langosteira	3 h 01 m 56 s	22:01:08
Ferrol	3 h 05 m 25 s	22:04:37
Gijón	3 h 29 m 33 s	22:28:45
Bilbao	3 h 34 m 21 s	22:33:33
Santander	3 h 21 m 57 s	22:21:09
Reykjavik	3 h 42 m 08 s	22:41:20
Peniche	3 h 18 m 08 s	22:17:20
Nazare	3 h 20 m 26 s	22:19:38
Leixoes	3 h 28 m 29 s	22:27:41
Ponta Delgada	2 h 52 m 32 s	22:51:44
Palmeira	5 h 51 m 21 s	00:50:34 (14th)

Derived from IGN run of model TTT SDK v3.3.

The 1-min filtered sea level time series are displayed in Figure 11 for the subset of these 50 stations where a signal, however small, was found. The red line indicates the time of the earthquake. Expected arrival times from the numerical simulation presented in Table 2 have been added by means of vertical black arrows.

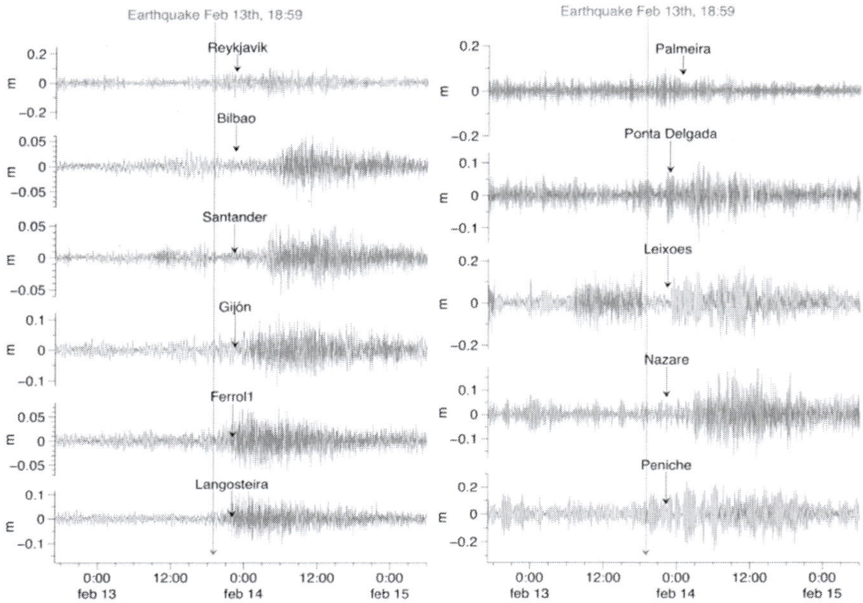

FIGURE 11. 1 min high-pass filtered sea level time series from those stations on figure **10**where a signal, even if small, might be present after the earthquake of february 13th 2015 showing the expected arrival times in table **2**, derived from the tsunami propagation model, at each tide gauge (black arrows)**.** The earthquake time is indicated by the red arrow.

These plots reveal, firstly, that there is no clear signal at most of the stations, if the times of earthquake and expected arrival of the wave are considered. The time series show a different behavior after the arrival time at all the stations except Palmeira, Ponta Delgada, and Reikjavik. In particular an increase of variability is observed immediately after or coincident with the arrival time only at Langosteira, Ferrol, Gijón, and Leixoes. Others show a signal several hours before or after the estimated arrival time.

It is difficult to determine clearly if a very small tsunami could have reached these tide gauges mainly for two reasons: (a) the uncertainty in the simulated arrival times could be large for those points where local bathymetry is not well resolved, (b) the local variability of each harbor and their relation to other oceano-meteorological reasons such as meteotsunamis or wind-waves.

Analysis of Langosteira Tide Gauge

It is interesting that, by chance or not, the increase of variability at this station coincides rather clearly with the expected arrival time from the model in this case, and that there is not significant energy during the day or few hours before the event.

Langosteira and Gijón 1 min high-pass filtered signals are displayed in Figure 12 along with their spectrogram and estimated amplitude. In this case the spectrogram was derived using a 512 points window (around 8 h) with a 128 points (2 h) overlap. It can be seen that the event recorded in Langosteira reflects not only a small increase of the magnitude of the signal, with waves reaching amplitudes of 10–15 cm, but also the appearance of higher frequency oscillations, with energy suddenly present at all the lower periods up to 2 or 2.5 min, as observed in the events of 2014 presented in previous section, along with the 15–25 min period signal already present before the event. Once again, it seems there is a kind of permanent background oscillation in this band of frequencies in this part of the Spanish coast, observed now also at Langosteira and Gijón tide gauges, similar to the one observed at Coruña during January–February 2014, and to the one found by Frère et al. (2014).

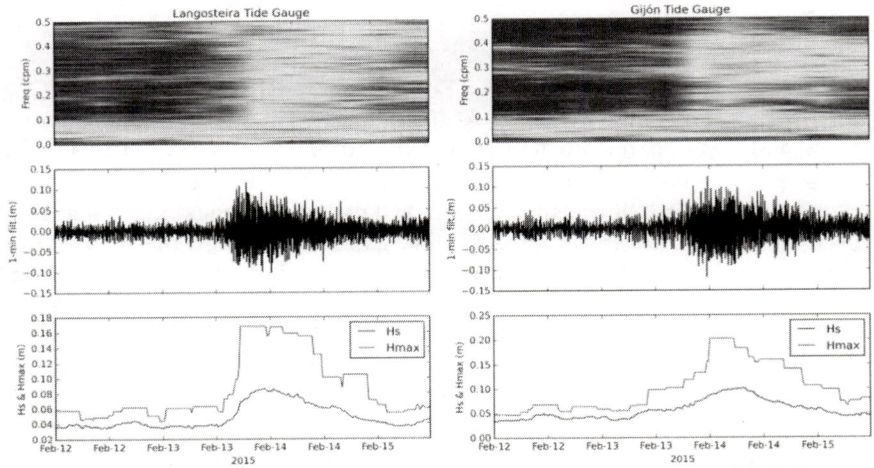

FIGURE 12. Top: spectrogram of 1 min high-pass filtered sea level data at Langosteira (left) and Gijón (right) tide gauges, showing the temporal evolution of the spectra along February 12–15th 2015. Applied to 512 points windows with 128 points overlap (colors in variance units). A constant signal with periods in the range of 15–25 min is present before the event at both stations; new signals for lower periods (until 2 or 2.5 min) appear when the event starts; middle and bottom: 1 min high-pass filtered time series and corresponding estimated "seiche" amplitude (half the significant wave height Hs and the maximum wave height Hmax) during this period (zero-crossing procedure).

The daily spectra of 1-min data for 1 day before (12th of February) and 1 day after the event (15th of February) are displayed in Figure 13 (left panel), in the same way described for Figure 9 for January–February 2014. This figure confirms the previous statement that a signal with a period between 15 and 25 min is always present (peak on 17.2 min); the higher frequency oscillations (up to 2 min) are mostly excited during the event (red line). Notice the coincidence of this signal with the ones observed in previous section for Coruña tide gauge (nearby Langosteira). In that case the excitation of the lower periods seemed to be associated to the storms and huge wind waves; it could be the case also in this event.

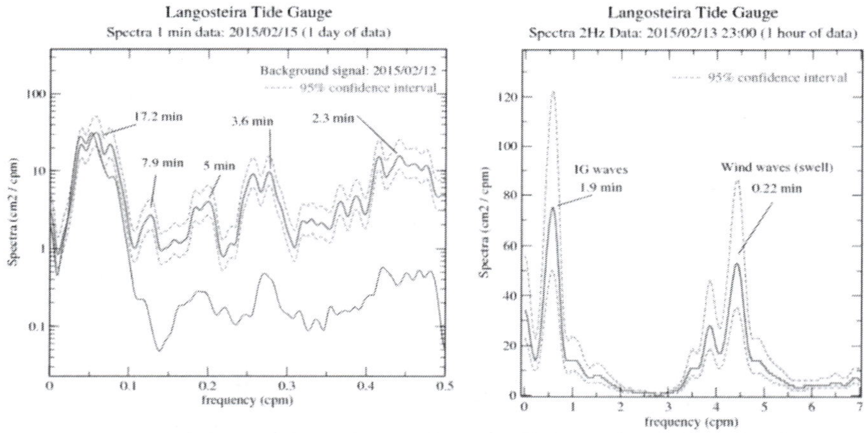

FIGURE 13. Left: spectral analysis of 1 min (left) and 2 Hz (right) data for Langosteira tide gauge on February 13th 2015. The 1-min and 2 Hz data have been high-pass filtered (periods <3 h and periods <45 min respectively). Red lines: spectra for 1 day/hour of maximum sea level oscillations (and 95% confidence intervals). Spectra from the background signal was added to the 1-min spectra (blue line) and a logarithmic vertical scale used in this case for easier comparison of the peaks: the energy at higher frequencies (e.g., periods of 3.6 and 2.3 min) increase significantly during this event. The 2 Hz spectrum reveals the energy content on the infragravity waves (peak very close to the Nyquist frequency of 1 min data) and on the swell bands. Spectra method: auto-covariance function smoothed with a Parzen lag window (40 degrees of freedom).

In the right panel of Figure 13 the spectral analysis of the 2 Hz data from Langosteira tide gauge is displayed (again in the same way as those in Figure 9). Here, the presence of a peak on the infragravity waves band is also clear, with a period of 1.9 min, clearly inside the infragravity frequency range. Wind waves swell are also present in this case with a peak period of 13.2 s (0.22 min), significantly lower, nevertheless, than in 2014 events, what would explain the lower period of this infragravity wave, most common according to the literature.

Although there were not extreme waves and storms during February 2015, their influence cannot be completely disregarded for Langosteira. Figure 14 shows the increase in significant wave height (up to 6.5 m) and mean period of the wind waves recorded at the nearby buoy of Estaca de Bares (offshore, north of Langosteira), during the appearance of the oscillations in this station. Based on the experience of other events, therefore, this sudden change in amplitude and period of the wind waves would be most likely the origin of the oscillations (infragravity waves) observed in this tide gauge coincidentally when a potential small tsunami could have reached the station. The 1 min meteorological data recorded at Vigo tide gauge during the same period also simply reflect an increase in the wind velocity at that time (related to wind waves on the other hand); the atmospheric pressure data does not show any relevant feature before February 13th, confirming the origin of the observed sea level oscillations in Langosteira in the infragravity waves.

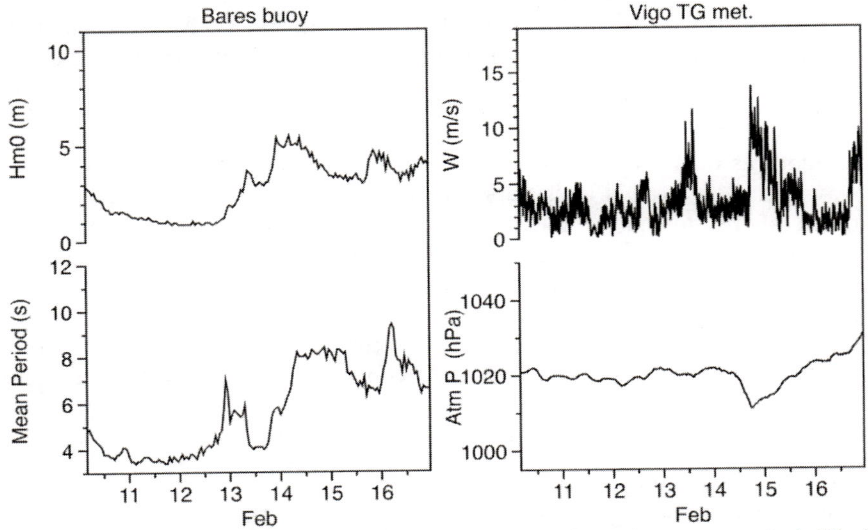

FIGURE 14. Left: significant wave height (Hm0) and mean period (Tm) recorded by Bares open waters buoy during the oscillations event of February 2015; right: mean wind velocity and atmospheric pressure recorded by the meteorological sensors of Vigo tide gauge for the same period (1-min sampling resolution).

The discussion above illustrates how considering these kind of signals as a small tsunami useful for validation of tsunami modeling without more careful analysis can be dangerous, and how the analysis of other environmental variables, as well as a better knowledge of the local sea level variability in the station, and the local response to the different forcings, is needed.

DISCUSSION

Nowadays we are in a fortunate new position to improve our understanding of sea level oscillations at higher frequencies (periods of the order of minutes) thanks to significant changes in observational sea level networks. Detection of small tsunami signals at tide gauges is of great interest for assessing the skill of existing tsunami warning systems and for validation of tsunami propagation models, and this need is leading to the development of new methodologies of sea level data processing and automatic detection algorithms such as the ones described in this paper. However, the tsunami footprint on sea level records resembles other more frequent types of phenomena, already known and studied from an academic point of view, but not usually considered within sea level warning systems.

The analysis of the two events presented in this paper has provided several lessons. First of all, it reveals the importance of considering these oscillations for a more precise design of sea level alert systems. Although wave-setup is already considered of relevance for sea level forecasts and this may be solved by means of coupled wind-wave and storm surge models, additional sea level oscillations in the frequency of the "tsunami" signals should also be taken into account in the future. In the examples described in this paper the forcing of these oscillations seems to be mainly wind waves, but in other regions the forcing may be different (meteotsunamis such as the "rissaga" events in the Mediterranean Sea, for example). Case studies like this one will contribute to a better physical knowledge of all these phenomena.

Another important lesson from this study it that detection of small tsunamis from tide gauges is uncertain due to the inherent sea level variability caused by all these oceano-meteorological agents, but also due to errors in tsunami propagation models or problems in the data (transmission fails, spikes, etc). This leads to a continuous need for refinement of automatic algorithms based on the analysis of events like the ones described here.

The storms of January and February 2014 represent an example of real extreme events where both wind-waves and sea level, with their different components combined, and even without a significant storm surge, may cause severe damage and flooding in harbors and along the coastline. Therefore, future extreme analysis should consider this random process to determine maximum (and minimum) sea levels. The low sampling interval allows an adequate measurement of the effect of these oscillations in sea level records, which is the main reason for the upgrade of the tide gauge networks that in the past usually worked with 10 or 15 min sampling intervals at best. The sea level signals observed along the Spanish coast in January and February 2014 were not that different from the ones of a small tsunami.

The tsunami automatic detection algorithm in PdE performed perfectly well during these storms, providing red alerts for oscillations. The multi-parameter alert system provided as well red alerts for wind wave heights in open waters (although the most anomalous feature of these storms were the long peak periods), and yellow or orange alerts for total sea level. However, the sea level forecast did not reach the level of alert obtained from observations as it was based on just the forecast of the tide and the storm surge. This confirms the interest of the multi-hazard approach and the need for taking into account these physical phenomena in the future developments of the sea level forecasting system.

Concerning the observational network, in this paper we present the development of a new procedure for characterization of these oscillations in terms of amplitude and period. Initially implemented for the 1 min data submitted in real time in PdE, and tested with success in these two examples, the procedure needs to be improved and adapted to automatic processing of the original 2 Hz raw data. This is important especially for the periods of the infragravity waves, very close to the Nyquist frequency of 1 min data, as we have seen here, and will become possible as the network communications and access to these data improve in the future. A prototype is in fact already in operation for one tide gauge from the REDMAR network.

The Atlantic Ridge earthquake on 13th February 2015 was not expected to generate a tsunami signal; however, the occurrence of small oscillations in the North and West coast of the Iberian Peninsula revealed the complexity of interpretation of tide gauge data and the need for analysis of sea level in combination with other environmental variables. Although the signal observed in a few tide gauges is coherent with the expected arrival time of a small tsunami, and the wind waves at that time were not extreme, a detailed analysis has shown that these oscillations could be most likely infragravity waves and not a tsunami signal (spectral analysis of the time series reveals similar behavior of this event and the previous one).

Although we point to the wind waves as the main mechanism of the oscillations observed in these two examples, we cannot discard completely the possibility of a combination of effects and the occurrence as well of a meteotsunami, especially during one of the events of 2014. Considerable research has been done about these oscillations associated with atmospheric pressure disturbances in the last decades; the availability of microbarographs like those already in operation in the REDMAR network is expected to be of help in the establishment of future meteotsunami warning systems. These data could be integrated into these systems as the sea level data are already included in the tsunami warning systems. Although ideally offshore bottom pressure sensors would be more useful in both cases, the reality is that nowadays the only data available in real time for most of the countries in Europe are provided by tide gauges at the harbors, so it makes sense that we should improve the tools to use these data in near-real time and to interpret them correctly. Future work should focus therefore on more detailed studies of the local response of small bays and harbors to all these external forcings. This local knowledge will allow the design of more reliable and refined alert systems. This is the objective of future projects in PdE, that is starting now to pay more attention to the tremendous amount of 2 Hz and 1 min data recorded inside the Spanish harbors.

Finally, tsunami warning centers should be aware of these limitations and the complexity of the sea level measurements: once again, multi-hazard experts, with a combined knowledge of tsunamis, oceanography and meteorology, and making use of information from the whole network of stations (and not relying just on local information) will ensure reliable interpretation of the detection networks and a better understanding of the sea level risks taking into account all the different phenomena and the catastrophic consequences that their superposition may generate.

AUTHOR CONTRIBUTIONS

BP: originator of the idea of the paper, manuscript writing; preparation of figures and tables, development of the algorithm for 1 min data processing, responsible of the sea level data processing in PdE including design of automatic quality control and alert issues. FM: implementation of an upgrade of the tsunami algorithm for tsunami detection in Java; generation of an offline Java script of this algorithm for easy application to the stations not included in the Portus system (Puertos del Estado). CG, JC, FS: tsunami wave arrival times at each station, description of the earthquake characteristics. EA: detailed review of the article, including re-writing of several sections and an improved description of the measurements during the storms of January–February 2014, with emphasis in the analysis of the storm surge, seiche and tide components.

FUNDING

PdE, Instituto Geográfico Nacional and CEA. SOPRANO project (Spanish national project: Plan Estatal de Investigación Científica) y Técnica y de Innovación 2013–2016 (Convocatoria 2014), Ministerio de Economía y Competitividad).

CONFLICT OF INTEREST STATEMENT

The authors declare that the research was conducted in the absence of any commercial or financial relationships that could be construed as a potential conflict of interest.

ACKNOWLEDGMENTS

We wish to thank first all the operators and maintenance technicians of Puertos del Estado networks of tide gauges and buoys (Sidmar and Oceanor personnel, Marta de Alfonso and María Isabel Ruiz). We thank as well the Sea Level Station Monitoring Facility of the IOC (UNESCO), the IBIROOS *In-Situ* Tac and the different national operators for the provision of tide gauge data of non-Spanish stations for the event of February 13th 2015. The Tsunami Travel Time software (TTT SDK v 3.3) was developed by Dr. Paul Wessel (Geoware, http://www.geoware-online.com), and is used by the NOAA Pacific Tsunami Warning Center and licensed to NOAA/ITIC for redistribution. Finally, we would like to express our gratitude to Dr. Angela Hibert, who kindly accepted to read the manuscript and make suggestions to improve the English writing. This study has been performed in the framework of the Spanish funded project SOPRANO.

REFERENCES

1. Alasset, P.-J., Hébert, H., Maouche, S., Calbini, V., and Meghraoui, M. (2006). The tsunami induced by the 2003 Zemmouri earthquake (MW= 6.9, Algeria): modelling and results. *Geophys. J. Int.* 166, 213–226.

2. Álvarez-Fanjul, E., de Alfonso, M., Ruiz, M. I., Lopez, J. D., and Rodriguez, I. (2002). "Real time monitoring of Spanish coastal waters: the deep water network," *Proceedings of the Third International Conference on EuroGOOS* (Athens).
3. \
4. Álvarez-Fanjul, E., Pérez-Gómez, B., and Rodríguez, I. (2001). Nivmar: a storm surge forecasting system for Spanish waters. *Sci. Mar.*65, 145–154.
5. BBC (2014). *UK Storms: Extreme Weather Caused Years of Erosion 21 February 2014. BBCNews Magazine.* Available online at: http://www.bbc.com/news/uk-26277373 (Accessed February 2016).
6. Beltrami, G. M., Di Risio, M., and De Girolamo, P. (2011). "Chapter 27: Algorithms for automatic, real-time tsunami detection in sea level measurements," in *The Tsunami Threat-Research and Technology*, ed N.-A. Moerner (InTech), 549–573.
7. Bressan, L., Zaniboni, F., and Tinti, S. (2013). Calibration of a real-time tsunami detection algorithm for sites with no instrumental tsunami records: application to coastal tide-gauge stations in eastern Sicily, Italy. *Nat. Hazards Earth Syst. Sci.* 13, 3129–3144.
8. Church, J. A., Gregory, J. M., Huybrechts, P., Kuhn, M., Lambeck, K., Nhuan, M. T., et al. (2001). "Changes in sea level," in *Climate Change: The Scientific Basis*, eds J. T. Houghton, Y. Ding, D. J. Griggs, M. Noquer, P. J. van der Linden, X. Dai, K. Maskell, and C. A. Johnson (New York, NY: Cambridge University Press), 639–694.
9. Defant, A. (1961). *Physical Oceanography, Vol. 2.* Oxford: Pergamon Press.
10. Dziewonski, A. M., Chou, T.-A., and Woodhouse, J. H. (1981). Determination of earthquake source parameters from waveform data for studies of global and regional seismicity. *J. Geophys. Res.* 86, 2825–2852.
11. Ekström, G., Nettles, M., and Dziewoński, A. M. (2012). The global CMT project 2004–2010: centroid-moment tensors for 13,017 earthquakes. *Phys. Earth Planet. Inter.* 200–201, 1–9.
12. El País (2014). *El Temporal Provoca Olas Gigantes de Hasta 13 Metros de Altura en el Cantábrico.* 3 February 2014. Available online at: http://elpais.com/elpais/2014/02/03/actualidad/1391412691_659227.html
13. Frère, A., Daubord, C., Gailler, A., and Hèbert, H. (2014). Sea level surges of June 2011 in the NE Atlantic Ocean: observations and possible interpretation. *Nat. Hazards* 74, 179–196.
14. Gómez-Lahoz, M., and Carretero-Albiach, J. C. (2005). Wave Forecasting at the Spanish Coasts. *J. Atmos. Ocean Sci.* 10, 389–405.
15. Gomis, D., Monserrat, S., and Tintoré, J. (1993). Pressure-forced seiches of large amplitude in inlets of the Balearic Islands. *J. Geophys. Res.* 98, 14437–14445.
16. Herbers, T. H. C., Elgar, S., and Guza, R. T. (1995). Generation and propagation of infragravity waves. *J. Geophys. Res.* 100, 24863–24872.
17. Hibiya, T., and Kajiura, K. (1982). Origin of the Abiki phenomenon (a kind of seiche) in Nagasaki Bay. *J. Oceanogr. Soc. Jpn.* 38, 172–182.
18. IOC UNESCO (2007). *North-East Atlantic, the Mediterranean and Connected Seas Tsunami Warning and Mitigation System, NEAMTWS, Implementation Plan*. Technical report 73, Intergovernmental Oceanographic Commission.
19. Jansà, A., Monserrat, S., and Gomis, D. (2007). The rissaga of 15 June 2006 in Ciutadella (Menorca), a meteorological tsunami. *Adv. Geosci.* 12, 1–4.
20. Jenkins, G. M., and Watts, D. G. (1968). *Spectral Analysis and its Applications.* San Francisco, CA: Holden Day.
21. Longuet-Higgins, M. S., and Stewart, R. W. (1962). Radiation stress and mass transport in gravity waves, with application to "surf beats." *J. Fluid Mech.* 13, 481–504.
22. Marcos, M., Monserrat, S., Medina, R., Orfila, A., and Olabarrieta, M. (2009). External forcing of meteorological tsunamis at the coast of the Balearic Islands. *Phys. Chem. Earth* 34, 938–947.

23. Monserrat, S., Vilibić, I., and Rabinovich, A. B. (2006a). Meteorological tsunamis: Atmospherically induced destructive ocean waves in the tsunami frequency band. *Phys. Chem. Earth* 34, 1035–1051.
24. Monserrat, S., Gomis, D., Jansá, A., and Rabinovich, A. B. (2006b). Ciutadella harbour, Menorca, Spain, 15 June 2006 rissaga. *Tsunami Newsletter* 38, 5–7.
25. Munk, W. H. (1949). Surf beats. *Eos. Trans. Am. Geophys. Un.* 30, 849–859.
26. Munk, W. H. (1962). *Long Waves, in The Sea*. New York, NY: J. Wiley.
27. Nomitsu, T. (1935). A theory of tsunamis and seiches produced by wind and barometric gradient. *Mem. Coll. Sci. Imp. Univ. Kyoto A* 18, 201–214.
28. Orlić, M. (1980). About a possible occurrence of the Proudman resonance in the Adriatic. *Thalassia Jugo-slavica* 16, 79–88.
29. Pattiaratchi, C. B., and Wijeratne, E. M. S. (2015). Are meteotsunamis an underrated hazard? *Philos. Trans. R. Soc. A* 373, 1–23.
30. PdE (2014). *Historical Information Available online at: Puertos del Estado website*: http://portus.puertos.es/Portus_RT/?locale=en(Accessed February 2016).
31. Pérez-Gómez, B., Alvarez-Fanjul, E., Pérez, S., de Alfonso, M., and Vela, J. (2013). Use of tide gauge data in operational oceanography and sea level hazard warning systems. *J. Oper. Oceanogr.* 6, 1–18.
32. Pugh, D. T. (2004). *Changing Sea Levels. Effects of Tides, Weather and Climate*. Cambridge, UK: Cambridge University Press.
33. Pugh, D. T., and Woodworth, P. L. (2014). *Sea Level Science: Understanding Tides, Surges, Tsunamis and Mean Sea Level Changes*. Cambridge, UK: Cambridge University Press.
34. Rabinovich, A. B. (2009). "Seiches and harbour oscillations," in *Handbook of Coastal and Ocean Engineering*, ed Y. C. Kim (Singapore: World Scientific Publishing Co), 193–236.
35. Rabinovich, A. B., and Eblé, M. C. (2015). Deep-ocean measurements of tsunami waves. *Pure Appl. Geophys.* 172, 3281–3312.
36. Rabinovich, A. B., and Monserrat, S. (1996). Meteorological tsunamis near the Balearic and Kuril Islands: descriptive and statistical analysis. *Nat. Hazards* 13, 55–90.
37. Rabinovich, A. B., Stroker, K., Thomson, R., and Davis, E. (2011). DARTs and CORK in cascadia basin: high-resolution observations of the 2004 Sumatra tsunami in the northeast Pacific. *Geophys. Res. Lett.* 38, L08607.
38. Rabinovich, A. B., Vilibić, I., and Tinti, S. (2009). Meteorological tsunamis: atmospherically induced destructive ocean waves in the tsunami frequency band. *Phys. Chem. Earth* 34, 891893.
39. Sahal, A., Roger, J., Allgeyer, S., Lemaire, B., Lavigne, F., Cedex, M., et al. (2009). The tsunami triggered by the 21 May 2003 Boumerdes-Zemmouri (Algeria) earthquake: field investigations on the French Mediterranean coast and tsunami modelling. *Nat. Hazards Earth Syst. Sci.* 9, 1823–1834.
40. Šepić, J., Vilibić, I., Lafon, A., Macheboeuf, L., and Ivanovic, Z. (2015b). High-frequency sea level 512 oscillations in the Mediterranean and their connection to synoptic patterns. *Prog. Oceanogr.* 137, 284–298.
41. Šepić, J., Vilibić, I., Rabinovich, A. B., and Monserrat, S. (2015a). Widespread tsunami-like waves of 23-27 June in the Mediterranean and Black Seas generated by high-altitude atmospheric forcing. *Sci. Rep.* 5:11682.
42. Sheremet, A. T., Staples, F., Ardhuin, S., and Suanez, Fichaut, B. (2014). Observations of large infragravity wave runup at Banneg Island, France. *Geophys. Res. Lett.* 41, 976–982.
43. Sibley, A., Cox, D., Long, D., Tappin, D., and Horseburgh, K. (2016). Meteorologically generated tsunami-like waves in the North Sea on 1/2 July 2015 and 28 May 2008. *Weather* 71, 68–74.
44. Tappin, D., Sibley, A., Horsburgh, K., Daubord, C., Cox, D., and Long, D. (2013). The English channel tsunami of 27 June 2011 – a probable meteorological source. *Weather* 68, 144–152.

45. Tintoré, J., Gomis, D., Alonso, S., and Wang, D. P. (1988). A theoretical study of large sea level oscillations in the Western Mediterranean. *J. Geophys. Res.* 93, 10797–10803.
46. Vela, J., Pérez-Gómez, B., González, M., Otero, L., Olabarrieta, M., Canals, M., et al. (2014). Tsunami Resonance in Palma Bay and Harbor, Majorca Island, as Induced by the 2003 Western Mediterranean Earthquake. *J. Geol.* 122, 165–182.
47. Vich, M. D. M., and Monserrat, S. (2009). Source spectrum for the Algerian tsunami of 21 May 2003 estimated from coastal tide gauge data. *Geophys. Res. Lett.* 36, 1–5.
48. Vilibić, I., Domijan, N., and Cupić, S. (2005). Wind versus air pressure seiche triggering in the Middle Adriatic coastal waters. *J. Mar. Syst.* 57, 189–200.
49. Vilibić, I., Monserrat, S., and Rabinovich, A. B. (2014). Meteorological tsunamis on the US East Coast and in other regions of the World Ocean. *Nat. Hazards* 74, 1–9.
50. Vilibić, I., Sepić, J., Rabinovich, A. B., and Monserrat, S. (2016). Modern approaches in meteotsunami research and early warning. *Front. Mar. Sci.* 3:57.
51. Wessel, P. (2009). Analysis of observed and predicted tsunami travel times for the pacific and Indian Oceans. *Pure Appl. Geophys.* 166, 301–324.
52. Woodworth, P. L., and Blackman, D. L. (2002). Changes in extreme high waters at Liverpool since 1768. *Int. J. Climatol.* 22, 697–714.
53. Woodworth, P. L., and Blackman, D. L. (2004). Evidence for systematic changes in extreme high waters since the mid-1970s. *J. Clim.* 17, 1190–1197.
54. Wu, J. K., and Liu, P. L.-F. (1990). Harbour excitations by incident wave groups. *J. Fluid Mech.* 217, 595–613.

CHAPTER 15

An Evaluation of Onshore Digital Elevation Models for Modeling Tsunami Inundation Zones

Jonathan Griffin[1,2*], Hamzah Latief[3], Widjo Kongko[4], Sven Harig[5], Nick Horspool[1], Raditya Hanung[3], Aditia Rojali[3], Nicola Maher[1], Annika Fuchs[5], Jakir Hossen[6], Supryiati Upi[7], Sri Edi Dewanto[7], Natalja Rakowsky[5] and Phil Cummins[1,6]

[1]*Community Safety Group, Geoscience Australia, Canberra, ACT, Australia*

[2]*Australia-Indonesia Facility for Disaster Reduction, Jakarta, Indonesia*

[3]*Faculty of Earth Sciences and Technology, Bandung Institute of Technology, Bandung, Indonesia*

[4]*Coastal Dynamics Research Centre, Agency for the Assessment and Application of Technology, Yogyakarta, Indonesia*

[5]*Computing Center, Alfred Wegener Institute, Helmholtz Center for Polar and Marine Research, Bremerhaven, Germany*

[6]*Research School of Earth Sciences, Australian National University, Canberra, ACT, Australia*

[7]*National Disaster Management Agency, Jakarta, Indonesia*

ABSTRACT

A sensitivity study is undertaken to assess the utility of different onshore digital elevation models (DEMs) for simulating the extent of tsunami inundation using case studies from two locations in Indonesia. We compare airborne IFSAR, ASTER, and SRTM against high resolution LiDAR and stereo-camera data in locations with different coastal morphologies.

Tsunami inundation extents modeled with airborne IFSAR DEMs are comparable with those modeled with the higher resolution datasets and are also consistent with historical run-up data, where available. Large vertical errors and poor resolution of the coastline in the ASTER and SRTM elevation datasets cause the modeled inundation extent to be much less compared with the other datasets and observations. Therefore, ASTER and SRTM should not be used to underpin tsunami inundation models. A model mesh resolution of 25 m was sufficient for estimating the inundated area when using elevation data with high vertical accuracy in the case studies presented here. Differences in modeled inundation between digital terrain models (DTM) and digital surface models (DSM) for LiDAR and IFSAR are greater than differences between the two data types. Models using DTM may overestimate inundation while those using DSM may underestimate inundation when a constant Manning's roughness value is used. We recommend using DTM for modeling tsunami inundation extent with further work needed to resolve the scale at which surface roughness should be parameterized.

Keywords: digital elevation model, tsunami inundation, LiDAR, IFSAR, SRTM

INTRODUCTION

Large and destructive tsunami occur infrequently and for any particular location typically have recurrence intervals that exceed human lifespans. Extreme events such as the 2011 Tohoku Tsunami have recurrence intervals on the order of 1000 years (Minoura et al., 2001) meaning only in exceptional circumstances do historical records capture previous events of similar magnitude. Therefore, hydrodynamic models of tsunami inundation from realistic but hypothetical scenarios are fundamental to understanding the threat posed to a particular section of coast from tsunami. The results of tsunami inundation models can inform disaster management activities including evacuation and contingency planning, impact and risk assessment and warning, as well as structural mitigation measures (e.g., protective barriers, vertical evacuation shelters, and building design standards). A key input to tsunami inundation models is elevation data, and hydrodynamic models are known to be sensitive to the resolution and accuracy of elevation data (Satake, 1995; Casas et al., 2006; Horritt et al., 2006; Sanders, 2007; Gallegos et al., 2009; Tang et al., 2009; Tsubaki and Kawahara, 2013). Considering on-shore elevation (topography) data, there are several widely available digital elevation models (DEMs) that vary in resolution, accuracy, extent of coverage, and cost. This study assesses different available DEMs in regards to their utility for accurate tsunami inundation modeling.

In order to assess the different elevation datasets the purpose for which the inundation model will be used must be considered. In this study we focus on one fundamental purpose: defining the extent of the coastal region that may be inundated in a particular tsunami scenario. This supports first-order disaster planning by allowing disaster managers to assess who and what may be inundated, and where safe places are for evacuating people to. Therefore, in our assessment estimates of inundation extent must be sufficiently robust for decision making in this context. It is acknowledged that other calculated quantities such as flow depth and velocity may be subject to greater levels of uncertainty due to uncertainties in elevation data, and can be important for informing other disaster management decisions beyond the focus of this study.

While it can be expected that the best data in terms of accuracy and resolution will support the most accurate tsunami inundation modeling results, it is important to assess whether lower quality data (that may be cheaper and/or available in more areas) can be used to produce sufficiently reliable tsunami inundation maps to inform first-order evacuation planning (Lewis et al., 2013). This study was originally conceived to inform a data acquisition program for tsunami hazard assessment in Indonesia. We model tsunami inundation for two locations in Indonesia using different DEMs for onshore elevation and different resolutions of the numerical mesh of solution. We then assess which elevation datasets and model resolutions give sufficiently accurate results for the purposes articulated above. We then compare different parameterizations of roughness in order to better understand sensitivity to this parameter and its dependence on the choice of elevation data.

Available Digital Elevation Models (DEMs)

In this study we consider a variety of DEMs that are either commercially or freely available and acquired using different remote sensing techniques, both airborne and satellite derived. Some DEMs are available as both Digital Surface Models (DSM), which usually involve processing first returns from active source techniques (LiDAR and IFSAR, defined below), and Digital Terrain Models (DTM) which attempt to remove buildings, vegetation, and other elements from the dataset to provide a bare-earth elevation dataset. The DEMs considered here are based on what was available in the two study locations at the time of this study. The methods used to derive the DEMs are described below and summarized in Table 1. Note that for all datasets actual vertical accuracies will vary from reported accuracies due to land cover and slope (Hodgson et al., 2003).

TABLE 1. A comparison of different topographic data types used in this study.

	Method	Horizontal resolution	Vertical accuracy	Advantages	Disadvantages
LiDAR	Laser pulse	0.5–2 m	0.15 m	Most accurate DEM DTM and DSM	Expensive, can only fly in dry season Large data volumes provide challenges for storage and processing
High resolution stereoscopic camera	Stereocamera	0.3 m	0.5 m	High resolution and accuracy Cheaper than LiDAR DTM and DSM	Still expensive New technique without commerical providers in study regions
Airborne IFSAR	IFSAR	5 m	1–3 m	Reasonable vertical accuracy Lower cost than LiDAR DTM and DSM	Less accurate in steep and densely vegetated areas
SRTM 90 m	IFSAR	90 m	5.6–9.0 m (absolute) 4.7–9.8 m (relative)	Almost worldwide coverage Free	DSM only Low vertical accuracy
Aster	Stereocamera	30 m	6–15 m (absolute)	Covers 99% globe Free Accuracy less affected by steep topography than other methods	DSM only Cloud cover problems Low vertical accuracy

Light Detection and Ranging (LiDAR) data are collected using an airborne laser transmitter-receiver that measures the time taken for a laser pulse to reflect from surfaces below in order to determine elevation (Hodgson et al., 2003). LiDAR has a horizontal resolution of 0.5–2 m and a stated vertical accuracy of 0.15 m. Clusters of reflections can be used to identify bare earth, vegetation canopies, and buildings. However, this data is expensive and requires cloud-free weather for collection—in tropical regions this can make acquisition during the wet season difficult or impossible.

The *Multifunctional camera—High resolution Stereoscopic Camera (HRSC)* system was developed by the German Aerospace Centre (DLR) and Remote Sensing Solutions (RSS) (Anko et al., 2008) and uses three camera angles to derive high resolution elevation and imagery data. This system was used in Padang, West Sumatra to develop a digital elevation model with 0.3 m resolution and 0.5 m stated vertical accuracy (Taubenböck et al., 2009).

Airborne Interferometric Synthetic Aperture Radar (IFSAR or INSAR) data have 5 m horizontal resolution and 1–3 m stated vertical accuracy. This uses an airborne X-band (3 cm wavelength) source and sensors to generate a high resolution DEM. X-band radar does not fully penetrate vegetation (Hodgson et al., 2003) and therefore the raw product uses first returns to generate a DSM. Auxiliary data and imagery collected during the survey are then used to manually identify bare earth locations to produce a DTM by interpolation between bare earth points. This method is therefore more effective in open areas but less accurate in densely vegetated areas and areas of high relief, as interpolation of the terrain surface between bare earth locations can smooth topography.

Advanced Spaceborne Thermal Emission and Reflection Radiometer (ASTER) data have been used to generate a DSM using space-borne stereoscopic infrared cameras. It has a horizontal resolution of 30 m and vertical accuracy of 6–15 m depending on the location, which much higher accuracies in bare-earth areas (Tachikawa et al., 2011). This data is free and has almost complete global coverage.

Shuttle Radar Topography Mission (SRTM) data have been used to generate a DSM with approximately 90 m horizontal resolution over much of the earth's surface (Farr and Kobrick, 2000). These data were collected using C-band IFSAR and aimed for a vertical accuracy better than 16 m (Rodriguez et al., 2005). However, Rodriguez et al. (2005) compared SRTM with known ground points and found that SRTM performed better than this, with an absolute height error of 5.6–9.0 m and a relative height error of 4.7–9.8 m for 90% of the data, varying across the earth's surface. Random vertical noise of the order of 6 m has been documented for SRTM (Lewis et al., 2013). SRTM is freely available at 90 m resolution between 60°N and 56°S latitude, or about 80% of the earth's surface (Cowan and Cooper, 2005). A 30 m DSM derived from X-band data is increasingly being made freely available in some regions but was not available for the regions in this study at the time of analysis.

Figure 1 shows a comparison of the different datasets for the Wuring Peninsula, Maumere, and illustrates how the different collection methods generate different elevation surfaces. Representation of the coast in ASTER and SRTM datasets is poor. The SRTM dataset has land protruding up to 100 m seaward of the coastline digitized from high resolution (1 m) imagery, with elevations in excess of 10 m seaward of the coast. Conversely, there are gaps in the ASTER coverage of the Wuring Peninsula. There are also differences in the representation of the coast between the high resolution datasets, LiDAR, and IFSAR. This can be seen in the intertidal region covered by mangroves in the bottom right of the image, which is included in the IFSAR DEM but has been excluded as part of the ocean by the LiDAR providers. Furthermore, the IFSAR DEM appears to capture reflections from some boats in the nearshore region, which must be removed before integration with bathymetry data. Plotting a cross-section of the data (Figure 2) highlights differences in vertical accuracy and horizontal resolution between the datasets and differences between DTMs and DSMs. DTMs can be seen to be smoother than DSMs, and the IFSAR DTM is smoother than the LiDAR DTM.

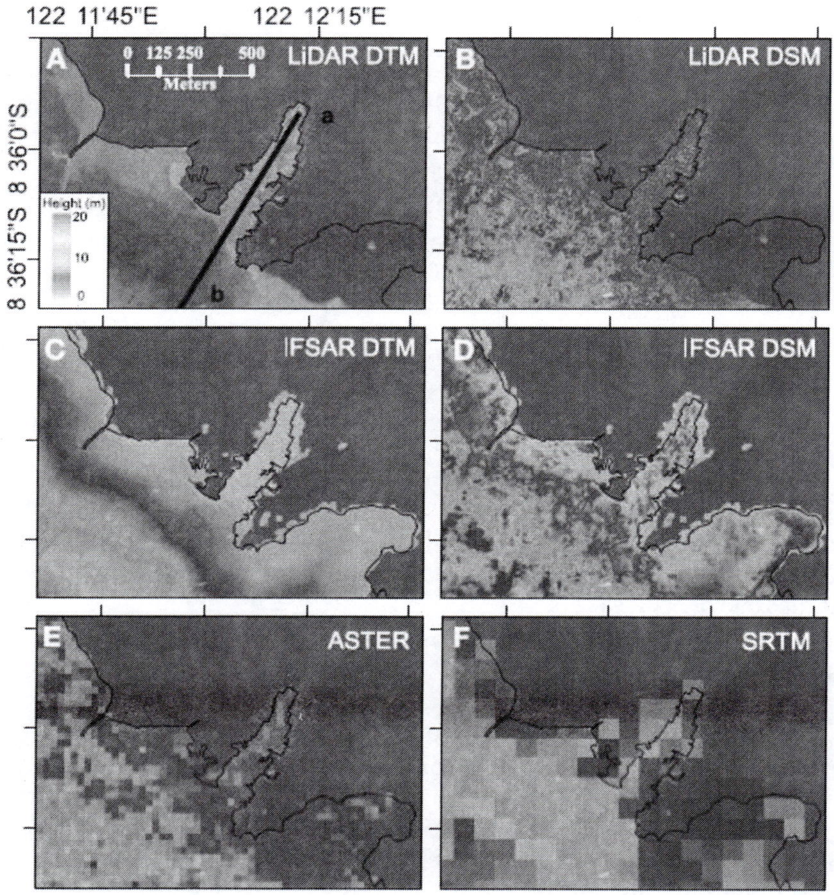

FIGURE 1. Comparison of elevation data for the Wuring Peninsula, Maumere for (A) LiDAR DTM; (B) LiDAR DSM; (C) IFSAR DTM; (D) IFSAR DSM; (E) ASTER30; and (F) SRTM90. Black line is the coastline digitized from high resolution imagery collected during the LiDAR survey. Line a-b in (A) defines the cross-section shown in Figure 2. See Figure 5 for location of Wuring Peninsula. Underlying imagery from Bing Imagery accessed through ArcGIS.

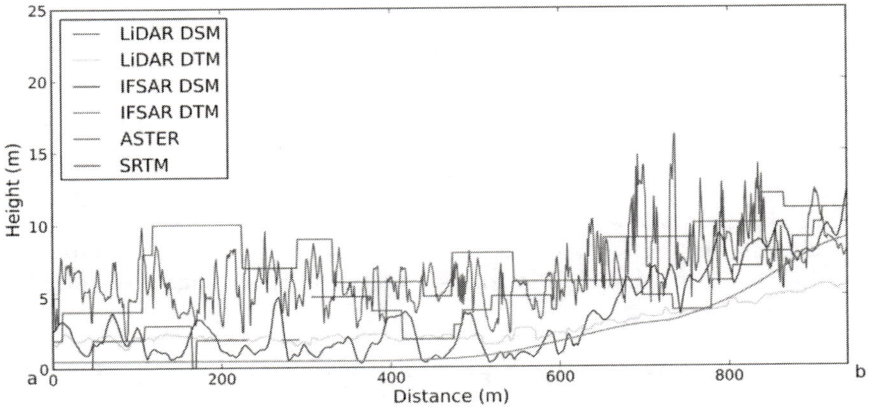

FIGURE 2. Cross-section of elevation profiles for the different DEM datasets along the line a-b shown in Figure 1A. Note that the ASTER cross-section has discontinuities where there is no data near the coast.

Tsunami Inundation Modeling

Tsunami propagation in the ocean can be described using shallow water theory and onshore inundation may be included by appropriate inundation schemes. For tsunami hazard assessment the 2D (depth integrated) non-linear shallow water (NLSW) equations are commonly implemented and solved numerically on a mesh or grid, and there are many software packages that implement NLSW solvers, such as TUNAMI (Goto et al., 1997), ANUGA (Nielsen et al., 2005), and TsunAWI (Rakowsky et al., 2013). While exact implementation varies, in two dimensions the NLSW equations can be expressed (e.g., Chaudhry, 2008) as two equations describing the conservation of momentum:

$$\frac{\partial uh}{\partial t} + \frac{\partial \left(u^2 h + \frac{gh^2}{2}\right)}{\partial x} + \frac{\partial uvh}{\partial y} = -gh\frac{\partial z}{\partial x} - ghS_{fx} \tag{1}$$

$$\frac{\partial vh}{\partial t} + \frac{\partial \left(v^2 h + \frac{gh^2}{2}\right)}{\partial y} + \frac{\partial uvh}{\partial x} = -gh\frac{\partial z}{\partial y} - ghS_{fy} \tag{2}$$

and one describing conservation of mass:

$$\frac{\partial h}{\partial t} + \frac{\partial uh}{\partial x} + \frac{\partial vh}{\partial y} = 0. \tag{3}$$

In the above equations, u and v are the flow velocities along horizontal x and y axes, t is time, h is the flow depth, z is the bed elevation, and g is the acceleration due to gravity. S_{fx} and S_{fy} are frictional forcing terms in the x and y directions respectively. Initial conditions of the deformation of the sea surface due to coseismic crustal deformation are used to force the model. The NLSW equations are also used to simulate fluvial flooding.

Sensitivity of Hydrodynamic Models to Elevation Data

Tsunami inundation is simulated through numerical solution of the NLSW equations (Equations 1–3) over a model of bathymetry and topography with appropriate extensions to model wetting and drying processes. In ANUGA, the conservative finite volume scheme allows discontinuities at the edges of all mesh triangles and therefore can simulate wetting and drying of mesh elements (Jakeman et al., 2010). In the finite element TsunAWI model wetting and drying is performed by an extrapolation scheme extending model quantities from the sea area to land (Lynett et al., 2002; Harig et al., 2008). The numerical solution is sensitive to both elevation data and the resolution of the mesh over which the equations are solved. Furthermore, elevation data must be interpolated to the model mesh introducing interpolation errors.

Horritt et al. (2006) found that model output for shallow fluvial flow was more sensitive to changes in model mesh resolution than changes in topographic data and suggested that mesh resolutions higher than the resolution of the underlying data may improve model performance. Sanders (2007) compared models of flow over a floodplain using LiDAR, airborne IFSAR, SRTM, and the United States Geological Survey's National Elevation Data (NED) and found that using SRTM resulted in an overestimation of the flood footprint by about 25%, which was considered reasonable for some applications. This performance is attributed to SRTM having better relative vertical accuracy in low relief environments such as floodplains. SRTM has also successfully been used to model storm surge inundation in the very low relief Bay of Bengal, after removal of vegetated pixels and smoothing to 900 m (Lewis et al., 2013). Conversely, Gallegos et al. (2009) found that neither SRTM nor the United States Geological Survey's NED were sufficiently accurate for reproducing observed flood footprints for a dam-break scenario in an urban area. Tsubaki and Kawahara (2013) have shown that the sensitivity of NLSW models to mesh resolution is dependent on slope, with low resolution meshes over-predicting the inundated area more significantly in steep areas compared with relatively flat areas.

Compared with fluvial flooding, there has been less work on the sensitivity of hydrodynamic models to elevation data for modeling tsunami inundation. Satake (1995) identified that tsunami models were more sensitive to bathymetry and elevation data than to a choice between use of the linear or NLSW wave equations for reproducing tsunami observations from tide gauges records. Tang et al. (2009) showed that using different elevation datasets gave substantially different inundation results for tsunami inundation models in Hawaii. Tsubaki and Kawahara (2013) found that elevation errors translate into errors in inundation depth of the same magnitude, while errors in flow velocity are larger than the magnitude of the elevation errors. However, if large scale topography is well represented then elevation uncertainties overlaying this may be less significant for estimating inundation extent, as the large scale topography controls the inundated area except in flatter areas.

Titov and Synolakis (1998) identified that a model resolution of 50 m was needed to accurately simulate tsunami propagation over typical coastal features however they do not provide comment regarding the accuracy of elevation data. Kazusa et al. (2004) in developing Cabinet Office recommendations for producing tsunami hazard maps in Japan advised that much higher resolution meshes must be developed to capture small coastal features such as streams, and that accurate modeling required elevation data with an accuracy of 1 m. In the flood context, Schubert and Sanders (2012) identify that appropriate mesh resolution will depend on if and how buildings are including the model.

Roughness

Friction at the boundary between the flow and the sea bottom/land reduces flow momentum. Features of the elevation surface with scales too small to be captured within the elevation model can exert a frictional force on the flow. For example, a surface covered with dense vegetation may retard flow more than a smoother surface such as an open sandy beach. Bed frictional forcing is parameterized through application of the Manning's formula:

$$S_{fx} = \frac{un^2\sqrt{u^2 + v^2}}{h^{4/3}} \tag{4}$$

$$S_{fy} = \frac{vn^2\sqrt{u^2 + v^2}}{h^{4/3}} \tag{5}$$

where n is the Manning's roughness coefficient that can be derived empirically or estimated based on measurements of flow across real or idealized surfaces.

Forcing due to bed roughness relative to other terms in the NLSW equations in the deep ocean is small (being proportional to $h^{-1/3}$) but becomes more significant in shallow water. Dao and Tkalich (2007) find that changing n from 0.025 to 0.015 in the open ocean does not change the quality of model fit with Jason-1 satellite observations of the 2004 Indian Ocean Tsunami, however friction values do change nearshore tsunami amplitudes significantly. They find a better match with observations using n = 0.025. Imamura et al. (2006), Latief and Hadi (2007), Imamura (2009) and Sugawara and Goto (2012) also suggest a value of n = 0.025 in the ocean and coastal waters while Satake (1995) uses n = 0.03 for coastal water. Fujima (2001) suggests friction values in the range 0.02–0.05 for modeling tsunami propagation. Kaiser et al. (2011) use a lower value of 0.011 in the sea, equivalent to a smooth surface, in order to fit observational data. Shimozono et al. (2012) use 0.02 in the ocean. Friction parameters are greater on land due to the resistance effects of vegetation and structures that are not included in the elevation model. Shimozono et al. (2012) use 0.05 for all on-land areas in reproducing the 2011 Tohoku tsunami. Kaiser et al. (2011)provide suggested values for Mannings n for a range of land cover types in coastal Thailand.

Figure 2 illustrates the difficulty in choosing appropriate roughness values, as different datasets capture different levels of detail of surface features. In urban areas, buildings may be represented either as elevation in a DSM or through parameterization of roughness using a DTM. Kaiser et al. (2011) compared these two possibilities and found that inundation depth and extent could reasonably be reproduced by using higher friction in urban areas, however friction values needed to be applied at the building scale to accurately reproduce flow velocities between buildings. Similar results have been found for flood modeling by Schubert and Sanders (2012), who compared four different methods for parameterizing buildings in finite-volume flow simulations including application of roughness coefficients at both the building and urban area scale. They found that calculation of flood extent and depth was not strongly sensitive to the roughness method while accurate simulation of flow velocity required building level parameterization.

STUDY AREAS

Tsunami inundation is modeled for two Indonesian communities: Maumere and Padang (Figure 3).

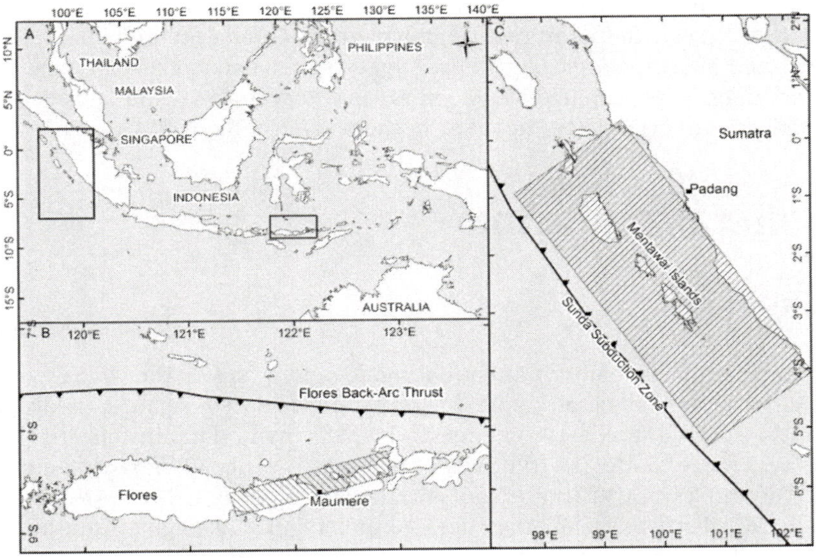

FIGURE 3. Regional setting (A) with inset maps showing the two study areas (B) Maumere, and (C) Padang. Hatched rectangles show the area of the fault rupture for each source model. Note that the source model for Maumere does not align with where the Flores Back-Arc Thrust is typically mapped. Fault traces are those used in the 2010 revision of Indonesia's national seismic hazard map (Irsyam et al., 2010).

Maumere is located on the north coast of Flores Island in the Indonesian province of East Nusa Tenggara. The population is 78,000 (BPS (Indonesian Central Agency for Statistics), 2014a) and is spread along the coastline, with high densities on the Wuring Peninsula and in the center of the city near the harbor. Coastal areas are either open beaches or fringed by mangroves, with some coral reefs extending offshore. This shallow platform drops off rapidly to depths of over 200 m within 0.7–2.0 km of the coastline. To the north of Maumere is the Flores Back-Arc Thrust, the main source of tsunami hazard for Maumere and northern Flores. In 1992 a moment magnitude (Mw) 7.8 earthquake generated a tsunami that inundated Maumere and northern Flores, killing about 2000 people (Tsuji et al., 1995).

Padang is a densely populated city of 877,000 people (BPS (Indonesian Central Agency for Statistics), 2014b) situated on a low lying coastal plain located at the foot of the Barisan Mountains in West Sumatra. Several beach ridges run parallel to the coast and it is bounded to the north, east, and south by rivers. Offshore the bathymetry slopes to depths of 200 m at about 20 km offshore. Padang was inundated by tsunami generated by earthquakes on the Mentawai Segment of the Sunda Subduction Zone in 1797 and 1833. Geological evidence suggests such events have a recurrence interval of approximately 200 years and therefore a similar event is anticipated to occur in the next few decades (Sieh et al., 2008).

MODEL DESCRIPTION

Software

A number of tsunami inundation models that solve the NLSW wave equations are available and have been validated against analytical solutions, wave tank experiments, and real tsunami data. In this study we use two models: ANUGA (finite volume; Nielsen et al., 2005; Jakeman et al., 2010); and TsunAWI (finite element; Rakowsky et al., 2013). ANUGA and TsunAWI discretise elevation data onto irregular triangular meshes. In TsunAWI elevation data is fitted to the mesh using a nearest-neighbor technique while ANUGA fits elevation data to the mesh using a global penalized least-squares fitting. ANUGA solves the NLSW equations in Cartesian coordinates while TsunAWI uses geographic coordinates and includes Coriolis forcing. Therefore, although both models are validated for their ability to realistically model tsunami, differences in the results are expected based on the use of different numerical schemes, forcing terms, and the way elevation data is represented in the model.

Tsunami Source Models

Several earthquake source models were published following the 1992 Flores earthquake and tsunami event (Imamura and Kikuchi, 1994; Beckers and Lay, 1995; Hidayat et al., 1995), however these were all limited to one or two subfaults. For this study a new source model is generated using 10 × 10 km subfaults by jointly fitting body wave, surface wave, and coseismic displacement data using the method of Thio et al. (2004), allowing much higher spatial resolution of the slip distribution. The two observed vertical coseismic displacements closest to the hypocenter have been down-weighted by a factor of 0.5 to reduce their dominance of the solution. The source model, waveform fits, and surface deformation are shown in

Figure 4. The seismic waveform data and coseismic uplift/subsidence constrained the fault position and geometry to be similar to but significantly different from those used in previous studies (Table 2). Furthermore, our model has a region of high slip (20 m) near the hypocenter, which was required to fit the magnitude and pattern of cosesmic uplift and subsidence along the northern coast of Flores observed after the earthquake. Note that our source model, along with those obtained in previous studies, agrees with the observed pattern of aftershocks but does not coincide with the location of the Flores Back-Arc Thrust as typically indicated on maps of the regional tectonics (e.g., Irsyam et al., 2010). Rather, it is partly onshore suggesting that the Flores Back-Arc region is more complicated tectonically than is typically portrayed.

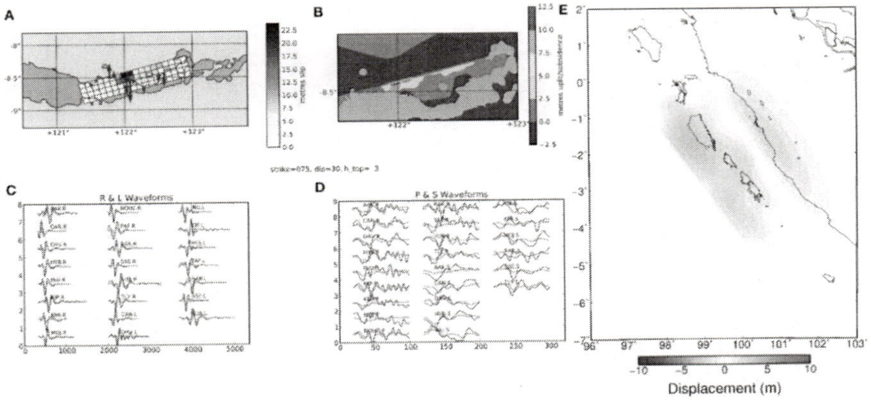

FIGURE 4. Source model for the 1992 Flores earthquake and tsunami. (A) Best fit model of slip distribution with observed (blue) and modeled (green) subsidence shown by vertical arrows; (B)resulting surface deformation; (C) Surface wave observations (blue) and model fit (red); **(D)** body wave observations (blue) and model fit (red). Surface deformation for the Padang consensus source mode **(E)**.

TABLE 2. Comparison of fault parameters for the 1992 Flores earthquake.

Study	Strike	Dip	Rake	Moment (10^{20} Nm)
Imamura and Kikuchi, 1994	61	32	64	6.4
Beckers and Lay, 1995	70	28	80–140	7.5–8.0
This study	75	40	70–135	4.8

The tsunami threat to Padang from the Sumatran Megathrust directly offshore has been well studied (Borrero et al., 2006; Natawidjaja et al., 2006; McCloskey et al., 2008; Sieh et al., 2008; Schlurmann et al., 2010). We use the consensus source model of Schlurmann et al. (2010) (Figure 4E) that estimates the accumulated slip on the megathrust since previous ruptures in 1797 and 1833 (Natawidjaja et al., 2006; Chlieh et al., 2008). Note that this model does not take into account reduction in accumulated stress along the southern-most section of this source due to the recent Mw 7.8 25 October 2010 Mentawai Islands earthquake and tsunami. It also does not consider the potential for slip to occur right at the trench (which is not recorded in the geodetic studies used to inform this model) and therefore potentially does not represent the worst case for tsunami generation (Hill et al., 2012; Griffin et al., 2015).

Model Setup

Figure 5 shows the mesh used for the ANUGA models for Maumere and Padang and the TsunAWI model for Maumere. Both models use unstructured triangular meshes, with internal polygons used to define the maximum allowable size of an individual mesh element within that region. The smallest polygon shows the area of interest (AOI) that is used in the comparisons of inundation extent for the different data types. For the comparison of the different elevation datasets the same computational mesh was used in each simulation. Bathymetry data is not varied between simulations. We use a 90 m commercial bathymetry dataset provided by TCarta Marine for depths shallower than 1000 m and the General Bathymetric Chart of the Oceans (GEBCO) in deeper areas. The mesh resolution within the AOI was then varied to compare the effect of changing mesh resolution.

Comparison of Inundated Area

In the absence of high quality observational data in the study areas, we define the benchmark model based on the highest resolution DTM available for each area with our best estimate of the appropriate friction coefficient from the available literature. That is, for Maumere we use the LiDAR DTM and for Padang the HRSC DTM. Following Kaiser et al. (2011) we apply a Manning's n of 0.06 for onshore regions of the Maumere model (medium density urban area) and 0.08 for onshore regions of the Padang model (high density urban area). Manning's n is 0.025 in offshore regions for the benchmark models.

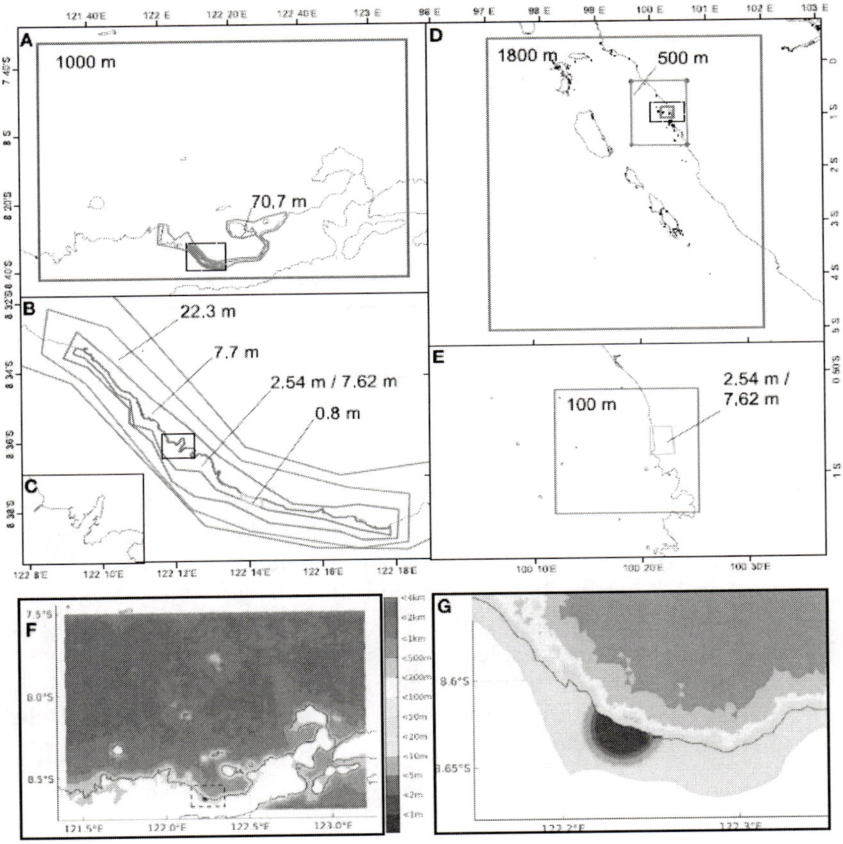

FIGURE 5. Polygons showing regions of different mesh resolution for the ANUGA models for: (A) north-eastern Flores, with the dashed black line showing the high resolution area shown in (B); (B) the coastline near Maumere, including the Wuring Peninsula, which is indicated by the dashed black rectangle; (C) the Wuring Peninsula, which is shown in Figures 1, 7, 12; (D) the Padang coastline, with the dashed black line showing the high resolution area shown in (E); (E) the coastline near Padang. The mesh resolution (m) in each polygon is indicated by the numbers in black. The blue rectangles in (A, D)outline the model boundaries for Maumere and Padang, respectively. Internal mesh resolution for TsunAWI model for Maumere is shown in (F) with dashed black line showing the high resolution area shown in (G).

The area of modeled inundation, $A(I_m)$, is compared against modeled inundation for a benchmark model, $A(I_b)$. This is quantified following Jakeman et al. (2010) using the measures:

$$P_{in} = \frac{A(I_b) - A(I_b \cap I_m)}{A(I_b)} \quad (6)$$

$$P_{out} = \frac{A(I_m) - A(I_m \cap I_b)}{A(I_b)} \quad (7)$$

Where P_{in} refers to the proportion of the area inundated in the benchmark model that is not captured by the model being tested, whereas P_{out} is the area of the inundated area of the model being tested that falls outside of the benchmark model area, relative to the benchmark model area. Note that as both measures have the same denominator, they are not symmetrical. P_{in} scales between 0 and 1 with zero indicating a perfect fit. P_{out} is not constrained in the same way; values approaching 0 indicate that the modeled area outside the benchmark region is small relative to the area of the benchmark region, but do not necessarily indicate a good model fit unless P_{in} is also close to 0 (e.g., considering the case of disjoint inundation area).

For the comparison datasets, we use a constant low Manning's n of 0.025 in both the ocean and on land to ensure sensitivities are due to elevation data and not varying friction values. We then compare increased on-land friction values. This allows us to compare the sensitivity to friction parameters with sensitivity to onshore elevation data.

The field survey of the 1992 Flores tsunami by Tsuji et al. (1995) provides tsunami run-up data along the northern coast of Flores, with GPS locations of observed run-ups and sketch maps provided. The sketch maps were georeferenced using marked GPS observation points and key buildings, roads, and other geographic features that are visible in BING Maps aerial imagery for the region using the ArcGIS software. GPS locations for some features from Tsuji et al. (1995) do not match the locations of the features as observed in imagery and we estimate horizontal errors of 100 m in the resulting georeferenced sketches. These sketches allow us to use the P_{in} and P_{out} measures to assess the performance of our models against observation data for the Wuring Peninsula, one of the most heavily impacted parts of Maumere during the 1992 event.

RESULTS

Comparison of Different Data Types

Figure 6 shows the values of P_{in} and P_{out} for inundation model results derived using each dataset for Maumere and Padang, calculated against our benchmark models. Figures 7, 8 plot inundation extents in the two locations. The results show that the IFSAR and LiDAR DTMs predict a greater inundation extent while all the other datasets, which are all DSMs, predict a smaller inundation extent. Differences between the DTM and DSM results for IFSAR and LiDAR are greater than differences between these two data types when comparing just DTMs or **just DSMs.**

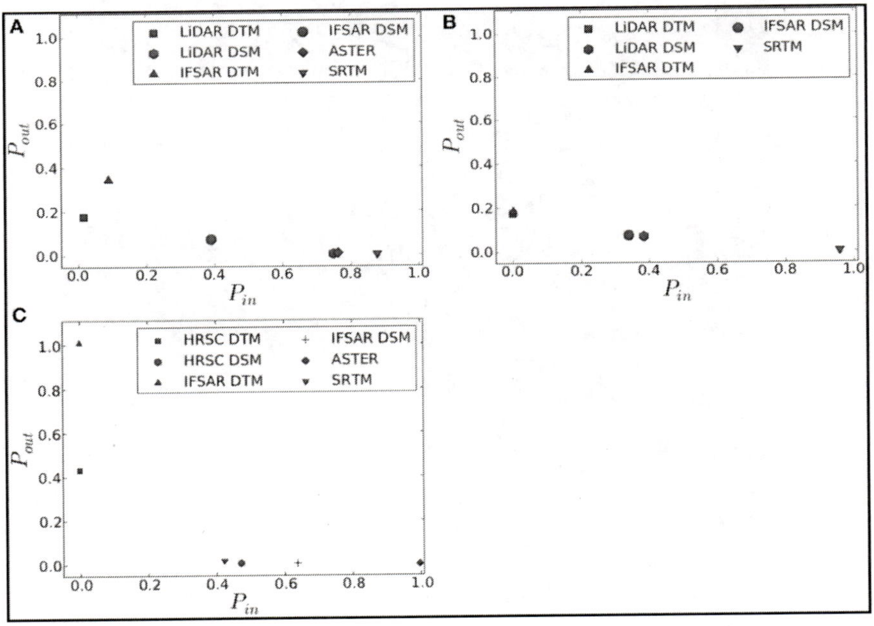

FIGURE 6. P_{IN} AND P_{OUT} VALUES FOR (A) MAUMERE, ANUGA MODEL; (B) MAUMERE, TSUNAWI MODEL; AND (C) PADANG, ANUGA MODEL, CALCULATED AGAINST THE BENCHMARK MODELS. Comparison models use Manning's $n = 0.025$ while benchmark models have onshore $n = 0.06$(Maumere) and 0.08 (Padang).

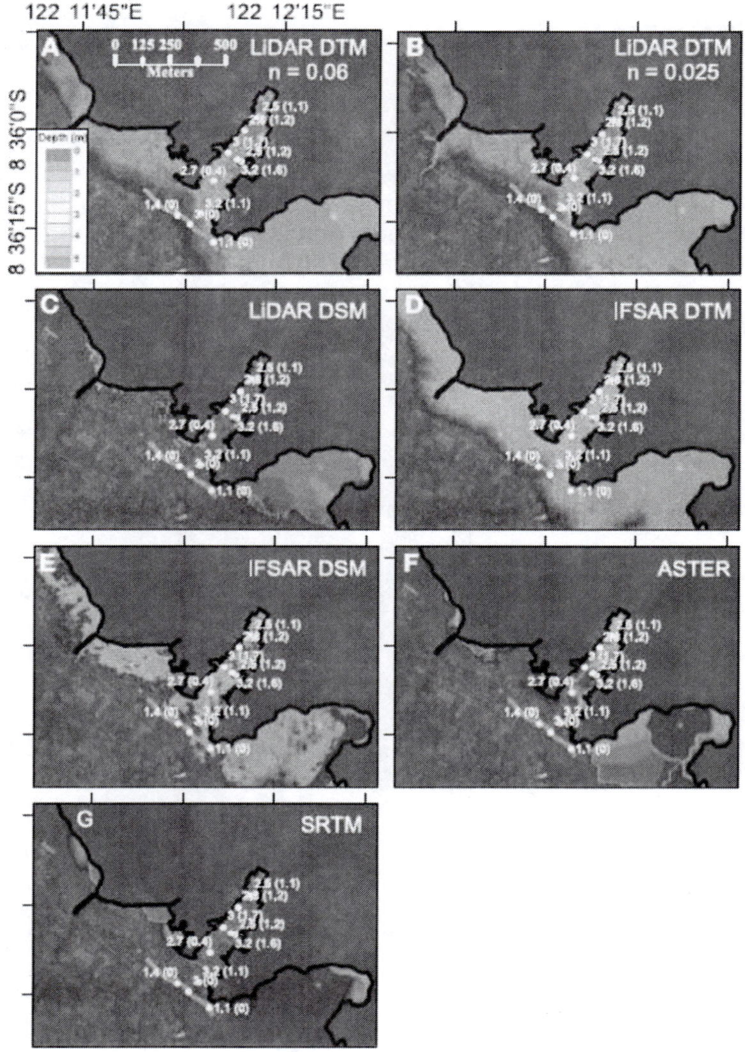

FIGURE 7. COMPARISON BETWEEN INUNDATION LIMIT FROM GEOREFERENCED OBSERVATIONAL SKETCH FROM **Tsuji et al. (1995)** (ORANGE LINE) AND ANUGA INUNDATION MODELS FOR WURING PENINSULA, FLORES, FOR (A) BENCHMARK MODEL USING LIDAR DTM (N = 0.06); AND COMPARISON MODELS (N = 0.025) FOR (B) LIDAR DTM; (C) LIDAR DSM; (D) IFSAR DTM; (E) IFSAR DSM; (F) ASTER; AND (G) SRTM. Mesh resolution is 7.62 m. White figures show tsunami heights in meters from Tsuji et al. (1995) observations with tsunami flow depths in parentheses.

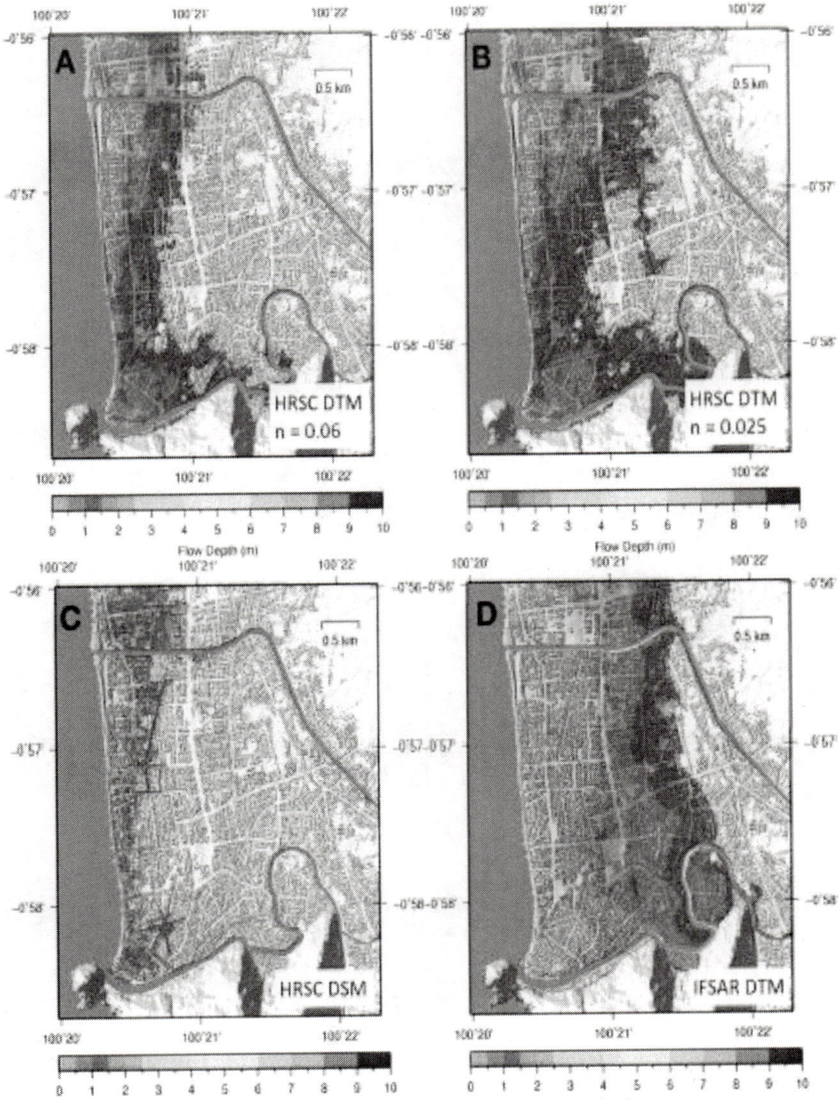
A
0.5 km
HRSC DTM
n = 0.06
B
0.5 km
HRSC DTM
n = 0.025
Flow Depth (m)
C
0.5 km
HRSC DSM
D
0.5 km
IFSAR DTM
100°20'
100°21'
100°22'
-0°56'
-0°57'
-0°58'
0 1 2 3 4 5 6 7 8 9 10

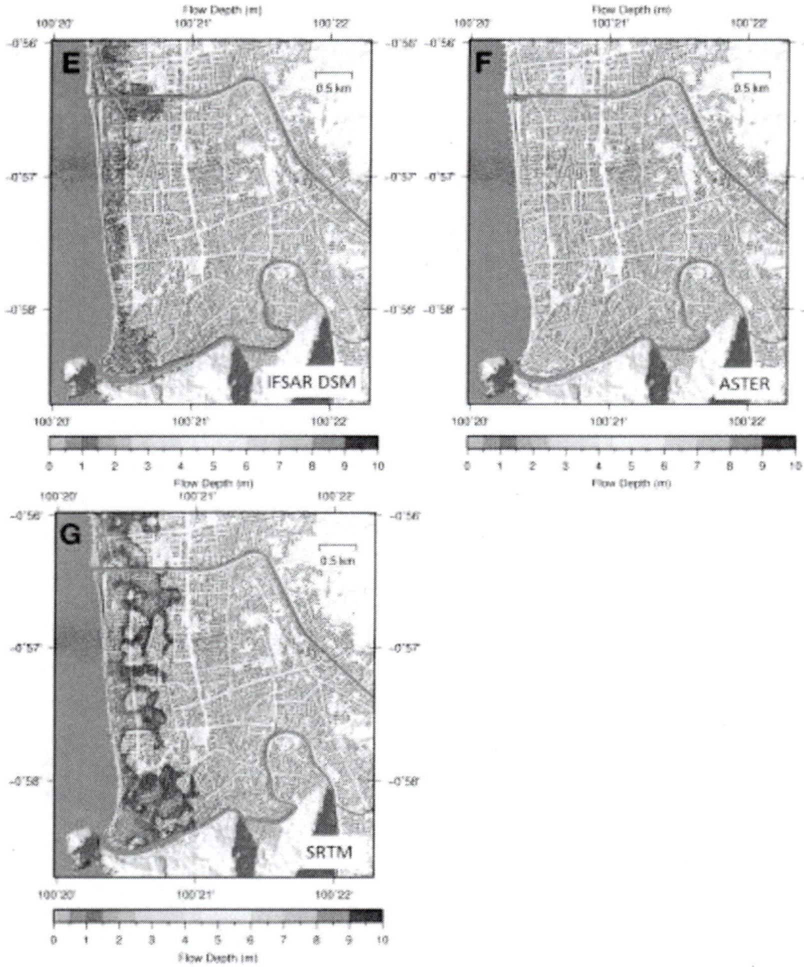

FIGURE 8. INUNDATION RESULTS FOR THE 7.62 M MESH RESOLUTION ANUGA MODEL FOR PADANG FOR (A) HRSC DTM BENCHMARK MODEL (N = 0.08); AND COMPARISON MODELS (N = 0.025) (B) HRSC DTM; (C) HRSC DSM; (D) IFSAR DTM; (E) IFSAR DSM; (F) ASTER; AND (G) SRTM.

Figure 7 compares modeled and observed inundation for the 1992 Flores tsunami for the Wuring Peninsula, Maumere, with P_{in} and P_{out} calculations shown in Figure 9. The LiDAR and IFSAR DTMs slightly overestimate the observed inundation (P_{out} values of 0.22 and 0.28 respectively) noting that we probably use unreasonably low Manning's roughness values. The IFSAR DSM also matches the observations reasonable well, underestimating and overestimating the inundation extent by a similar amount in different areas. Inundation models using the LiDAR DSM, ASTER, and SRTM DSMs produce much less inundation than that observed (P_{in} values of 0.56, 0.39, and 0.77 respectively), with almost no inundation occurring in the SRTM case.

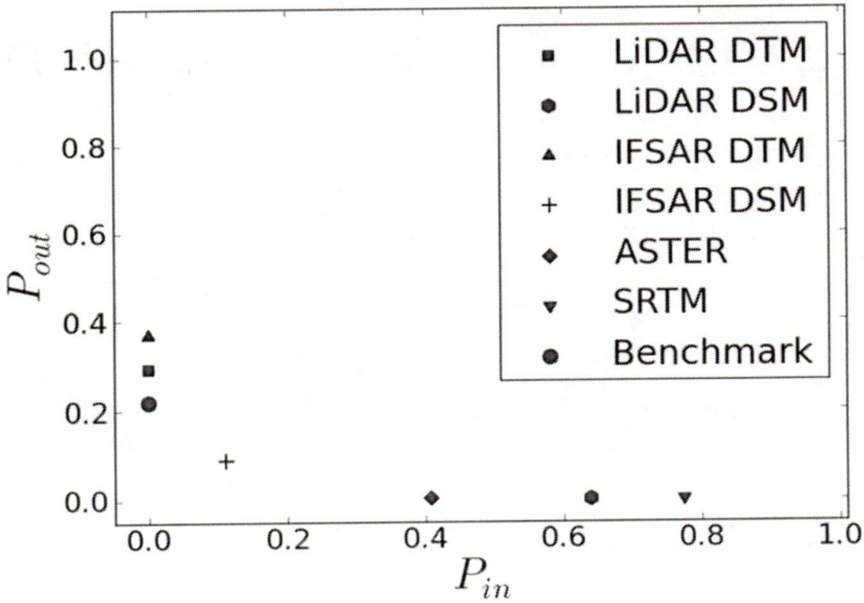

FIGURE 9. P_{IN} AND P_{OUT} VALUES FOR MODELED RESULTS COMPARED WITH OBSERVATIONS OF TSUNAMI INUNDATION FROM **Tsuji et al. (1995)** FOR THE WURING PENINSULA (AREA OF SKETCH SHOWN IN FIGURE **6**) FOR ANUGA SIMULATIONS.

Comparison of Different Mesh Resolutions

A comparison of the same model (ANUGA) and elevation data (IFSAR DTM) using different mesh resolutions for Maumere is shown in Figure 10. The inundated area is very similar for all resolutions with only the 68.63 m model showing clear deviations as a result of the coarser resolution. This suggests that in this example, which is for a fairly uniform coastal geomorphology, the IFSAR data is sampling the topography at a higher horizontal resolution than is needed to simulate inundation.

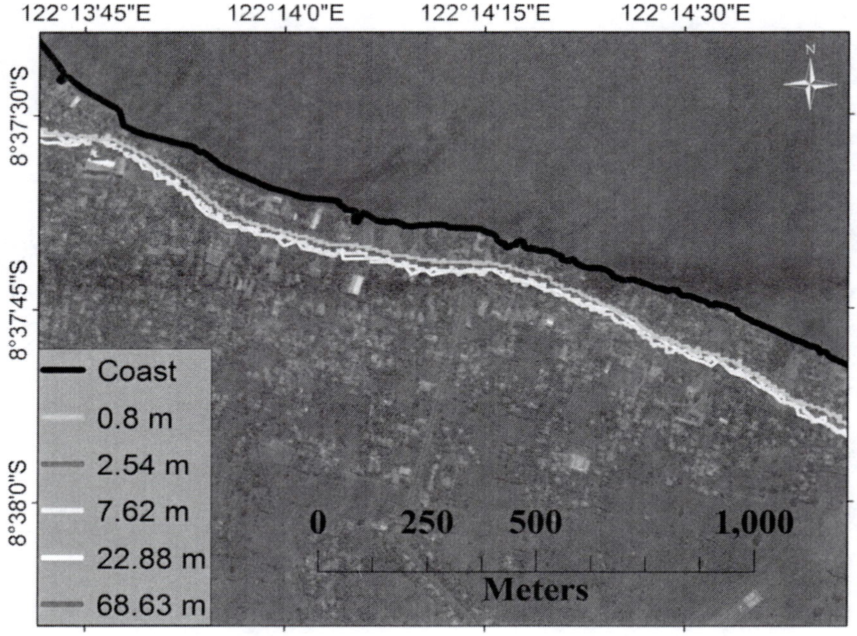

FIGURE 10. COMPARISON OF MODELED TSUNAMI INUNDATION DEPTH FOR ANUGA MODELS USING THE IFSAR DTM FOR DIFFERENT MESH RESOLUTIONS FOR PART OF MAUMERE. See Figure 5B for location.

The effect of increasing model resolution is different for DTM and DSM. Figure 11 shows P_{in} and P_{out}values for LiDAR DTM and DSM, with the benchmark model in this case being the highest resolution model where convergence errors are expected to be at a minimum. All models use constant Manning's $n = 0.025$. For the DTM, as model resolution is decreased the model predicts a greater inundation area. For DSM as model resolution is increased the P_{in} values increase. This is likely because a finer resolution is need to resolve flow between buildings, trees, and other surface features. As a coarser resolution is used, interpolation of elevation to the model mesh may smooth these features toward average elevation values that are above the level of the terrain surface resulting in reduced inundation in some areas. The TsunAWI model results show less sensitivity to mesh resolution than the ANUGA results.

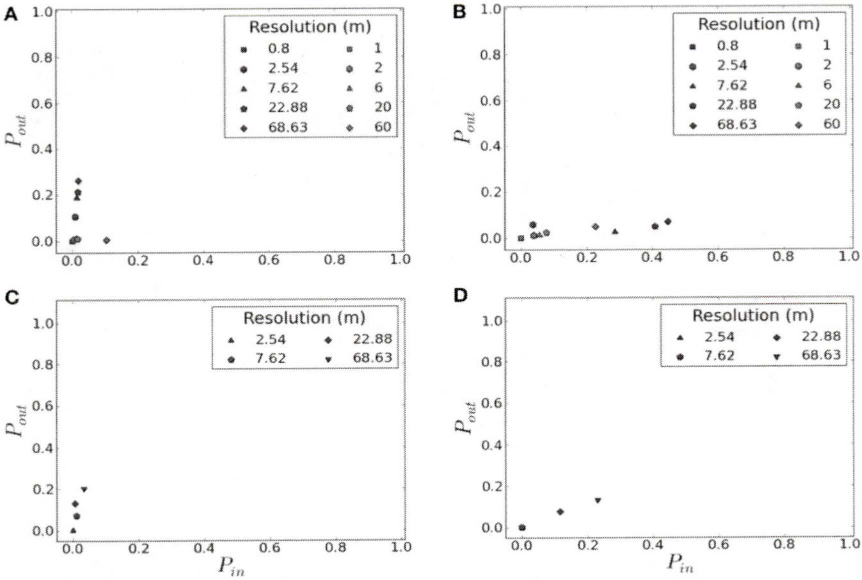

FIGURE 11. P_{IN} AND P_{OUT} VALUES FOR DIFFERENT MESH RESOLUTIONS FOR (A) LIDAR DTM AND (B) LIDAR DSM FOR MAUMERE; (C) HRSC DTM AND (D) HRSC DSM FOR PADANG. The benchmark model is the model with the highest mesh resolution where convergence errors are assumed to be most reduced. ANUGA results are in blue, TsunAWI results are in red.

Sensitivity to Onshore Friction Values

Figures 12, 13 show different onshore friction values for the Wuring Peninsula, Maumere, and Padang, respectively. For Maumere, changing the onshore friction value does not significantly alter the inundation extent, whereas for Padang the effect is dramatic. In the case of Maumere, the extent of inundation is controlled more strongly by the topography. Conversely, for Padang the elevation is very flat and does not provide a topographic barrier to tsunami propagation, therefore friction is more dominant in controlling inundation extent. For Padang the inundation extent using HRSC DTM and Manning's $n = 0.01$ (Figure 13A) is similar to that obtained using IFSAR DTM and Manning's $n = 0.025$ (Figure 8D).

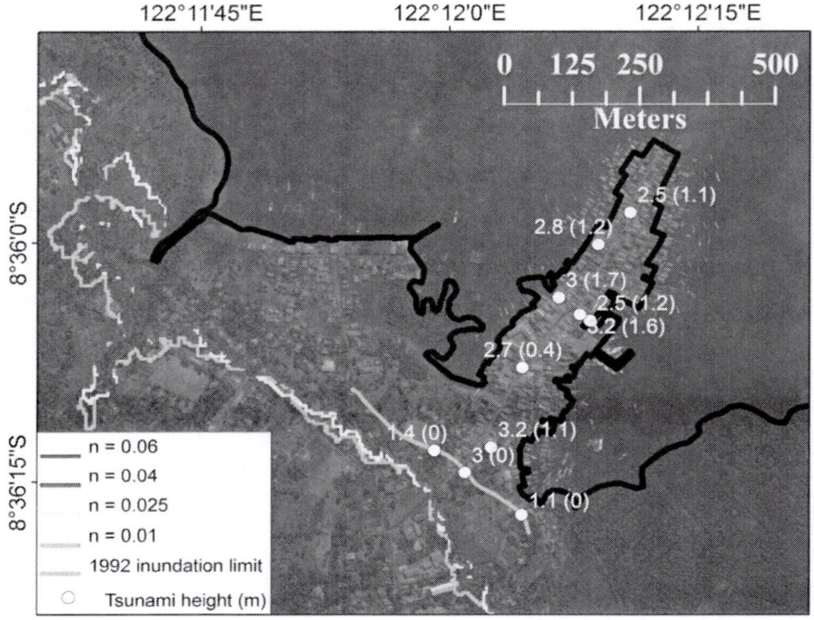

FIGURE 12. COMPARISON BETWEEN INUNDATION LIMIT (ORANGE LINE) FROM GEOREFERENCED OBSERVATIONAL SKETCH FROM **Tsuji et al. (1995)** AND ANUGA INUNDATION MODELS FOR WURING PENINSULA, FLORES, FOR LIDAR DTM USING ONSHORE MANNING**'S** *N*-VALUES OF 0.01, 0.025, 0.04, AND 0.06. Mesh resolution is 7.62 m. White numbers give tsunami heights from the field survey with tsunami flow depths in parentheses.

DISCUSSION

The results presented in this paper clearly demonstrate that the present generation of freely available global DEMs (i.e., ASTER and SRTM90) are not sufficiently accurate to simulate tsunami inundation with confidence. These datasets are limited by their vertical accuracy, with coastal elevation errors that may exceed the amplitude of the tsunami wave being simulated. Unfortunately tsunami hazard assessments continue to use these datasets based on the argument that they are the "best available" when more accurate datasets are not available for part or all of the study region (e.g., DLR and GTZ, 2010; Latief, 2012; Norwegian Geotechnical Institute (NGI) and Geoscience Australia (GA), 2015). We therefore advocate for the development of a global, freely available, high vertical accuracy digital elevation model that would have many benefits to society, one of which would be informing tsunami hazard assessment and evacuation planning.

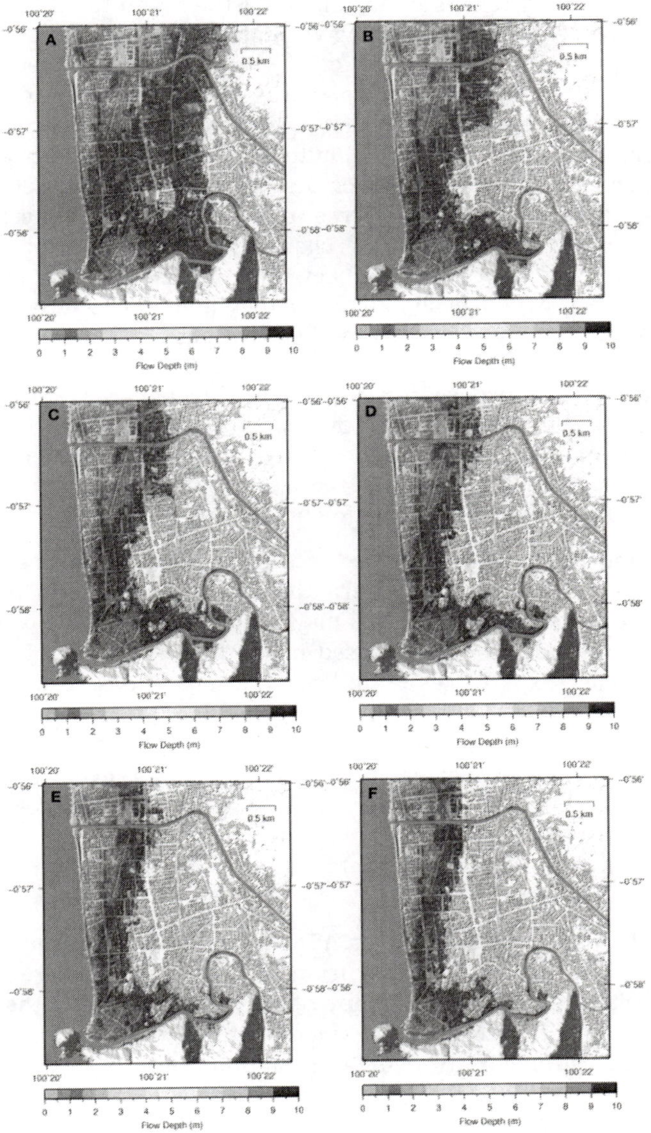

FIGURE 13. COMPARISON BETWEEN ANUGA INUNDATION MODELS FOR PADANG FOR LIDAR DTM USING ONSHORE MANNING'S *N*-VALUES OF (A) 0.01; (B) 0.025; (C) 0.04; (D) 0.06; (E) 0.08, AND (F) 0.1. Mesh resolution is 7.62.

Simulations of the 2011 Tohoku Tsunami (Løvholt et al., 2012) and 2007 Tropical Cyclone Sidr storm surge (Lewis et al., 2013) events have demonstrated good results by making adjustments to the SRTM elevation model by removing cells containing vegetation and interpolating elevation values from adjacent cells. Lewis et al. (2013) also smoothed the SRTM dataset to a resolution of 900 m, which was sufficient for reproducing storm surge inundation extent in the very low relief Bay of Bengal. However, generalization of these results is limited by their dependence on manual identification of vegetation cover using additional datasets and the presence of nearby vegetation-free cells in order to interpolate the elevation surface. Calibration of channel depth (Lewis et al., 2013), Mannings *n* (Løvholt et al., 2012) Lewis et al., 2013), and the tsunami source model (Løvholt et al., 2012) further limit the generality of these results. We are not aware of any method that has been applied on a global scale that improves the vertical accuracy of the SRTM elevation model to be sufficient for the purposes articulated in this paper. Nevertheless, the promising results shown by (Løvholt et al., 2012) and Lewis et al. (2013) suggest further research should examine whether additional datasets such as imagery can be used to significantly improve the vertical accuracy of the global SRTM elevation model.

In cases where ASTER or SRTM are presently the "best available" elevation datasets we recommend that alternative strategies are used to underpin tsunami evacuation planning. This may include conservative estimation of the potential inundation zone based on historical records of tsunami or recent tsunami experiences from neighboring regions. Tsunami hazard maps derived using ASTER or SRTM may be dangerously misleading, and this may be amplified as the use of computational models may imply a greater confidence in the results than is warranted, even if appropriate caveats are given. Therefore, we argue that it is better not to have a modeled tsunami inundation map and instead develop alternative strategies than to use a map that may severely underestimate tsunami inundation extents.

The key limitation in the use of ASTER and SRTM datasets is vertical accuracy. This is in part derived from the fact that they are DSMs, and therefore may represent the canopy of coastal vegetation rather than the surface, along with uncertainties inherent in the methods used to derive elevation measurements. Horizontal resolution is less important in the case studies presented here, if DTMs are used. Using a mesh resolution of 25 m (i.e., similar to the horizontal resolution of ASTER data) is sufficient for modeling inundation extent; higher resolutions are only required if there is a requirement to resolve flow through and around individual features that may be captured in a DSM such as seawalls, inlets or buildings. For example, for hydrodynamic modeling of flood flow between buildings Gallegos et al.

(2009) suggest a mesh resolution of approximately 5 m. These higher resolutions may be required for engineering purposes or detailed risk assessment, however in order to define evacuation areas a 25 m model resolution using an accurate DTM is sufficient.

Modeled inundation extents obtained using airborne IFSAR are comparable with those using higher accuracy LiDAR and HRSC datasets. Differences in modeled inundation between DTM and DSM for IFSAR and LiDAR/HRSC are greater than differences between the two data types. In the DSM models, features such as buildings and vegetation become solid parts of the elevation data and water must flow over or around them. Clearly this does not represent reality; although some structures may resist flow in a real tsunami, water can flow through buildings and vegetation, and some structures can be expected to be destroyed. A DTM, on the other hand, removes all of these features, and therefore does not capture any of the flow resistance that would be offered by these structures in a real tsunami. Flow resistance due to these features can be modeled by the inclusion of spatially varying friction parameters (Manning's roughness) and this has been done, for example, by Kaiser et al. (2011) for Patong, Thailand. This is much more important in flat areas, such as Padang, where friction becomes significant relative to the effect of topography. To avoid underestimating inundation extent as may happen with DSMs, we suggest that the best approach for assessing inundation extent is to use DTMs using appropriate roughness parameters to capture the effect of surface features. Further research is needed to determine the spatial scale at which friction needs to be parameterized through detailed comparison with observational run-ups. However, in any particular study it must be considered whether high resolution information on flow between buildings or other structures is required in which case more explicit inclusion of a building model derived from a DSM may be important (e.g., Schubert and Sanders, 2012).

Inundation distances are generally a little further for IFSAR DTM compared with LiDAR DTM. This can be partly explained by differences in actual elevation values in the IFSAR DTM, but an additional factor is that the LiDAR DTM is rougher than the IFSAR DTM (Figure 2). Along with the smoothing provided by the coarser horizontal resolution, the method used to extract the DTM from raw observations is different for each of the datasets. The LiDAR DTM is derived from the lowest return signals recorded whereas the IFSAR DTM is based on manual identification of bare earth points and then interpolation between these points, which can smooth over topographic features in between if they are obscured by the canopy. Furthermore, the model mesh resolution will also determine which topographic features are captured explicitly and which become "sub-grid" features captured by the roughness coefficient. Therefore, we conclude that it will not be possible to develop definitive recommendations regarding roughness coefficients for particular land cover types that are independent of the particular DEM and mesh resolution used in a tsunami inundation model.

The study presented here is limited in scope to considering onshore elevation data with the same "best available" bathymetry model being used in each case. This allows us to test sensitivity to changing the onshore elevation data type but leaves open the criticism that the "best available" bathymetry data used in this study may not be sufficiently accurate for modeling tsunami inundation. Comparison of our model results with observations of the 1992 Flores tsunami (Figure 7) shows that the bathymetry data used is sufficiently accurate to achieve robust results in this study but nevertheless we recommend a similar study be undertaken to assess the sensitivity of inundation results to different bathymetry datasets, particularly in the nearshore environment.

CONCLUSIONS

Different digital elevation data types have been tested for modeling the extent of tsunami inundation at two different locations in Indonesia. Based on the results of this study the following statements can be made:

- Tsunami inundation models developed using DEMs that are currently freely available at a global scale (i.e., ASTER and SRTM) have the potential to dangerously underestimate the inundation extent. These datasets should not be used to assess tsunami inundation zones using hydrodynamic models.
- Tsunami inundation models developed using airborne IFSAR DTMs may give inundation extents that are similar to those obtained using higher accuracy data, and observations where available, for model mesh resolutions between 0.8 and 68.63 m. These models overestimate inundation extent and therefore are suitable for use in the absence of higher accuracy data.
- Comparing results using airborne IFSAR with LiDAR and HRSC datasets, differences in inundation extent between DSMs and DTMs for the same dataset are greater then differences in inundation extent between the datasets. Tsunami inundation models developed using DSMs underestimate inundation extent while those developed using DTMs overestimate inundation extent where we have observations. It is recommended to use DTMs for modeling inundation extent.

- Further research is needed to determine at what scale spatially variable roughness should be applied, and the dependence of roughness values on the roughness of the underlying DEM.
- The importance of roughness is dependent on the topography of the coastal region. Changing Manning's roughness coefficients from 0.01 to 0.06 has a negligible effect on inundation extent for Maumere while halving the inundation extent in Padang.
- A mesh resolution of 25 m is suitable for estimating inundation extent using a DTM. Higher resolutions are needed to accurately model flow between and around individual structures if a DSM is used, or to better capture flow up rivers and inlets.

CONFLICT OF INTEREST STATEMENT

The authors declare that the research was conducted in the absence of any commercial or financial relationships that could be construed as a potential conflict of interest.

ACKNOWLEDGMENTS

The authors would like to thank DLR for sharing the HRSC digital elevation model for Padang. Funding for other elevation data purchases was provided by the Australian Department of Foreign Affairs and Trade and the Indonesian National Disaster Management Agency through the Australia-Indonesia Facility for Disaster Reduction (AIFDR). Funding for participation in this project by the Alfred Wegener Institute was provided by PROTECTS (Project for Training, Education and Consulting for Tsunami Early Warning Systems) by the German Federal Ministry for Education and Research (BMBF), grant 03TSU07. The authors thank Kristy Van Putten, Ole Nielsen, Trevor Dhu (AIFDR) and Leharne Fountain (Geoscience Australia) for assistance with presenting the results of this work. The manuscript was substantially improved by reviews from Rikki Weber and Gareth Davies (Geoscience Australia), Matthew Lewis (Bangor University) and Dai Yamazaki (JAMSTEC). Imagery is courtesy of Bing Imagery provided through ArcGIS software. Some data processing and figure preparation was done using the Generic Mapping Tools (Wessel and Smith, 1998). This paper is published with the permission of the CEO, Geoscience Australia.

REFERENCES

1. Anko, B., Heiko, H., Karsten, S., Michael, S., and Jurgen, W. (2008). *"MFC—a modular line camera for 3D world modulling,"* in *RobVis LNCS*, Vol. 4931, eds G. Sommer and R. Klette (Berlin; Heidelberg: Springer), 319–326.
2. Beckers, J., and Lay, T. (1995). Very broadband seismic analysis of the 1992 Flores, Indonesia, earthquake (Mw = 7.9). *J. Geophys. Res.* 100, 18179–18193.
3. Borrero, J. C., Sieh, K., Chlieh, M., and Synolakis, C. E. (2006). Tsunami inundation modeling for western Sumatra. *Proc. Natl. Acad. Sci. U.S.A.* 103, 19673–19677.
4. BPS (Indonesian Central Agency for Statistics). (2014a). *Sikka Regency in Figures.* Jakarta: BPS Catalogue.
5. BPS (Indonesian Central Agency for Statistics). (2014b). *Sumatra Barat in Figures.* Jakarta: BPS Catalogue.
6. Casas, A., Benito, G., Thorndycraft, V. R., and Rico, M. (2006). The topographic data source of digital terrain models as a key element in the accuracy of hydraulic flood modelling. *Earth Surf. Proc. Landf.* 31, 444–456.
7. Chaudhry, M. H. (2008). *Open Channel Flow.* New York, NY: Springer.
8. Chlieh, M., Avouac, J. P., Sieh, K., Natawidjaja, D. H., and Galetzka, J. (2008). Heterogeneous coupling of the Sumatran megathrust constrained by geodetic and paleogeodetic measurements. *J. Geophys. Res. Solid Earth* 113:B05305.
9. Cowan, D., and Cooper, G. (2005). The Shuttle Radar Topography Mission—a new source of near-global digital elevation data. *Explor. Geophys.* 36, 334–340.
10. Dao, M., and Tkalich, P. (2007). Tsunami propagation modeling: a sensitivity study. *Nat. Hazards Earth Sys. Sci.* 7, 741–754.
11. DLR (German Aerospace Centre) GTZ (German Technical Cooperation Agency). (2010). *Tsunami Hazard Maps for Kabupaten Cilacap.* Technical Documentation. Available online at: http://www.gitews.org/tsunami-kit/en/E1/further_resources/hazard_maps /central_java/Technical%20Dokumentation%20Tsunami% 20Hazard %20Map%20Cilacap.pdf
12. Farr, T. G., and Kobrick, M. (2000). Shuttle Radar Topography Mission produces a wealth of data. *Eos Trans. Am. Geophys. Union* 81, 583–585.
13. Fujima, K. (2001). *"Long wave propagation on large roughness,"* in *Proceedings International Tsunami Symposium* (Seattle, WA), 7–22.
14. Gallegos, H. A., Schubert, J. E., and Sanders, B. F. (2009). Two-dimensional, high-resolution modeling of urban dam-break flooding: a case study of Baldwin Hills, California. *Adv. Water Resour.* 32, 1323–1335.
15. Goto, C., Ogawa, Y., Shuto, N., and Imamura, F. (1997). *IUGG/IOC Time Project: Numerical Method of Tsunami Simulation with the Leap-Frog Scheme.* Paris: IOC Manuals and Guides No. 35; UNESCO.
16. Griffin, J. D., Pranantyo, I. R., Kongko, W., Haunan, A., Robiana, R., Miller, V., et al. (2015). *Assessing Tsunami Hazard Using Heterogeneous Slip Models in the Mentawai Islands, Indonesia. Geohazards in Indonesia.* London, UK: Geological Society of London Special Publication.
17. Harig, S., Chaeroni, C., Pranowo, W. S., and Behrens, J. (2008). Tsunami simulations on several scales. *Ocean Dyn.* 58, 429–440.
18. Hidayat, D., Barker, J. S., and Satake, K. (1995). Modeling the seismic source and tsunami generation of the December 12, 1992 Flores Island, Indonesia, earthquake. *Pure Appl. Geophys.* 144, 537–554.
19. Hill, E. M., Borrero, J. C., Huang, Z., Qiu, Q., Banerjee, P., Natawidjaja, D. H., et al. (2012). The 2010 Mw 7.8 Mentawai earthquake: very shallow source of a rare tsunami earthquake determined from tsunami field survey and near-field GPS data. *J. Geophys. Res. Solid Earth* 117:B06402.

20. Hodgson, M. E., Jensen, J. R., Schmidt, L., Schill, S., and Davis, B. (2003). An evaluation of LIDAR-and IFSAR-derived digital elevation models in leaf-on conditions with USGS Level 1 and Level 2 DEMs. *Remote Sens. Environ.* 84, 295–308.
21. Horritt, M. S., Bates, P. D., and Mattinson, M. J. (2006). Effects of mesh resolution and topographic representation in 2D finite volume models of shallow water fluvial flow. *J. Hydrol.* 329, 306–314.
22. Imamura, F. (2009). *"Tsunami modelling: calculating inundation and hazard maps,"* in *Tsunamis. The Sea, Ideas and Observations on Progress in the Studies of the Sea*, Vol. 15, eds E. Bernard and A. Robinson (Cambridge, MA: Harvard University Press), 321–332.
23. Imamura, F., and Kikuchi, M. (1994). Moment realease of the 1992 Flores Island earthquake inferred from tsunami and teleseismic data. *Sci. Tsunami Hazards* 12, 67–76.
24. Imamura, F., Yalciner, A., and Ozyurt, G. (2006). *Tsunami Modeling Manual.* Available online at: http://www.tsunami.civil.tohoku.ac.jp/hokusai3/E/projects/manual-ver-3.1.pdf
25. Irsyam, M., Sengara, W., Aldiamar, F., Widiyantoro, S., Triyoso, W., Natawidjaja, D. H., et al. (2010). *Development of Seismic Hazard Maps for Indonesia for Revision of Seismic Hazard Map in SNI 03-1726-2002.* Bandung: Bandung Institute of Technology.
26. Jakeman, J., Nielsen, O., Van Putten, K., Mleczko, R., Burbidge, D., and Horspool, N. (2010). Towards spatially distributed quantitative assessment of tsunami inundation models. *Ocean Dyn.* 60, 1115–1138.
27. Kaiser, G., Scheele, L., Kortenhaus, A., Løvholt, F., Römer, H., and Leschka, S. (2011). The influence of land cover roughness on the results of high resolution tsunami inundation modeling. *Nat. Hazards Earth Sys. Sci.* 11, 2521–2540.
28. Kazusa, S., Takeda, M., Kageyama, T., Hosomi, Y., and Sunohara, Y. (2004). *Tsunami and Storm Surge Hazard Map Manual.* Study Committee on Tsunami and Storm Surge Hazard Maps, Cabinet Office (Disaster Management). Available online at: http://www.icharm.pwri.go.jp/publication/pdf/2004/tsunami_and_storm_surge_hazard_map_manual.pdf
29. Latief, H. (2012). *"Kaijian risiko tsunami di Provinsi Sumatera Barat dan Upaya Mitigasinya,"* in *Proceedings of the 37th HAGI Annual Convention and Exhibition* (Palembang).
30. Latief, H., and Hadi, S. (2007). *"Coastal protection in the aftermath of the Indian Ocean tsunami: what role for forests and trees?"* in *The Role of Forests and Trees in Protecting Coastal Areas against Tsunamis*, eds S. Braatz, S. Fortuna, J. Broadhead, and R. Leslie (Khao Lak: Food and Agricultural Society of the United Nations, Regional Office for Asia and the Pacific).
31. Lewis, M., Bates, P., Horsburgh, K., Neal, J., and Schumann, G. (2013). A storm surge inundation model of the northern Bay of Bengal using publicly available data. *Q. J. R. Meteorol. Soc.* 139, 358–369.
32. Løvholt, F., Kaiser, G., Glimsdal, S., Scheele, L., Harbitz, C. B., and Pedersen, G. (2012). Modeling propagation and inundation of the 11 March 2011 Tohoku tsunami. *Nat. Hazards Earth Sys. Sci.* 12, 1017–1028.
33. Lynett, P. J., Wu, T.-R., and Liu, P. L.-F. (2002). Modeling wave runup with depth-integrated equations. *Coas. Eng.* 46, 89–107.
34. McCloskey, J., Antonioli, A., Piatanesi, A., Sieh, K., Steacy, S., Nalbant, S., et al. (2008). Tsunami threat in the Indian Ocean from a future megathrust earthquake west of Sumatra. *Earth Planet. Sci. Lett.* 265, 61–81.
35. Minoura, K., Imamura, F., Sugawara, D., Kono, Y., and Iwashita, T. (2001). The 869 Jogan tsunami deposit and recurrence interval of large-scale tsunami on the Pacific coast of northeast Japan. *J. Nat. Disaster Sci.* 23, 83–88.
36. Natawidjaja, D. H., Sieh, K., Chlieh, M., Galetzka, J., Suwargadi, B. W., Cheng, H., et al. (2006). Source parameters of the great Sumatran megathrust earthquakes of 1797 and 1833 inferred from coral microatolls. *J. Geophys. Res.* 111, 1–37.

37. Nielsen, O., Roberts, S., Gray, D., McPherson, A., and Hitchman, A. (2005). *"Hydrodynamic modelling of coastal inundation,"* in *MODSIM* (Melbourne, VIC).
38. Norwegian Geotechnical Institute (NGI) Geoscience Australia (GA). (2015). *Tsunami Methodology and Result Overview in UNISDR Global Assessment Report on Disaster Risk Reduction 2015: Background Papers for Global Risk Assessment*, Geneva.
39. Rakowsky, N., Androsov, A., Fuchs, A., Harig, S., Immerz, A., Danilov, S., et al. (2013). Operational tsunami modelling with TsunAWI—recent developments and applications. *Nat. Hazards Earth Sys. Sci.* 13, 1629–1642.
40. Rodriguez, E., Morris, C. S., Belz, J. E., Chapin, E. C., Martin, J. M., Daffer, W., et al. (2005). *An Assessment of the SRTM Topographic Products*. Technical Report JPL D-31639, Jet Propulsion Laboratory.
41. Sanders, B. F. (2007). Evaluation of on-line DEMs for flood inundation modeling. *Adv. Water Resour.* 30, 1831–1843.
42. Satake, K. (1995). Linear and nonlinear computations of the 1992 Nicaragua Earthquake Tsunami. *Pure Appl. Geophys.* 144, 3–4.
43. Schlurmann, T., Kongko, W., Goseberg, N., Natawidjaja, D. H., and Sieh, K. (2010). *"Near field tsunami hazard map Padang, West Sumatra: utilizing high resolution geospatial data and reasonable source scenarios,"* in *Proceedings of the ICCE No.32* (Shanghai).
44. Schubert, J. E., and Sanders, B. F. (2012). Building treatments for urban flood inundation models and implications for predictive skill and modeling efficiency. *Adv. Water Resour.* 41, 49–64.
45. Shimozono, T., Sato, S., Okayasu, A., Tajima, Y., Fritz, H. M., Liu, H., et al. (2012). Propagation and inundation characteristics of the 2011 Tohoku tsunami on the central Sanriku coast. *Coast. Eng. J.* 54:1250004.
46. Sieh, K., Natawidjaja, D. H., Meltzner, A. J., Shen, C.-C., Cheng, H., Li, K.-S., et al. (2008). Earthquake supercycles inferred from sea-level changes recorded in the corals of West Sumatra. *Science* 322, 1674–1678.
47. Sugawara, D., and Goto, K. (2012). Numerical modeling of the 2011 Tohoku-oki tsunami in the offshore and onshore of Sendai Plain, Japan. *Sediment. Geol.* 282, 110–123.
48. Tachikawa, T., Kaku, M., Iwasaki, A., Gesch, D., Oimoen, M., Zhang, Z., et al. (2011). *ASTER Global Digital Elevation Model Version 2 -Summary of Validation Results*. Joint Report of NASA Land Processes Distributed Active Archive Center and the Join Japan-US ASTER Science Team.
49. Tang, L., Titov, V. V., and Chamberlin, C. D. (2009). Development, testing, and applications of site–specific tsunami inundation models for real–time forecasting. *J. Geophys. Res.* 114:C12025.
50. Taubenböck, H., Goseberg, N., Setiadi, N., Lämmel, G., Moder, F., Oczipka, M., et al. (2009). "Last-Mile" preparation for a potential disaster—interdisciplinary approach towards tsunami early warning and an evacuation information system for the coastal city of Padang, Indonesia. *Nat. Hazards Earth Sys. Sci.* 9, 1509–1528.
51. Thio, H. K., Graves, R. W., Somerville, P. G., Sato, T., and Ishii, T. (2004). A multiple time window rupture model for the 1999 Chi-Chi earthquake from a combined inversion of teleseismic, surface wave, strong motion, and GPS data. *J. Geophys. Res.* 109:B08309.
52. Titov, V. V., and Synolakis, C. E. (1998). Numerical modeling of tidal wave runup. *J. Waterway Port Coast. Eng.* 124, 157–171.
53. Tsubaki, R., and Kawahara, Y. (2013). The uncertainty of local flow parameters during inundation flow over complex topographies with elevation errors. *J. Hydrol.* 486, 71–87
54. Tsuji, Y., Matsutomi, H., Imamura, F., Takeo, M., Kawata, Y., Matsuyama, M., et al. (1995). Damage to coastal villages due to the 1992 Flores Island earthquake tsunami. *Pure Appl. Geophys.* 144, 481–524.

55. Wessel, P., and Smith, W. H. F. (1998). New, improved version of generic mapping tools released. *Eos Trans. Am. Geophys. Union* 79, 579.

Index